Cholera Toxins

Keya Chaudhuri • S.N. Chatterjee

Cholera Toxins

Springer

Dr. Keya Chaudhuri
Indian Institute of Chemical Biology
4 Raja S.C. Mullick Road
Kolkata-700032
India
keya.chaudhuri@gmail.com

Prof. S.N. Chatterjee
Bidhan Nagar
Sector 1 Block-CF-69
Kolkata-700064
India
sncac.08@gmail.com

ISBN 978-3-540-88451-4 e-ISBN 978-3-540-88452-1

Library of Congress Control Number: 2008937891

Cover design: WMXDesign GmbH, Heidelberg, Germany

Printed on acid-free paper

9 8 7 6 5 4 3 2 1

springer.com

Preface

To start with, we feel that we should explain why the book has been entitled *Cholera Toxins*. In fact, the enterotoxin secreted by *Vibrio cholerae*, which is primarily responsible for causation of the disease, is conventionally known as or referred to as cholera toxin, or CT. By using the word "toxins" (in its plural form), we wanted to cover all of the different types of toxins—and not just CT—produced by *V. cholerae*. We could have used the title *Toxins of Vibrio cholerae*, but we believe that *Cholera Toxins* is simpler and equally as expressive. However, due to its relative importance, the story of CT covers most of this book. Also, compared to all other toxins of *V. cholerae*, CT has been investigated more extensively.

This book was jointly written by us. It is not a multiauthor book in which each expert writes one chapter. In that respect our task is harder. On the other hand, it has given us the unique opportunity to present the entire subject in the way that we conceived it. Besides, our objective is to cater to the needs of not only active research scientists but also students from different disciplines—microbiology, molecular physiology and pharmacology, basic medicines, etc.—and as such, we have attempted to present the subject in a way that will be appreciated by general readers. Further, we have provided some information that students and predoctoral researchers may find useful at the end of the book. We also have taken the liberty of listing all of the references alphabetically at the end of the book and not at the end of each chapter, primarily to avoid repetitions and also for the benefit of young researchers. Similarly, instead of adding concluding paragraphs at the end of each chapter, we have inserted a concluding chapter as the last chapter of the book, in which we have tried to present a glimpse of our introspection on the future prospects of cholera research—at least the way we look at it—keeping in mind, in particular, the ultimate needs of mankind in general.

One of us (SNC) had the unique opportunity while working at the School of Tropical Medicine, Calcutta, to interact directly with S.N. De (Sambhu Nath De) at the neighboring institution, Calcutta Medical College, where De made his outstanding contribution to cholera research. Incidentally, De happened to be the thesis examiner of one of SNC's students, J. Das. The interactions with De often became very interesting and lively in the presence of D. Barua, who was at that time in the Department of Bacteriology at the School of Tropical Medicine, and J.K. Sarkar, Professor of Virology. In fact, those interactions motivated him (SNC) to initiate studies on the cellular and molecular biology of *Vibrio cholerae*, an area that was practically untrodden by medical or other biomedical scientists in the late 1950s or early 1960s. It is amazing to discover how with the advent of various sophisticated techniques and instruments, the sciences of molecular biology, genetics, etc. have contributed toward unraveling the nature of the disease cholera and the mechanism of action of the causative organism, *V. cholerae*. In this context, it would be interesting to place on record how De worked in his laboratory with practically no such instruments except for some optical microscopes. The term "cholera toxin" has perhaps been implanted in the brain of one of us (SNC) since that time, eventually leading to the idea of writing this book. Both of us have in fact spontaneously agreed to state that this book should be considered our humble tribute to S.N. De.

The first three chapters of this book should be considered an introduction to the subject, and are meant for general readers and students only. Chapter 1 gives an idea of the basic nature of the disease cholera and how it has created havoc in different parts of the world. Different bacteria produce different types of toxins, which vary in terms of their modes of action, sites of action, etc. This basic or background knowledge, presented in Chapter 2, will, it is hoped, help general readers to assess the nature and function of cholera toxins in the proper perspective. Chapter 3 similarly provides basic information about the organism that produces all of the toxins discussed in this book.

Chapters 4–6 deal with the endotoxins or lipopolysaccharides of *V. cholerae*, and are to a large extent reproductions of the three review articles written by us for *Biochimica Biophysica Acta—Molecular Basis of Disease*, with some updating performed wherever needed.

Unraveling the structure of the cholera toxin molecule and its interactions has played a big role in elucidating its mode of action that leads to the causation of the disease. Accordingly, Chapter 7 elucidates the primary, secondary, tertiary, and quaternary structures of the cholera toxin (CT) molecule, as derived by different physicochemical techniques and particularly X-ray crystallography. This chapter also elucidates the structure of the CT-GM1 complex, explaining the details of the binding of CT to the receptors at the surfaces of the epithelial cells, and also how the CTA1 fragment is activated upon binding with ARF6 complexed with GTP.

The genetics of the biosynthesis of CT and its regulation by the different environmental conditions prevailing within the intestinal lumen are illustrated in Chapters 8 and 9. The mechanism of how CT is secreted by the vibrios to make it available for interaction with the epithelial cells lining the intestinal lumen remained elusive for quite some time until the discovery of the specialized secretion or translocation channels in the inner and outer membranes of the *V. cholerae* organism, as described in Chapter 10 of this book.

The next chapter, Chapter 11, unfolds a beautiful facet of the subject of molecular physiology and describes the fascinating story of the retrograde voyage of the cholera toxin molecule via lipid-raft-mediated endocytosis to the lumen of the endoplasmic reticulum, where the A1 fragment dissociates from the parent holotoxin, is unfolded, and is translocated by the PDI via the sec61 channel of the ER membrane to the cytosol. The A1 fragment is then activated by interacting with the

ARF6 molecule complexed with GTP, leading to the activation of the membrane-bound adenylate cyclase, which in turn catalyzes the conversion of the ATP molecule to cyclic AMP (cAMP). This leads to the heavy drainage of intracellular Cl^-, water, and other electrolytes to the lumen and the massive diarrhea that are characteristic of cholera.

Chapter 12 gives an account of the immune response of the host to CT and the present status of the cholera vaccines that have been constructed by various laboratories and are at different stages of trial. Since it is widely acknowledged that antibacterial immunity is superior to antitoxin immunity, the immune response to CT is briefly discussed. Although not very relevant, this chapter also presents some aspects of the adjuvant properties of CT. Since antibacterial immune response mainly arises upon exposure to the LPS coating the outer envelope of the bacteria, a lengthy discussion on the current status, the problems, and the prospects of developing an effective cholera vaccine are included in the concluding parts of the chapters dealing with LPS of *V. cholerae* due to their immediate relevance.

Then, to do justice to the title of this book, one fairly big chapter—Chapter 13—is devoted to elucidating the nature, discovery, mechanism of action, etc., of ten exotoxins other than CT produced by *V. cholerae*, among which the zonula occludens toxin (Zot) occupies a significant position.

In Chapter 14, the concluding chapter of this book, we attempt to provide a brief introspection on the achievements made in this and related areas of research, the targets that are yet to be achieved in spite of the best efforts of scientists, and what we are expected to do, bearing in mind our primary obligations to the society at large. Certainly, the wealth of knowledge acquired is our strength; sooner or later, it will enable us to produce an effective solution to the cholera problem.

This book is based on works published by many scientists over several decades. The authors have tried their best to cite these works and include them in the list of references given at the end of the book, and we sincerely apologize for any unintentional omission.

It is now our noble duty to gratefully acknowledge and extend our most sincere thanks to the many illustrious scientists and publishers who have ungrudgingly and promptly helped us in our efforts to realize this book. In particular, we would like to express our sincere thanks and indebtedness to (i) J.J. Mekalanos, Harvard Medical School, Boston, (ii) J.B. Kaper, University of Maryland School of Medicine, Baltimore, (iii) W.G. Hol, Biomolecular Structure Center, University of Washington, Seattle, (iv) A. Fasano, University of Maryland School of Medicine, Baltimore, (v) P. Watnick, Childrens' Hospital Boston, Boston, (vi) P. Alzari, Institut Pasteur, Paris, (vii) U.H. Stroeher, University of Adelaide, Australia, (viii) W.F. Wade, Dartmouth Medical School, Lebanon, (ix) O.C. Stine, University of Maryland, Baltimore, (x) G.B. Nair, National Institute of Cholera and Enteric Diseases, Kolkata, and (xi) Lisa Craig, Simon Fraser University, Canada, who spontaneously provided us with valuable materials for reproduction in our book or granted us permission to reproduce figures or diagrams from their published works. We are particularly grateful to Drs. Fasano, Hol, and Craig for providing much valuable input into our book. We are equally thankful to the editors and publishers of various journals or online data banks, including those associated with (i) the *Proceedings of the National Academy of Sciences USA*, (ii) *Science*, (iii) American Society of Microbiology, (iv) Elsevier, Amsterdam, (v) BMC, (vi) pLoS, and (vii) RCSB-PDB, for granting permission to reproduce specific figures or diagrams from their publications in our book. Specific acknowledgements of the abovementioned scientists and publishers have also been recorded gratefully at the appropriate positions in this book. We are equally thankful to Drs. Pat Crowley and Andy Deelen from Elsevier Publishing Ltd. for promptly helping us by providing information on the rights of authors for Elsevier, as given on the Elsevier website. Last but not the least, we offer our grateful thanks to Springer-Verlag, Heidelberg, Germany, the publisher of this book, for kindly agreeing to publish this volume, and to their editors, particularly Drs. Sabine Engelmann and Anette Lindqvist, for providing us with many valuable suggestions and guidance throughout. We are equally thankful to the members of the Molecular and Human Genetics Division, Indian Institute of Chemical Biology, Kolkata, to Drs. Arunava Bandopadhyay, Rajdeep Chowdhury, and Mr. Arun Roy, for providing technical or other help, and particularly to Mr. Avirup Dutta for his continuous technical support.

Finally, KC would like to acknowledge the technical help and suggestions provided, at times, by her two beloved daughters, Kamalika and Sibapriya, the encouraging patience exhibited and the cooperation extended by her husband, Durgadas, and the inspiration provided by her mother. While appreciating the interesting tips offered, at times, by his son, Saugata, from a great distance, SNC would like to put on record his gratefulness to and appreciation of his wife, Amala (who has made a significant contribution to biophysics), who, in spite of her failing health and serious illness, has provided many critical suggestions and all the support and encouragement needed for the successful completion of our project.

Kolkata, India
August 2008

Keya Chaudhuri
S.N. Chatterjee

Contents

Chapter 1
An Introduction to Cholera

Abstract A brief discussion of the nature of the disease cholera, the seven pandemics experienced, and also the large-scale epidemic caused by the *Vibrio cholerae* serogroup O139 around 1992 is presented for the benefit of general readers, including students of microbiology.

1.1 Nature of the Disease

Cholera (often called Asiatic cholera) is a disease that begins with a sudden onset of massive diarrhea accompanied by the loss of a profuse amount of protein-free fluid along with electrolytes, bicarbonates and ions within a short time (maybe several hours after onset), as well as vomiting, resulting in hypovolemic shock and acidosis. If the patient is left untreated without prompt attention, the resulting dehydration produces tachycardia, hypotension, and vascular collapse. Under such conditions, the disease becomes fatal in most cases.

1.2 Epidemics and Pandemics

Cholera often occurs in outbreaks or epidemics and pandemics. A cholera pandemic is a cholera epidemic that can last many years or even a few decades at a time, and that spreads to many countries and across continents and oceans. It is generally recognized that seven distinct pandemics of cholera have occurred and that the first one took place in the early nineteenth century. There are practically no authentic records of what happened prior to that date. A brief account of the different pandemics of cholera is presented in Table 1.1.

In each of the first seven pandemics, cholera spread to other continents from Asia. These pandemics affected many countries and extended over many years. Except for the seventh pandemic, which originated in the island of Sulawesi in Indonesia, the pandemics arose in the Indian subcontinent, usually from the Ganges Delta of Bengal. The first pandemic, according to records, began in 1817, when cholera

K. Chaudhuri and S.N. Chatterjee, *Cholera Toxins*.

Table 1.1 A brief account of the pandemics of cholera witnessed so far

Serial number	Period (calendar years)	Affected areas; origin
1	1817–1823	India, China, Caspian Sea; Bengal (undivided), India
2	1829–1851	Europe, London, Paris, Russia, Quebec, Ontario, New York, Pacific coast of North America; India
3	1852–1859	Mainly Russia; India
4	1863–1879	Europe, Africa, North America; India
5	1881–1896	South America, mainly Argentina, Chile, Peru among other countries; India
6	1899–1923	Russia; India
7	1961–1970s[a]	Bangladesh, India, U.S.S.R., North Africa, Italy, Japan, and South Pacific; Indonesia

The contents of the table are based on data collected from different sources (Politzer 1959, Barua and Burrows 1974, Blake 1994, Kaper et al. 1995)

The years marking the beginning or end of any pandemic are approximate and can be debated

[a] In 1961, which is known to be the year in which the seventh pandemic started, the El Tor biotype spread out of Indonesia. It is also believed in many quarters that the seventh pandemic continued until the beginning of 1991

spread out of India. During the second pandemic, cholera reached the British Isles in the early 1830s, and O'Shaughnessy (in 1831–1832) first demonstrated that the characteristic rice water stools of patients contained salts and alkali, i.e., were very high in electrolyte content. John Snow from London made the fundamental epidemiologic observation in the period 1847–1854 that cholera is transmitted through water. The second pandemic also originated in India but took a devastating toll on both Europe and America. The third pandemic of cholera swept England and Wales in 1848–1849. In the course of field investigations carried out in Egypt during the fifth pandemic, Robert Koch isolated a bacterium from the rice water stools of cholera patients, which he referred to as "comma bacilli" because of their shape. Koch subsequently moved to Calcutta, where in 1884 he again isolated the same bacterium from cholera patients (Koch 1884). The fifth pandemic extensively affected South America. The sixth pandemic extensively affected populations in the Near and Middle East, as well as the Balkan peninsula. The seventh pandemic of cholera exhibited many important and notable characteristics. This pandemic, unlike the earlier ones, originated in the island of Sulawesi in Indonesia. It was the most extensive one in terms of geographic spread and duration. Further, unlike the earlier pandemics, the causative agent of the seventh pandemic was *V. cholerae* O1 biotype El Tor. The El Tor strain possessed several characteristics that conferred upon it a high degree of epidemic virulence, allowing it to spread across the world. First, the ratio of cases to carriers was much less than in the cholera due to the classical biotype. Second, the duration of carriage after infection was longer for the El Tor strain than for the classical one. Third, the El Tor strain survived for longer periods in the extraintestinal environment. Thus, the El Tor biotype virtually replaced the classical biotype, which was previously responsible for the annual cholera epidemics in India and Bangladesh.

During the period 1992–1993, epidemic cholera was reported at several sites, including Madras and Calcutta, in India and in Bangladesh. Surprisingly, a new serogroup, *V. cholerae* O139, not *V. cholerae* O1, was the causative agent. From these initial sites, the organism spread throughout India, Pakistan, Nepal, China, Thailand, Kazakhstan, Afghanistan, and Malaysia. The O139 was derived genetically from the El Tor pandemic strain, but it had acquired a new antigenic structure such that there was no existing immunity to the disease and so it caused widespread havoc and fatal cases. However, the strain was properly diagnosed soon afterwards (Ramamurthy et al. 1993) and the disease was brought under control. Also, this particular strain exhibited quiescent periods with the re-emergence of the O1 serotype. Whether the 1992–1993 cholera epidemics could be termed the eighth pandemic remains a debatable proposition.

1.3 Causation of the Disease

It is now known that *V. cholerae* is the causative agent of cholera. Similar but milder diarrhea is also known to be caused by other vibrios, enterotoxic *Escherichia coli*, etc. Humans are apparently the only natural hosts for the cholera vibrios. Cholera is acquired through the ingestion of water or food contaminated with the feces of an infected individual. Because *V. cholerae* is sensitive to acid, most cholera-causing bacteria die in the acidic environment of the stomach. However, when a person has ingested food or water containing large amounts of cholera bacteria, some organisms will survive in the intestine. In the small intestine, the rapidly multiplying bacteria produce a toxin, now known as the cholera toxin (CT), which causes a large volume of water and electrolytes to be secreted into the bowels and then to be abruptly eliminated as watery diarrhea. Vomiting may also occur. In serious cases, the infection can produce a dangerous state of dehydration (even within several hours) due to the loss of a large volume of fluid and electrolytes. Immediate replacement of the lost fluid and electrolytes is necessary to prevent kidney failure, coma, and death.

1.4 Treatment Basics

The key to the treatment of cholera lies in preventing dehydration by replacing the fluids and electrolytes lost through diarrhea and vomiting. The treatment of cholera has now been greatly simplified by the discovery that rehydration can be accomplished orally (Phillips 1964, Sack et al. 1970, Cash et al. 1970). A simple and inexpensive oral replacement fluid or oral rehydration solution (ORS) containing appropriate amounts of water, sugar, and salts has been formulated from clinical studies, is recommended by the WHO, and is currently being used worldwide. This simple, cost-effective treatment is now available even in the remotest villages. In cases of severe dehydration, replacement fluid must be given intravenously. To shorten the duration of illness and reduce fluid loss, an antibiotic (tetracycline is used in most cases) can be administered orally. This is the basic idea, and is a greatly simplified view of the treatment of cholera.

Chapter 2
Bacterial Toxins: A Brief Overview

Abstract A basic idea of the different types of toxins produced by bacteria in general and the classification of them according to their (i) modes of production, (ii) sites of action, (iii) modes of action, etc. are presented as a relevant introduction to the subject matter of this book, cholera toxins.

2.1 Introduction

This chapter presents a general introduction to the various types of toxins produced by different bacteria. This will enable the readers to gain a useful perspective on the toxins produced by *Vibrio cholerae* compared to those produced by other bacteria.

2.2 Different Types of Toxins

The different types of toxins produced by a bacterial cell are broadly classified as

1. Endotoxins
2. Exotoxins
3. Enterotoxins

Endotoxins are part of the outer membrane or the cell wall of Gram-negative bacteria. The word endotoxin is in fact properly reserved for the lipopolysaccharide (LPS) complex associated with the outer membrane of Gram-negative bacteria in general, irrespective of whether the organism is a pathogen or not. These toxins are generally available for action only after the death and lysis of the bacteria to which they belong.

Exotoxins are soluble proteins excreted by a microorganism, including bacteria, fungi, algae, and protozoa. Both Gram-positive and Gram-negative bacteria produce exotoxins. They are highly potent and can cause major damage to the host by different mechanisms. The exotoxins may also be released during lysis of the cell.

K. Chaudhuri and S.N. Chatterjee, *Cholera Toxins*.

Table 2.1 Some of the general properties of typical endotoxins and exotoxins

Property	Endotoxin	Exotoxin
Chemical nature	Lipopolysaccharide	Protein
Relationship to cell	Part of outer membrane	Extracellularly secreted
Denatured by boiling	No	Usually yes
Antigenic	Yes	Yes
Forms toxoid	No	Yes
Potency	Relatively low	Relatively high
Specificity	Low degree	High degree
Enzyme activity	No	Usually yes
Pyrogenicity	Yes	Occasionally

Enterotoxins are a type of exotoxin released by a microorganism in the intestine. They are frequently cytotoxic and kill cells by altering the permeability of the epithelial cells of the intestinal wall. They are often pore-forming toxins, secreted by bacteria, that assemble to form pores in cell membranes leading to cell death. There are other enterotoxins which act differently, i.e., *V. cholerae* enterotoxin, whose mode of action will be discussed later. Examples of organisms secreting enterotoxins are: *Escherichia coli, Clostridium perfringens, V. cholerae*, and *Yersinia enterocolitica*. Table 2.1 presents some of the basic properties that differentiate the endotoxins and the exotoxins produced by different bacterial cells.

2.3 Bacterial Endotoxins: A General Introduction

Endotoxins are associated with Gram-negative bacteria and form part of their cell wall structures. More specifically, endotoxins refer to the lipopolysaccharide complex associated with the outer membranes of Gram-negative bacteria (Fig. 2.1) such as *E. coli, V. cholerae, Salmonella* species, etc., irrespective of whether they are pathogens or not. Most of the endotoxin generally remains associated with the cell wall until the bacteria disintegrate.

The lipolysaccharides (LPS) are complex amphiphilic molecules with a molecular weight of around 10kDa and a chemical composition that varies widely both between and among bacterial species. However, the LPS of Gram-negative bacteria have the general chemical structure shown in Fig. 2.2. LPS thus consists of three component regions, lipid A, core polysaccharide (comprising an inner and an outer core) or core-PS, and an O-polysaccharide (O-specific chain) or O-PS. Lipid A is the lipid component of LPS, which is hydrophobic and anchors to the outer membrane of bacteria. The structure of lipid A is highly conserved among Gram-negative bacteria. It consists of a phosphorylated *N*-acetylglucosamine (NAG) dimer with six or seven fatty acids (FA) attached. The core antigen region or core-PS is attached to the 6-position of one NAG molecule of lipid A and consists of a short chain of sugars. Two unusual sugars, heptose and 2-keto-3-deoxyoctolusonic

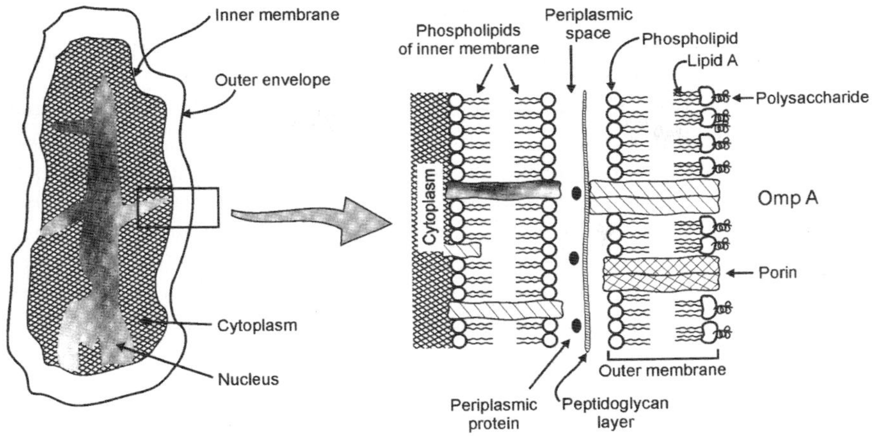

Fig. 2.1 Schematic diagram showing the organization of the inner and outer membrane (*right*) of a Gram-negative bacterium (*left*). The right hand picture is a highly amplified view of the portion within the box in the left hand one

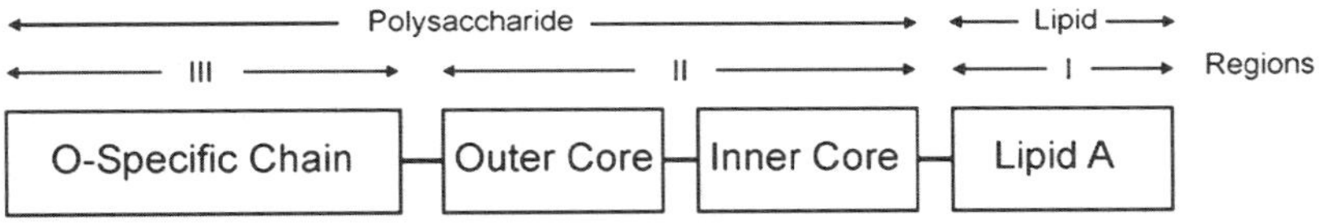

Fig. 2.2 The general structure of the lipopolysaccharide of a Gram-negative bacterium consisting of three different sectors, lipid A (I), the core polysaccharide (II) (comprising inner and outer cores), and the O-specific polysaccharide chain (III)

acid (Kdo), are usually present in the LPS. Kdo generally connects the lipid A to the core-PS of the LPS molecule. The O-PS (somatic O-antigen polysaccharide) is attached to the outer side of the core-PS and consists of repeating oligosaccharide subunits made up of 3–5 sugars. The individual chains vary in length, ranging up to 40 repeat units. The O-polysaccharide is much longer than the core polysaccharide and maintains the hydrophilic domain of the LPS molecule. A major antigenic determinant (antibody combining site) of the Gram-negative cell wall resides in the O-polysaccharide.

Endotoxins, although antigenic, cannot be converted into toxoids. Since the LPS is located on the outer face of the bacterial outer membrane, it mediates contact with the environment. It is essential for the proper functioning of the outer membrane, and because it is a structural component of the cell, it is likely to play several roles in the pathogenesis of Gram-negative bacterial infections. Acting as a permeability barrier, it is permeable only to low molecular weight, hydrophilic molecules. It acts as a barrier to lysozyme and many antimicrobial agents. It also impedes destruction of the bacterial cells by serum components and phagocytic cells. Furthermore, LPS plays an important role as a surface structure in the interaction of the pathogen with its host. It may be involved in adherence or resistance to phagocytosis, or antigenic shifts that determine the course and outcome of an infection.

The O-specific chain is the portion of LPS that is mostly responsible for its immune recognition. Minor variations in the structure of the O-polysaccharide make enormous differences to the virulence of bacterial infections. While it is the function of the immune system to recognize the O-polysaccharide and produce defenses against the invading bacteria, the lipid A fragments are the components that produce all of the harm caused by the infection. Loss of O antigen results in loss of virulence, suggesting that this portion is important during a host–parasite interaction. Loss of the more proximal parts of the core, which occurs in deep-rough mutants, makes the strains sensitive to a range of hydrophobic compounds, including antibiotics, detergents, bile acids, etc. This area contains a large number of charged groups and is thought to be important in maintaining the permeability properties of the bacterial outer membrane. Both lipid A (the toxic component of LPS) and the polysaccharide side chains (the nontoxic but immunogenic portion of LPS) act as determinants of virulence in Gram-negative bacteria.

Against this background information on the nature and properties of endotoxin (LPS) of Gram-negative bacteria in general, the characteristics and functions of the endotoxin (LPS) of *V. cholerae* will be presented in the following chapters.

2.4 Bacterial Exotoxins: A General Introduction

The exotoxins may be classified in accordance with their site or mode of action. Their different modes of action include: (i) damaging cell membranes; (ii) inhibiting protein synthesis; (iii) activating second messenger pathways; (iv) inhibiting the role of neurotransmitters; (v) activating the host immune response, etc. A general classification of exotoxins that takes both their sites and their modes of action into account is presented below.

2.4.1 Extracellularly Acting Toxins

These toxins may be further classified in accordance with their site of action:

2.4.1.1 Membrane-Damaging Toxins

(i) **Hemolysins.** These toxins cause lysis of red blood cells with resultant anemia. There are many different types of hemolysins but their net action is the same.
(ii) **Phospholipase.** This enzyme attacks any cell with phospholipids in its membrane. The result is widespread cell lysis.

(iii) **Lecithinase.** This is also known as phospholipase C, and it is an enzyme which breaks down lecithin in the human cell plasma membrane, resulting in cell lysis. It is especially active on red blood cells.

(iv) **Pore-forming toxins.** These toxins disrupt the selective influx and efflux of ions across the plasma membrane by inserting a transmembrane pore. This group of toxins includes the RTX (repeats in toxin) toxins from Gram-negative bacteria, streptolysin O produced by *Streptococcus pyogenes* and the *Staphylococcus aureus* α toxin.

2.4.1.2 Non-Membrane-Damaging Toxins

Some of the exotoxins belonging to this class are:

(i) **Hyaluronidase.** This is an enzyme that catalyzes the breakdown of hyaluronic acid, the substance that cements human cells together. This allows the bacterial cells to spread through tissue, causing a condition known as cellulitis.

(ii) **Fibrinolysin.** This toxin catalyzes the conversion of plasminogen to the fibrinolytic enzyme plasmin. In *S. aureus*, the gene for fibrinolysin is on a bacteriophage and is expressed during lysogeny.

(iii) **Lipase.** Production of excessive amounts of lipase allows bacteria to penetrate fatty tissue with the consequent formation of abscesses.

(iv) **Collagenase.** This enzyme catalyzes the degradation of collagen, a scleroprotein found in tendons, nails, and hair.

2.4.2 Intracellularly Acting Toxins

These may be further classified in accordance with their mode of action:

2.4.2.1 ADP-Ribosyl Transferases

These toxins promote the breakdown of nicotinamide adenine dinucleotide (NAD) into nicotinamide and adenine diphosphate ribose (ADPR) and the covalent binding of the ADPR to various proteins, thus inactivating the bound protein. Some examples are given below.

(i) **Diphtheria toxin.** *Corynebacterium diphtheriae* lysogenized with the β-phage produce a toxin which is a bipartite molecule, composed of a B subunit which mediates binding to a specific human cell surface receptor and an A subunit which possesses enzymatic (ADP-ribosyl transferase) activity. The reaction substrate is human elongation factor-2 (EF-2), an essential part of the protein synthesis machinery. The toxin catalyzes the reaction:

$$EF2 + NAD \rightarrow ADPR\text{-}EF2 + nicotinamide + H^+$$

This results in the inhibition of protein synthesis and cell death.

(ii) **Cholera toxin (CT).** *V. cholerae* growing in the intestine secretes an exotoxin composed of five B subunits and one A subunit. On exposure to small bowel epithelial cells, each B subunit binds to a receptor on the gut epithelium. Following binding, the toxin migrates through the epithelial cell membrane. The A1 subunit is an ADP-ribosyl transferase that catalyzes in the cell cytosol the transfer of ADPR from NAD to a guanine triphosphate (GTP) binding protein that regulates adenylate cyclase activity. The ADP-ribosylation of GTP binding protein inhibits the GTP turn-off reaction and causes a sustained increase in adenylate cyclase which results in excess secretion of isotonic fluid into the intestine with resulting diarrhea.

(iii) **Labile toxin (LT)** of *E. coli*. This toxin is identical to cholera toxin. Its production by the bacterial cell is mediated by a plasmid.

2.4.2.2 Nonribosylating Toxins

(i) **Shiga toxin.** Species of *Shigella* carry the gene for shiga toxin on the chromosome. This toxin has one A subunit and five B subunits. The A subunit can be divided into A1 and A2 subunits. The A1 moiety binds to the 60S human ribosome, which inhibits protein synthesis.

(ii) **Botulinum toxin.** *C. botulinum* produces an endopeptidase that blocks the release of acetylcholine at the myoneural junction, resulting in muscle paralysis. The botulinum toxin cleaves synaptobrevin, thus interfering with vesicle formation.

(iii) **Pyocyanin.** *Pseudomonas aeruginosa* produces this non-enzyme protein which binds to the flavoproteins of the cytochrome system and interferes with terminal electron transport, causing an energy deficit and cell death.

(iv) **Adenylate cyclase.** *Bordetella pertussis* produces a calmodulin-independent adenylate cyclase that inhibits and/or kills white blood cells.

2.4.3 Toxins Whose Mechanisms of Action Are Not Well Defined

Some of the other toxins whose mechanisms of action are less understood are cited below:

(i) **Trachea toxin**. *B. pertussis* tracheal cytotoxin kills cilia-bearing cells

(ii) **β-Toxin.** *C. difficile* produces a β-toxin which causes necrotic enteritis

(iii) Erythrogenic toxin. *S. pyogenes* produces this toxin, which is similar to toxic shock syndrome toxin produced by *S. aureus* and is mediated through the induction of interleukin-1, causing various symptoms, including fever

This chapter presents only a few typical examples of the toxins belonging to each class. In this context, the different toxins produced by *V. cholerae* are listed below.

2.5 Toxins of *Vibrio cholerae*

The different types of toxins produced by *V. cholerae* (Kaper et al. 1995) include:

(i) Cholera endotoxin (LPS)
(ii) Cholera enterotoxin or cholera toxin (CT)
(iii) Zonula occludens toxin (Zot)
(iv) Accessory cholera enterotoxin (Ace)
(v) Hemolysin/cytolysin
(vi) RTX toxin
(vii) Chinese hamster ovary (CHO) cell elongation factor (Cef)
(viii) New cholera toxin (NCT)
(ix) Shiga-like toxin (SLT)
(x) Thermostable direct hemolysin (TDH)
(xi) Heat-stable enterotoxin of nonagglutinable vibrios (NAG-ST)
(xii) The toxin WO-7

Of all the toxins of *V. cholerae* listed above, endotoxin (LPS) and cholera toxin (CT) are the two most important ones from a functional point of view. However, details of the properties and functions of all of the toxins referred to above will be discussed in the following chapters of this book.

Chapter 3
Vibrio cholerae, the Causative Organism

Abstract *Vibrio cholerae*, the causative organism of the disease cholera and the producer of several different toxins, belongs to the family *Vibrionaceae*, the genus *Vibrio* and the species *V. cholerae*. The basis of its further subclassification or subtyping into serogroups, serotypes, biotypes, and phage types has been elucidated. The molecular subtyping of *V. cholerae* strains in accordance with (i) plasmid profile analysis, (ii) restriction fragment length polymorphism (RFLP) of *ctx* and rRNA genes including their flanking regions, (iii) zymovar analysis of the enzymes produced, (iv) pulsed field gel electrophoresis of isolated DNA fragments, etc., has produced results of great epidemiological significance. This chapter further presents the basic ultrastructural features of *V. cholerae* cells and cholera bacteriophages, their important growth characteristics, and a general outline of the genetics of the bacteria to an extent relevant to the subject matter of this book, thereby providing a relevant introduction to the subject for the general reader.

3.1 Introduction

Vibrio cholerae is the causative organism of the disease cholera. The organism produces many different types of toxins, the nature and function of which will be discussed thoroughly in this book. However, a proper understanding of all of these aspects by a general reader requires an understanding of the basic nature of the causative organism, *V. cholerae*. Accordingly, this chapter presents a brief account of the basic bacteriology and cellular and molecular biology of the organism *V. cholerae*.

3.2 Classification and Identification

The classification of *V. cholerae* is given below:

Kingdom: Bacteria
Phylum: Proteobacteria

K. Chaudhuri and S.N. Chatterjee, *Cholera Toxins*.

Class: Gamma Proteobacteria
Order: Vibrionales
Family: *Vibrionaceae*
Genus: *Vibrio*
Species: *V. cholerae*
Binomial name: *Vibrio cholerae*

The vibrios belong to the family *Vibrionaceae*, along with *aeromonas, phobacterium*, and *plesiomonas* species. The genus *Vibrio* includes many different species, of which few are pathogenic. The classic example of a pathogenic species of this genus is *V. cholerae*, with others being *V. parahaemolyticus, V. alginolyticus, V. mimicus, V. vulnificus*, and certain nonagglutinable vibrios (previously termed "NAG vibrios") (Sakazaki 1968). Sakazaki (1992) made a significant contribution towards the classification and characterization of vibrios and defined the practical basis for the differential diagnosis of the genus *Vibrio* and related genera, viz., *Aeromonas, Plesiomonas*, and the *Enterobacteriaceae*. Among the vibrios, *V. cholerae* is a well-defined species due to the biochemical tests and DNA homology studies that have been performed on it (Baumann et al. 1984). *V. cholerae* and the other species belonging to the *Vibrio* and related genera can be differentiated for all practical purposes by a variety of simple tests (Table 3.1).

Table 3.1 Differentiation of *V. cholerae* from related species (reproduction of Table 1 of Chatterjee and Chaudhuri 2003)

Tests	*V. cholerae*	Other *Vibrio* species	Enterobacteriaceae
Oxidase[1]	+	+[a]	–
String test[2]	+	+/–	–
Acid from mannitol[3]	+	+/–	+/–
Acid from sucrose[3]	+	+/–	+/–
Lysine decarboxylase[4]	+	+/–	+/–
Ornithine decarboxylase[4]	+	+/–	+/–
Growth in 0% NaCl[5]	+	–[b]	–
Mol% G + C[6]	47–49	38–51	38–60

Positive response means: [1]test for the presence in bacteria of a certain oxidase that will catalyze the transport of electrons between donors in bacteria and a redox dye which is reduced to a deep purple color; [2]a mucoid string is formed when an inoculating loop is drawn slowly away from a drop of 0.5% aqueous solution of sodium deoxycholate in which a 24 h growth of the organism is suspended; the string is formed because the organisms are lysed, DNA is released and the mixture is made viscous; [3]ability of the organism to ferment the particular sugar added to the growth medium and produce acid which changes the color of an indicator (phenol red, bromothymol blue, etc.); [4]ability of the organism to decarboxylate a particular amino acid added to the growth medium with the liberation of carbon dioxide and the change of color of the medium to violet; [5] ability of the organism to grow in the absence of NaCl in the medium; [6] DNA base composition given in terms of mol% G + C

[a]Except for *V. metchnikovii*

[b]Except for *V. mimicus*

3.3 Subtyping

3.3.1 Serogroups

The species *V. cholerae* is, however, not homogeneous in many respects, and important distinctions within the species are made on the basis of serogroup, production of cholera enterotoxin, and potential for epidemic spread. The serogrouping is done on the basis of the heat-stable O-antigen of the bacteria. Around 200 serogroups (O1 to O200) of *V. cholerae* have already been identified, and more may surface in the future. Previously, all major epidemics of cholera were caused by the *V. cholerae* strains belonging to the same serogroup, O1. This idea is no longer valid, as a new serogroup, O139, emerged in 1992 to cause an epidemic of the disease in the Indian subcontinent. Only those strains of the two serogroups, O1 and O139, which produce cholera toxin (CT) are associated with epidemic cholera, but there are other strains of these serogroups which do not produce CT, do not produce cholera and are not involved in the epidemics. Although some strains belonging to serogroups other than O1 and O139 have produced occasional outbreaks of cholera, they have not been associated with any large epidemic or extensive pandemic so far. In respect to the third classification parameter, the potential for epidemic spread, the actual determinant of this potential is still not known, and under the circumstances, the possession of O1 or O139 antigen may be considered at least a marker of such a potential.

3.3.2 Serotypes

The O1 serogroup is further subdivided into three serotypes or subtypes—Inaba, Ogawa, and Hikojima—where the names denote their historical origins. Note that the Hikojima subtype is not recognized by all workers and is rare and unstable. The serotyping is based on the three antigenic forms of *V. cholerae* O1. The O antigen of *V. cholerae* O1 consists of three factors—A, B and C. The A factor may be the D-perosamine homopolymer, but the nature of the B and C factors is unknown. DNA sequence analysis of genes encoding the O1 antigen (*wbe*) reveals that the sequences for Inaba and Ogawa antigens are nearly identical. The shift from Inaba to Ogawa can result from a variety of changes, even from a single base change in the *wbeT* gene that creates a premature stop codon in this gene. A recent study of the serotype shift indicated that one Ogawa strain may have arisen from two mutations: one deletion mutation producing the initial Ogawa to Inaba shift, and a second insertion mutation which restored the original *wbeT* reading frame, thereby restoring the Ogawa serotype.

3.3.3 Biotypes

V. cholerae O1 strains are divided into two biotypes, classical and El Tor. This subdivision is based on several tests, as described in Table 3.2. Another approach to biotyping has recently been described on the basis of observed differences in DNA sequence between genes encoding the toxin-coregulated pilus (TCP) from classical and El Tor strains. Among other recent findings on the differentiation of classical and El Tor biotypes, differences in the restriction fragment length polymorphisms (RFLP) in rRNA genes of classical and El Tor strains are worth mentioning here (Koblavi et al. 1990, Popovic et al. 1993). The El Tor biotype was discovered as the causative agent of the seventh pandemic of cholera (Gallut 1971). However, the properties of this El Tor biotype are not considered sufficiently distinctive to make it a separate species (Hugh 1965a, b). A summary of the different serogroups, serotypes, and biotypes of *V. cholerae* is presented in Table 3.3.

Table 3.2 Tests generally used for biotyping of *V. cholerae* (reproduction of Table 2 of Chatterjee and Chaudhuri 2003)

Tests used	Responses of the two biotypes	
	Classical	El Tor
Hemolysis of sheep erythrocytes[a]	–	+/–
Agglutination of chicken erythrocytes	–	+
Voges–Proskauer reaction	–	+
Inhibition by polymyxin B (50 μg disk)	+	–
Lysis by Group IV cholera phage[b]	+	–
Lysis by FK cholera phage[b]	+	–

[a]This test is of limited value since both hemolytic and nonhemolytic El Tor strains have been isolated

[b]This test is dependable provided well-characterized bacteriophages are available from the WHO reference centers

Table 3.3 Classification of *V. cholerae* species into serogroups, serotypes, and biotypes (reproduction of Table 3 of Chatterjee and Chaudhuri 2003)

Serogroups	CT production (no. of strains, %)	Epidemic spread	Serotypes (no.)	Biotypes no. (names)
O1	+ (>95)[a]	+	Inaba, Ogawa, Hikojima (3)	2 (classical, El Tor)
O139	+ (>95)	+	Nil	1
Other non-O1	– (>95)[a]	–	Nil	1

[a]More than 95% of the strains produce (+) or do not produce (–) cholera toxin (CT) and cause (+) or do not cause (–) epidemic spread of the disease cholera

3.3.4 Phage Types

As described in Table 3.2, Mukerjee's typing phages, and particularly the Group IV phage, have been used to differentiate the classical and El Tor biotypes of *V. cholerae* O1. While the classical biotype is sensitive, the El Tor is resistant to the Gr. IV phage and also to the FK phage of Takeya (Mukerjee 1963a, Takeya and Shimodori 1963). This typing scheme is of limited use since (i) well-characterized vibrio phages are not always available except from a small number of reference centers and research laboratories all over the world, (ii) a consensus typing scheme is lacking, and (iii) multiple typing schemes are also limited by the fact that two or three phage types account for up to 80% of all *V. cholerae* strains examined.

3.4 Molecular Subtyping

The following methods are often used, either singly or in combination, for molecular subtyping of *V. cholerae* strains isolated from environments undergoing cholera outbreaks or pandemics. These methods have proven useful for classifying or characterizing the strains and have provided valuable information of epidemiological significance.

(i) Plasmid profile analysis
(ii) Restriction fragment length polymorphism (RFLP) of *ctx* genes
(iii) RFLP analysis of rRNA genes
(iv) DNA or gene probes
(v) Multilocus enzyme electrophoresis (MEE) or zymovar analysis
(vi) Pulsed field gel electrophoresis

3.4.1 Plasmid Profile Analysis

This method characterizes *V. cholerae* strains in terms of the presence or absence of different plasmids. However, this is of limited value, since many strains (particularly many of those isolated during the seventh pandemic) appear to be plasmidless.

3.4.2 RFLP Analysis of ctx Genes

In this method, the strains of *V. cholerae* are characterized in terms of the sequence divergence in genes flanking the *ctx* locus using the *ctx* gene probes (Kaper et al. 1982, 1986). A significant finding of epidemiological interest revealed that the US

Gulf Coast isolates in the period between 1973 and 1981 were shown to possess two small Hind III restriction fragments of sizes 6 and 7 kb which contained *ctx* sequences. No similar finding was obtained for any of the strains isolated from other countries. This finding established the clonal nature of the strains isolated from the US Gulf Coast.

3.4.3 RFLP Analysis of the rRNA Genes

Ribotyping is a sensitive and reproducible subtyping method based on rRNA gene restriction fragment length polymorphisms (RFLP). In the bacterial genome, rRNA sequences are highly conserved and are present in multiple copies. In ribotyping, restriction digestion of chromosomal DNA is followed by Southern blotting with suitably labeled 23S or 16S rRNA probes in order to generate DNA banding patterns, which allow subtype differentiation of bacterial isolates beyond the species and subspecies levels.

RFLP analysis of the *V. cholerae* strains isolated during epidemics or pandemics exhibited great divergence in DNA sequences flanking the rRNA genes (Koblavi et al. 1990, Wachsmuth et al. 1994). The rRNA genes of (particularly) the El Tor isolates exhibited greater diversity in the RFLP analysis than in multilocus enzyme electrophoresis (MEE) analysis (see Sect. 3.4.5). It was observed that the difference in results obtained using the two techniques were due to a higher mutation rate in the DNA sequences flanking the rRNA genes than in genes encoding the enzymes studied with the MEE technique. A standardized ribotyping scheme was thus developed for epidemiological purposes. In a study of 214 human and environmental isolates from 35 countries and 14 US states collected over 60 years, the *V. cholerae* DNA was digested with the restriction enzyme *Bgl*I and the digest was hybridized with *Escherichia coli* 16S and 23S rRNA. The rRNA gene was then subjected to RFLP analysis and the resulting fragments were grouped into seven different ribotypes among classical strains and 20 ribotypes among El Tor ones. The RFLP analysis of rRNA gene is preferred to other subtyping schemes when maximum divergence is sought among *V. cholerae* isolates in epidemiological investigations.

3.4.4 DNA or Gene Probes

In the gene probe method, the target gene or genes are detected in the isolate by hybridization with a probe. The DNA or gene probe is essentially a fragment of DNA of variable length containing the target gene or an internal portion of the gene or synthetic oligonucleotides complimentary to the target gene sequence. The probes are labeled with radioisotope or with immunological markers, such as digoxigenin. The labeled probe is then denatured into single DNA strands and hybridized to the target DNA (Southern blotting), which is immobilized on a membrane or in situ.

A DNA probe that specifically detects *ctxA* gene of cholera toxin was developed by Kaper et al. (1982, 1986), and has been widely used for the epidemiological analysis of cholera cases (Yam et al. 1989, Desmarchelier and Senn 1989, Nair et al. 1988, Minami et al. 1991). This probe consisted of a 554 bp *Xba*I-*Cla*I fragment containing 94% of the gene encoding the A subunit ligated to *EcoR*I linkers and cloned into pBR325 (Kaper et al. 1982, 1986).

The practical usefulness of the DNA or gene probes resulted from the discovery and application of the polymerase chain reaction (PCR) method, which can selectively and repeatedly replicate selected segments from a complex DNA mixture (Olsvik et al. 1993). The PCR method of amplifying rare sequences from a mixture has vastly increased the sensitivity of genetic tests including the use of DNA or gene probes. DNA probes are particularly useful for identifying strains of *V. cholerae* that contain genes for cholera toxin (*ctx*) by studying hybridization. An oligonucleotide probe labeled with alkaline phosphatase was used to screen colonies grown on nonselective media inoculated with stool samples from volunteers experimentally infected with toxigenic *V. cholerae* O1. In this colony hybridization assay, the colonies were transferred directly to nylon membranes, where lysis of cells, denaturation of DNA, neutralization, and hybridization were carried out (Wright et al. 1992, Yoh et al. 1993). Any sample containing more than 10^3 *V. cholerae* bacteria per gram of stool could be successfully detected by this probe. DNA probes have been extremely useful for distinguishing strains of *V. cholerae* that contain genes (*ctx*) encoding cholera toxin from those that do not contain these genes. This is of great epidemiological importance, since many of the environmental isolates often do not contain the *ctx* genes. Another application of the DNA probes was to examine the sequence divergence within the structural gene (*ctxB*) for the CTB subunit. For this, the PCR-generated amplicons of the *ctxB* sequences were subjected to automatic sequencing to examine the divergence among them. The *ctxB* sequences were found to be highly conserved (99%) (Mekalanos et al. 1983) and could be divided into three genotypes. Genotype 1 was found in the classical and El Tor strains from the US Gulf Coast, genotype 2 in El Tor strains from Australia, and genotype 3 in strains from the seventh pandemic and the recent Latin American outbreak. These findings are of considerable epidemiological significance.

3.4.5 Multilocus Enzyme Electrophoresis (MEE) Types

MEE or zymovar analysis of many bacterial species has shown divergence among bacterial isolates of the same species (Selander et al. 1986, Wachsmuth et al. 1994). For *V. cholerae* strains, variations in electrophoretic mobility of several enzymes have been found to divide the species into multiple electrophoretic types (ETs) and to distinguish classical and El Tor strains. This technique was used to show that all strains of *V. cholerae* O1 isolated from humans in Australia belonged to the same

ET, regardless of their ability to produce CT. This technique is believed to be useful when investigating the origins of new outbreaks of disease. Zymovar analysis or MEE is applied here to investigate the genetic variation of *V. cholerae* strains and to characterize strains or groups of strains of medical and epidemiological interest. Fourteen loci were analyzed in 171 strains of non-O1 non-O139, 32 classical and 61 El Tor from America, Africa, Europe, and Asia (Freitas et al. 2002). The mean genetic diversity was 0.339. It is shown that the same O antigen (both O1 and non-O1) may be present in several genetically diverse (different zymovars) strains. Conversely, the same zymovar may contain more than one serogroup. It is confirmed that the South American epidemic strain differs from the seventh pandemic El Tor strain in locus LAP (leucyl leucyl aminopeptidase). Here it is shown that this rare allele is present in one *V. mimicus* and four non-O1 *V. cholerae*. Nontoxinogenic O1 strains from South India epidemic share zymovar 14A with the epidemic El Tor from the seventh pandemic, while another group have diverse zymovars. The sucrose-negative epidemic strains isolated in French Guiana and Brazil have the same zymovar as the current American epidemic *V. cholerae*.

3.4.6 Pulsed Field Gel Electrophoresis (PFGE)

This technique is a distinct improvement over the simpler and routinely employed agarose gel electrophoresis for the separation of DNA fragments of different lengths. In this technique, the electric fields are applied in pulses and in different directions, resulting in more efficient separation of DNA fragments of different lengths. This technique has found very useful application in the analysis of cellular chromosomes, which range from the smallest yeast chromosome ($\sim 5 \times 10^5$ bp) to the largest animal and plant chromosomes (~ 2 or 3×10^8 bp).

3.5 Growth Requirements and Characteristics

The cholera vibrios are Gram-negative, slightly curved rods whose motility depends on a single polar flagellum. Their nutritional requirements are simple. Fresh isolates are prototrophic (i.e., they grow in media containing an inorganic nitrogen source, a utilizable carbohydrate, and appropriate minerals). In adequate media, they grow rapidly, with a generation time of less than 30 min. Although they reach higher population densities when grown with vigorous aeration, they can also grow anaerobically. Vibrios are sensitive to low pH and die rapidly in solutions below pH 6.0; however, they are quite tolerant of alkaline conditions. This tolerance has been exploited in the choice of media used for their isolation and diagnosis. The growth of *V. cholerae* is stimulated by the addition of NaCl (1%). Unlike other vibrio species, *V. cholerae* is able to grow in nutrient broth without added NaCl. Most vibrios have relatively simple growth factor requirements and will grow in synthetic media with glucose as a sole source of carbon and energy. A typical growth medium utilized in recent times for specific purposes is Luria–Bertani (LB) broth containing 10 g tryptone, 5 g NaCl and 5 g yeast extract per liter adjusted to pH 7.4. Cultures are usually grown with shaking at 37 °C (Yoon and Mekalanos 2008). In some cases, cells are grown in LB broth, pH 6.5, 30 °C, without shaking for 12–16 h for toxin-coregulated pilus (TCP) expression. Growth conditions favorable for toxin expression will be discussed in Chap. 9.

V. cholerae is facultatively anaerobic, asporogenous, and capable of respiratory and fermentative metabolism. Further, *V. cholerae* is oxidase-positive, reduces nitrate, and is motile via a single polar sheathed flagellum. The O group 1 cholera vibrios almost always fall into the Heiberg I fermentation pattern; that is, they ferment sucrose and mannose but not arabinose, and they produce acid but not gas. *V. cholerae* also possesses lysine and ornithine decarboxylase, but not arginine dihydrolase. Freshly isolated agar-grown vibrios of the El Tor biotype, in contrast to classical *V. cholerae*, produce a cell-associated mannose-sensitive hemagglutinin that is active on chicken erythrocytes. This activity is readily detected in a rapid slide test.

3.6 Ultrastructure of the Cell

V. cholerae bacteria are slightly curved (often comma-shaped) rods of average diameter 0.55 μm and length 1.8 μm. The bacterium possesses one single polar flagellum as the organ of motility. The flagellum has a core covered by a sheath (Figs. 3.1 and 3.2d). The diameters of the core and the sheath are about 150 and 300 Å, respectively (Das and Chatterjee 1966a, Sur et al. 1974). The single polar flagellum lies attached to a structure in the bacterial protoplasm called the basal granule, which is about 650 Å in size (Fig. 3.1). The bacterial protoplasm is bounded by a membrane called the plasma membrane. The plasma membrane exhibits a trilamellar structure and has an overall thickness of about 75 Å. On the outer side of the plasma membrane there is the periplasmic space of thickness 50–100 Å. The periplasmic space is again bounded by another layer of membranous structure, the cell wall, which consists of a thin layer of peptidoglycan covered by another trilamellar structure at the outermost periphery of the cell. The trilamellar structure of the cell wall is about 100 Å thick (Fig. 3.2c). Several different kinds of fimbriae or pili have been observed to radiate from the surface of the bacterial cell, which will be discussed separately later. The protoplasm of the bacterial cell consists of a considerably electron-transparent central zone (Fig. 3.2a, b), the nucleus, which contains intertwined fibrils of DNA (Fig. 3.2a) and is surrounded by an electron-dense region of the cytoplasm. Often densely packed ribosomal granules can be discerned in the cytoplasm. The electron density of the cytoplasm is mostly due to the presence of plenty of ribosomal particles. The nuclear zone is often spanned by bridges of cytoplasmic material. Electron microscopy revealed that the mechanism of division of *V. cholerae* cells is similar to that of any other Gram-negative bacterium, i.e., elongation to double the original length and then division through the simple pinching-off process (Fig. 3.3).

3.6.1 Pili (Fimbriae)

The presence of different types of pili on the surfaces of *V. cholerae* cells has been recorded by electron microscopy, and not all of these have been characterized in terms of function (Fig. 3.4). The presence of a small number of pili of diameter 60–80 Å on El Tor vibrios was recorded as early as 1964 (Fig. 3.5) (Barua and Chatterjee 1964). Subsequently, pili were detected on all of the mannose-sensitive hemagglutinating strains of *V. cholerae* El Tor, although the number of pili per cell was quite small. A direct correlation between the presence of pili and pellicle formation on the surface of static aerobic liquid cultures was also recorded (Adhikari and Chatterjee 1969). A completely different type of pili about 70 Å in diameter and laterally associated in bundles was detected by electron microscopy and found to be a colonization factor of *V. cholerae* O1 (Fig. 3.6) (Craig et al. 2003). A transposon,

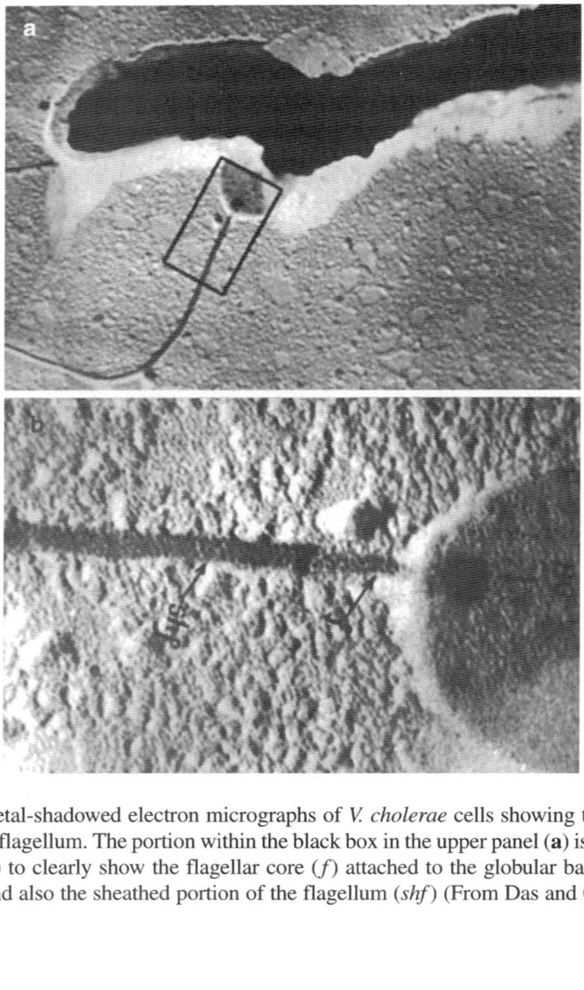

Fig. 3.1 a–b Metal-shadowed electron micrographs of *V. cholerae* cells showing the basal granule and the sheathed flagellum. The portion within the black box in the upper panel (**a**) is enlarged further at the bottom (**b**) to clearly show the flagellar core (*f*) attached to the globular basal granule (*BG*) within the cell and also the sheathed portion of the flagellum (*shf*) (From Das and Chatterjee 1966)

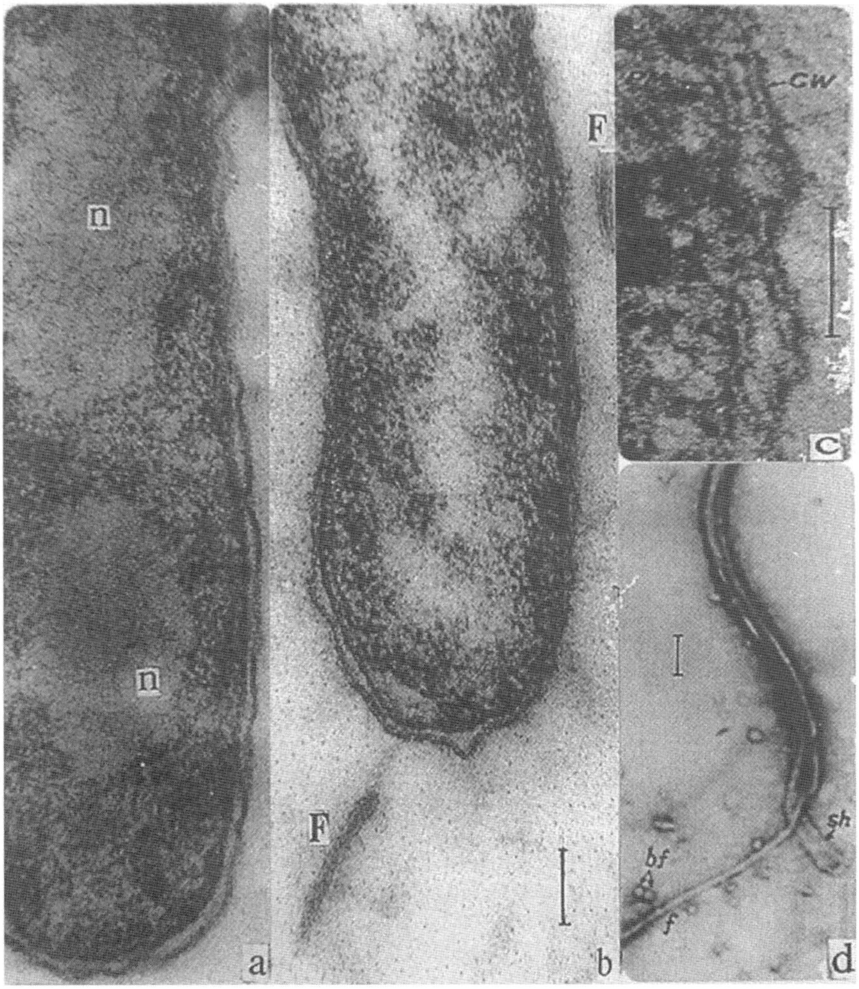

Fig. 3.2 a–d Electron micrograph of thin sections of *V. cholerae* cells. **a** Portion of a cell showing the electron-transparent nuclear zones (*n*) containing intertwined fibrils of DNA; **b** the sheathed flagellum (*F*) in thin section; **c** trilamellar structures of the plasma membrane (*PM*) and the cell wall (*CW*) in thin section; **d** negatively stained picture of lysed flagellum clearly showing the core (*f*), the released portion of the sheath (*sh*), and the broken flagellar cores (*bf*) (From Das and Chatterjee 1966, Chatterjee 1990)

Tn*phoA*, was used to identify bacterial virulence factors. When Tn*phoA* inserts in-frame into a gene encoding a secreted or membrane-spanning protein, a fusion protein is formed that expresses alkaline phosphatase activity. A Tn*phoA* mutant was found in colonies expressing alkaline phosphatase activity which exhibited greatly decreased colonization ability when tested in an infant mouse model. This mutant was defective in producing a pilus associated with hemagglutination in the

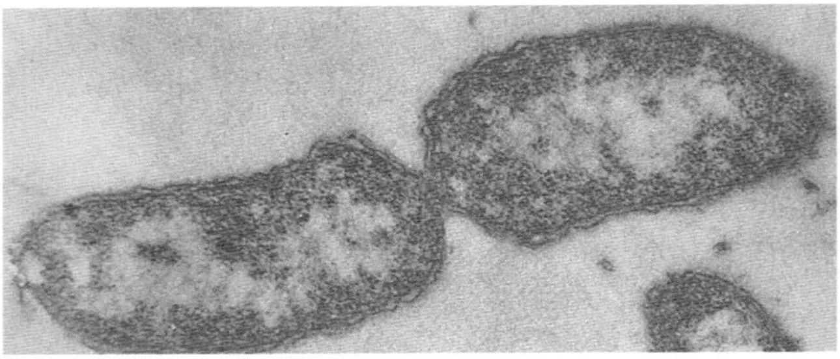

Fig. 3.3 Electron micrograph of a dividing cell of *V. cholerae* in thin section (From Chatterjee, unpublished)

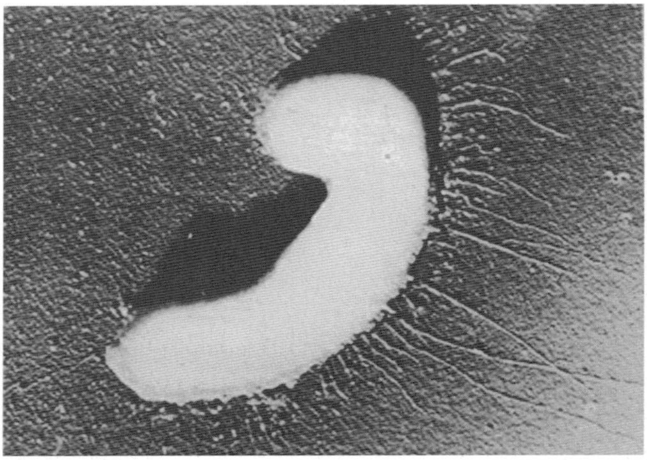

Fig. 3.4 Metal-shadowed electron micrograph of a comma-shaped *V. cholerae* cell, showing several fairly straight atypical pili radiating from its convex surface (From Adhikari and Chatterjee 1969, Chatterjee 1990)

presence of fucose. The expression of this type of pilus was found to be associated with the expression of cholera toxin, and hence this pilus was named toxin-coregulated pilus (TCP). It was observed that multiple types of pili can be expressed by the same strain of *V. cholerae*. Further aspects of the role of TCP in the colonization process and acquisition of toxin gene (*ctx*) by the vibrios will be discussed later.

3.7 Cholera Bacteriophages

A brief introduction to the different phages infecting *V. cholerae* cells is useful, since phages often reveal a lot about the lives and activities of bacteria. Mukerjee (Mukerjee 1963a, b; Mukerjee and Takeya 1974) were the first to classify the phages infecting *V. cholerae* classical strains into four different groups (Groups I–IV) on the

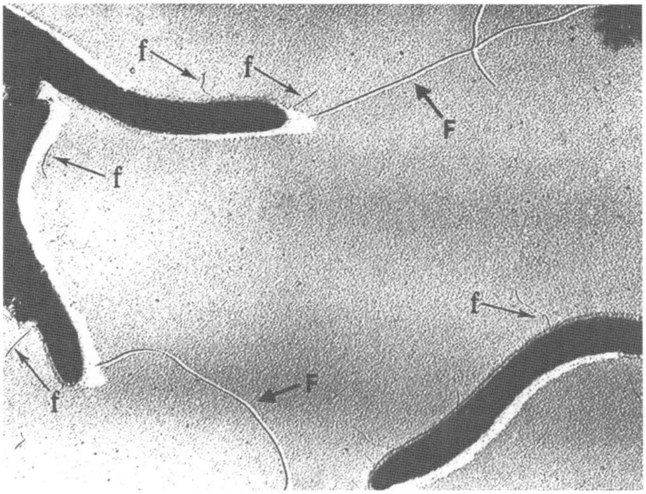

Fig. 3.5 Metal-shadowed electron micrograph of several *V. cholerae* El Tor cells bearing a very small number of pili per cell. Note that the pili (*f*) are much thinner than the flagella (*F*), and are much shorter and straight (From Adhikari and Chatterjee 1969)

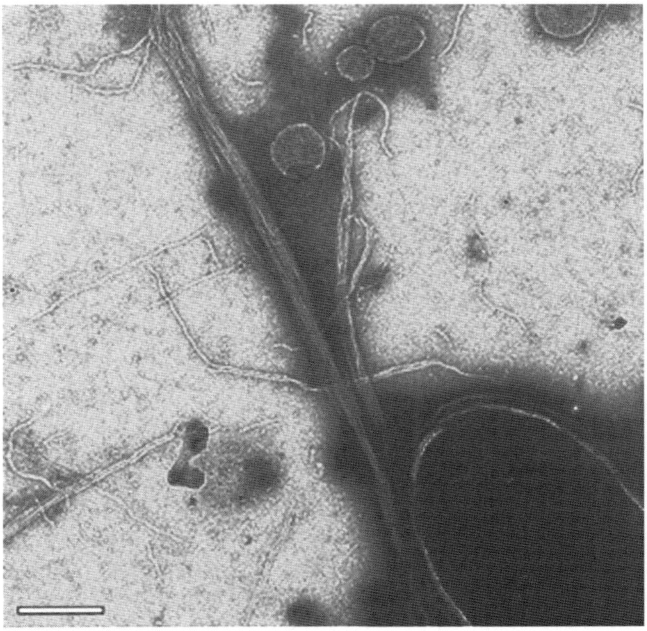

Fig. 3.6 Electron micrograph (negative staining) of toxin-coregulated pili (TCP) as a bunch of fibrils; the bar represents 200 μm. (Photograph obtained through the kind courtesy of M. Lim and L. Craig, Simon Fraser University, Burnaby, BC, Canada)

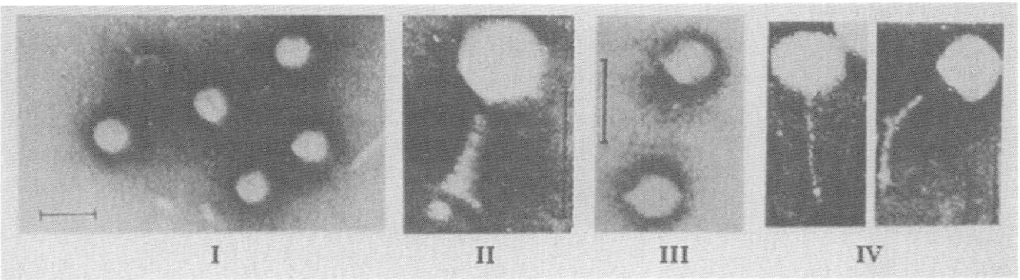

Fig. 3.7 Electron micrograph (negative staining) of Mukerjee's four groups (I–IV) of cholera bacteriophages. Note that phages belonging to any particular group have a distinct morphology (From Chatterjee et al. 1965, Das and Chatterjee 1966)

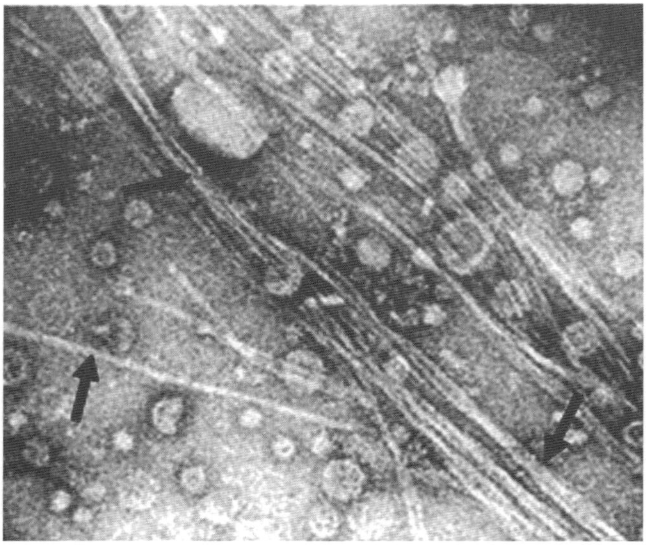

Fig. 3.8 Electron micrograph (negative staining) of filamentous CTXφ phages (*arrows*). (Figure 2 of Waldor and Mekalanos (1996), reproduced with kind permission from J.J. Mekalanos and also from the Editor of *Science*)

basis of their serological properties, plaque morphology, thermal death points, etc. Electron microscopy revealed significant characteristics of these phages, and the phages belonging to any of Mukerjee's four groups had distinct morphologies (Fig. 3.7) (Chatterjee et al. 1965, Das and Chatterjee 1966b,Maiti and Chatterjee 1971, Chatterjee and Maiti 1984). This was practically the first indication of the fact that electron microscopic morphology can be used as an important parameter in phage or viral taxonomy, which was subsequently utilized extensively for the classification of viruses by the International Committee for the Taxonomy of Viruses (ICTV). Mukerjee's classification was found to be very useful for differentiating the two biotypes, classical and El Tor, of *V. cholerae* strains, and for tracing the routes of the spread of infection (by classical strains in particular) during an epidemic. Subsequent attempts to classify El Tor phages and type the El Tor strains did not find success. Takeya (Takeya and Shimodori 1963, Mukerjee and Takeya 1974) made a notable contribution to the discovery and properties of the lysogenic cholera phage, Kappa. The taxonomic classification and the properties of vibriophages have been reviewed by Ackermann (Ackermann and Eisenstark 1974, Ackermann and DuBow 1987a, b, Ackermann 1992) and by Chatterjee and Maiti (1984).

Subsequently, several filamentous vibriophages were discovered. Of these, the one named CTXφ (Fig. 3.8) has been found to play a very significant role in the life cycle of *V. cholerae* strains (Waldor and Mekalanos 1996). The genome of this phage is a single-stranded DNA of size 6.9 kb, which contains several *V. cholerae* toxin genes, including the genes (*ctxA* and *ctxB*) of the cholera toxin (CT). The *V. cholerae* strain acquires these toxin genes only after it is infected by the phage CTXφ and the entire phage genome is integrated into the *V. cholerae* genome. Under some circumstances, the phage genome can also stay as a plasmid without being integrated into the bacterial genome. Toxin-coregulated pili (TCP) (Fig. 3.6), which occur in bunches, serve as the receptor for CTXφ through which the phage genome is inserted into the bacterial cell. Further details of the role of CTXφ in the toxinogenicity and other aspects of *V. cholerae* will be discussed in Chaps. 8 and 9 of this book.

3.8 Genetics of *V. cholerae*: An Outline of Some Relevant Areas

The genetics of *V. cholerae* did not receive any significant attention until the mid-1980s (Guidolin and Manning 1987). The few exceptions included the discovery of the conjugal system in *V. cholerae* by Bhaskaran (1958). It was shown that the conjugation in *V. cholerae* was mediated by the P factor, which, unlike the F factor of *E. coli*, could not integrate into the chromosome, and hence could not induce Hfr donors (Datta et al. 1973). Few plasmids were reported to be present in *V. cholerae*. The Bhaskaran strain 162 was shown to contain two plasmids besides P (Datta et al. 1973). It was also reported that there was a gradual increase in the incidence of R-factor-mediated drug resistance, and that this was almost exclusively due to plasmids of the *IncC* incompatibility group (Prescott et al. 1968). Parker et al. (1979) used P factor crosses to obtain a linear linkage of the *V. cholerae* classical strain 162, containing both selected and unselected markers.

V. cholerae DNA (classical, Ogawa 154) exhibited a melting temperature (T_m) of 89 °C in 1× SSC, pH 7.1, an average G + C content of 48.0 ± 0.1 (see Table 3.4 for recent values of the two individual chromosomes of some vibrio strains discovered later), and a unimodal distribution of G-C base pairs (Ghosh et al. 1976). Several studies showed the activities of different repair genes in *V. cholerae* DNA: (i) excision repair mediated by uvr-endonuclease enzymes (Banerjee and Chatterjee 1981), (ii) photorepair mediated by the enzyme photolyase and gene *phr* (Chanda and Chatterjee 1976, Samad et al. 1987) and *recA* (Goldberg and Mekalanos 1986) mediated induction of "SOS" repair genes (Rahman et al. 1993) leading to (a) filamentation (Fig. 3.9) of the cells (Raychaudhuri et al. 1970, Chatterjee and Maiti 1973), (b) prophage induction (Mandal and Chatterjee 1987), (c) mutation (Banerjee and Chatterjee 1984, Bhattacharyya et al. 1991) and (d) Weigle reactivation and mutagenesis (Bhattacharyya et al. 1991). *V. cholerae* DNA was also shown to contain functional genes, *ada* and *alk*, leading to adaptive repair response against both alkylating and oxidative DNA damage (Basak and Chatterjee 1994, Basak et al. 1992), and also the gene, *dam*, involved in DNA mismatch repair processes (Bandyopadhyay and Das 1994). In fact, *V. cholerae* has genes encoding several DNA-repair and DNA-damage response pathways, including homologs of many of the genes involved in the SOS response in *E. coli* (Heidelberg et al. 2000).

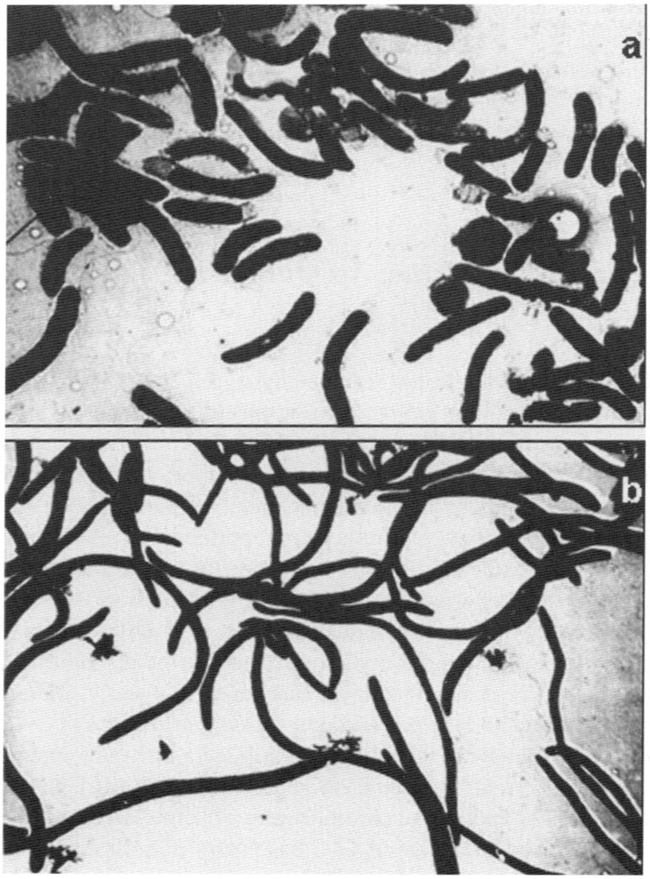

Fig. 3.9 a–b *V. cholerae* cells underwent filamentation (**b**) subsequent to SOS repair induced by the action of the synthetic nitrofuran drug, furazolidone. The normal or untreated cells (**a**) are shown on the *top*. Metal-shadowed electron micrographs are shown (Ray-Chaudhuri et al. 1970)

The genetics of the virulence of *V. cholerae* formed the next important area of research. The major virulence-associated genes that encode colonization factors and cholera toxin (CT) are parts of larger genetic elements composed of clusters of genes (Pearson et al. 1993). Although the manifestation of diarrhea is the result of the production of CT encoded by *ctxAB* genes, the pathogenesis of cholera is dependent on the synergistic action of a number of other genes, including the genes for one or more colonization factors (Kaper et al. 1995). There are at least two regions of the *V. cholerae* chromosome in which genes encoding virulence factors are clustered (Trucksis et al. 1993). These include the CTX element, which was subsequently shown to be the genome of a filamentous bacteriophage (CTXφ) (Waldor and Mekalanos 1996) and the TCP-accessory colonization factor (Acf) gene cluster referred to as the TCP pathogenicity island (Kovach et al. 1996). The characteristics of the pathogenicity island include the presence of groups of virulence genes, a transposase gene, specific (att-like) attachment sites flanking each end of the island, and an integrase with homology to a phage integrase gene (Faruque et al. 1998).

The toxinogenic *V. cholerae* carries one or more copies of CT genes (*ctxAB*) (Mekalanos 1983, Faruque et al. 1998, Basu et al. 1998, Mukhopadhyay et al. 1998). The A- and B-subunits of CT are encoded by two separate but overlapping ORFs. Besides, there are genes for several other types of toxin, including the (i) zonula occludens toxin (Zot), which increases the permeability of the small intestinal mucosa (Fasano et al. 1991), and (ii) the accessory cholera enterotoxin (Ace), which is capable of inducing fluid accumulation in rabbit ligated ileal loops (Trucksis et al. 1993).

The next significant contributions to the genetics of *V. cholerae* were that (i) its genome was shown to contain two chromosomes (Trucksis et al. 1998), a larger (replicon 1) and a smaller (replicon 2) one, of characteristic features given in Table 3.4, and (ii) the complete genome of the *V. cholerae* El Tor strain N16961 was sequenced (Heidelberg et al. 2000). The following features of the distribution of genes between the two chromosomes in El Tor strain N16961 were recorded. Most genes required for growth and viability are located on chromosome I. There is pronounced asymmetry in the distribution of genes known to be essential for growth and virulence between the two chromosomes. Several repair genes have been found to be located on chromosome II, indicating that this chromosome is required for full DNA repair capability, although significantly more genes encoding DNA replication and repair are encoded by chromosome I (Heidelberg et al. 2000). Also, more genes encoding DNA transcription, translation, cell wall biosynthesis and a variety of central catabolic and biosynthetic pathways are encoded by chromosome I. Similarly, most genes essential in bacterial pathogenicity (those encoding the toxin-coregulated pilus, cholera toxin, LPS, and the extracellular protein secretion machinery) are also located on chromosome I. In contrast, chromosome II contains a larger fraction of the hypothetical genes and genes of unknown function. The small chromosome also carries a gene capture system (the integron island) and host "addiction" genes that are typically found on plasmids, thereby indicating that the small chromosome may have originally been a megaplasmid that

Table 3.4 Some general features of the genomes of three *Vibrio* species (updated version of the Table 1 of Chatterjee and Chaudhuri 2004)

General features → / *Vibrio* spp. ↓		Size (bp)	(G + C) %	No. of proteins	ORF size[a]	Percent coding[b]	Hypothetical proteins[c]	No. of rRNA operons	No. of tRNAs
V. cholerae El Tor N16961	ChrI	2961149	47.7	2742	952	87	941	8	94
	ChrII	1072315	46.9	1093	918	84	570	0	4
V. cholerae O395[d]	ChrI	1108250	46.9	1133	847	86	560	0	4
	ChrII	3024069	47.8	2742	972	88	860	8	92
V. parahaemolyticus	ChrI	3288558	45.4	3080	926.9	86	1,076	10	112
	ChrII	1877212	45.4	1752	931.3	86	770	1	14
V. vulnificus Y016	ChrI	3354505	46.4	3259	912	87	1,089	8	100
	ChrII	1857073	47.2	1696	989	89	710	1	12

[a]Average size of all ORFs present in the complete genome

[b]Percentage coding = [Σ size of each ORF (bp)/genome size (bp)] × 100

[c]ORFs coding for proteins that are homologous to some proteins present in other organisms, but their functions remain unknown

[d]Data obtained from NCBI genome database at http: //www.ncbi.nlm.nih.gov/genomes/lproks.cgi

was possibly captured by an ancestral *Vibrio* species (Heidelberg et al. 2000). The origin of chromosome II, however, continues to be the subject of further investigations. Recently, the origins of replication of the two chromosomes have been defined and novel replicon-specific requirements for each chromosome as well as factors required for the replication of both chromosomes are found (Egan and Waldor 2003). The genetics of virulence and its regulation will be discussed in more detail in Chaps. 8 and 9 of this book.

Chapter 4
Endotoxin of *Vibrio cholerae*: Physical and Chemical Characterization

Abstract The lipopolysaccharide (LPS)—the endotoxin of *Vibrio cholerae*—plays an important role in eliciting the antibacterial immune response of the host and in classifying the vibrios into some 200 or more serogroups. This chapter presents an account of our current knowledge of the physical and chemical characteristics of the three constituents, lipid A, core polysaccharide (core-PS) and O-antigen polysaccharide (O-PS), of the LPS of *V. cholerae* of different serogroups, including the disease-causing ones, O1 and O139. The structure and occurrence of the capsular polysaccharide (CPS) on *V. cholerae* O139 are discussed. Similarities and dissimilarities between the structures of LPS of different serogroups, and particularly between O22 and O139, are analyzed with a view to learning their roles in the causation of the epidemic form of the disease through the avoidance of the host's defense mechanism and in the evolution of newer pathogenic strains in future. An overview of emerging research trends involving the use of immunogens prepared from synthetic oligosaccharides that mimic terminal epitopes of the O-PS of *V. cholerae* O1 in the development of a conjugate anticholera vaccine is also provided.

4.1 Introduction

V. cholerae is the causative organism of the disease cholera, which is both an endemic and epidemic disease, and is practically the only bacterial pandemic disease known so far. Although *V. cholerae* LPS has been investigated since the late 1960s, research on these molecules gained momentum in the late 1980s and has been progressing at an accelerating pace since the emergence of the non-O1 strain, *V. cholerae* O139, which caused an outbreak of the epidemic in the Indian subcontinent around 1992. An account of our current knowledge about the *V. cholerae* LPS in relation to its role in the causation and spread of as well as the defense against cholera is presented in this and the following two chapters of this book. This chapter presents the physical and chemical characterization of the different constituents (lipid A, Kdo, core-PS, and O-PS) of the LPS molecules belonging to *V. cholerae* of different serogroups, serotypes, and biotypes, which will form the background material when we explore

K. Chaudhuri and S.N. Chatterjee, *Cholera Toxins*.

topics such as the genetics of the biosynthesis and the biological functions of *V. cholerae* LPS in subsequent chapters (Chaps. 5 and 6) of this book. This chapter also presents a brief account of the physical and chemical characterization of the capsular polysaccharide (CPS) produced by many strains of *V. cholerae* under different conditions, in order to elucidate its differences from the LPS.

4.2 LPS of *V. cholerae*: Site of Occurrence

LPS was extracted from the surfaces of *V. cholerae* cells by the phenol-water method of Westphal et al. (1952). Aggregated, thin, sheet-like structures of LPS alongside the vibrios could be seen (Fig. 4.1). After extensive purification, the LPS preparation presented much smaller, thin, sheet-like structures of elongated or rounded shapes (Fig. 4.2). The purified LPS preparation was antigenically active, as detected by the spectrophotometric estimation of protein in the precipitin (collected by centrifugation) formed on incubation with O-antiserum prepared against heat-killed vibrios, and acted as receptor of cholera phage φ149 (Maiti et al. 1977, Chatterjee and Maiti 1984). Electron microscopy revealed the presence of a sheath on the flagellum of the vibrios. The sheath could be separated from the core of the flagellum and had a thickness that was almost identical to that of the trilamellar structure of the cell wall. Immunoelectron microscopy of the vibrios revealed the presence of O-antigen and hence LPS (Fig. 4.4; see legend for explanatory notes) on the flagellar sheath and vibrio surface (Fuerst and Perry 1988).

4.3 Extracellular LPS

Actively growing cells of *V. cholerae* O1 were shown to exhibit a novel excretory mechanism (Fig. 4.3), which included: (A) bulging out of the cell wall outer membrane in localized areas; (B) the formation of a constricted neck of the bulged-out portion, and; (C) the pinching off of the bulged-out portion into the external medium in the form of membrane vesicles (Chatterjee and Das 1966, 1967, Chatterjee et al. 1974, Chatterjee and Sur 1974). This was subsequently found to be a mechanism of release of membrane vesicles by Gram-negative bacteria in general (Li et al. 1998, Kuehn and Kesty 2005). The membrane vesicles released by *V. cholerae* were isolated from the culture filtrate, photographed by electron microscopy, and usually exhibited sizes of between 400 and 600 Å. These extracellular vesicles of *V. cholerae* were shown to contain LPS and somatic antigens (Chatterjee et al. 1974) besides other macromolecular constituents presumably derived from the periplasmic space of the bacterial cells. The membrane vesicles of other Gram-negative bacteria (Li et al. 1998, Kadurugamuwa et al. 1996) were shown to contain endotoxins (Devoe and Gilchrist 1973), DNA-binding proteins (Dorward and Garon 1989) and several other chemicals, and to exhibit several functions, including

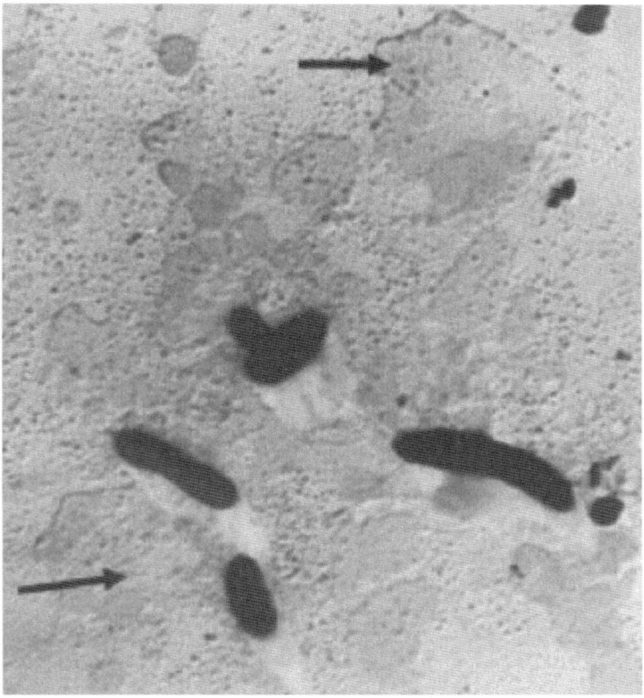

Fig. 4.1 Electron micrograph (metal-shadowed) of the *V. cholerae* O1 cells immediately after treatment with phenol (Westphal et al. 1952). The LPSs released appeared as aggregated, thin, sheet-like structures of varying sizes and shapes. (From Chatterjee et al. 1974, Chatterjee and Chaudhuri 2003)

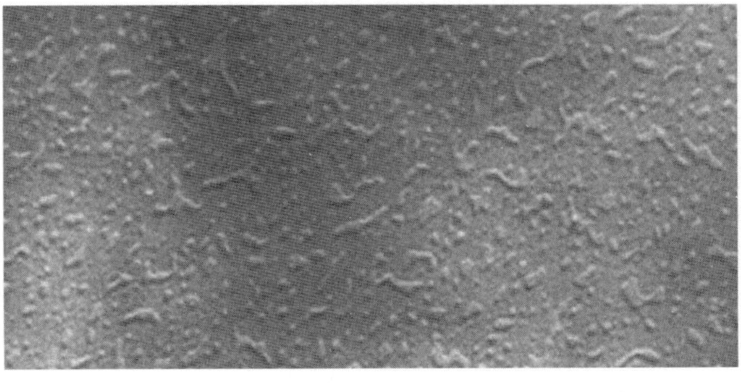

Fig. 4.2 Electron micrograph (metal-shadowed) of isolated *V. cholerae* LPS after extensive dialysis and purification. (From Chatterjee et al. 1974, Chatterjee and Chaudhuri 2003)

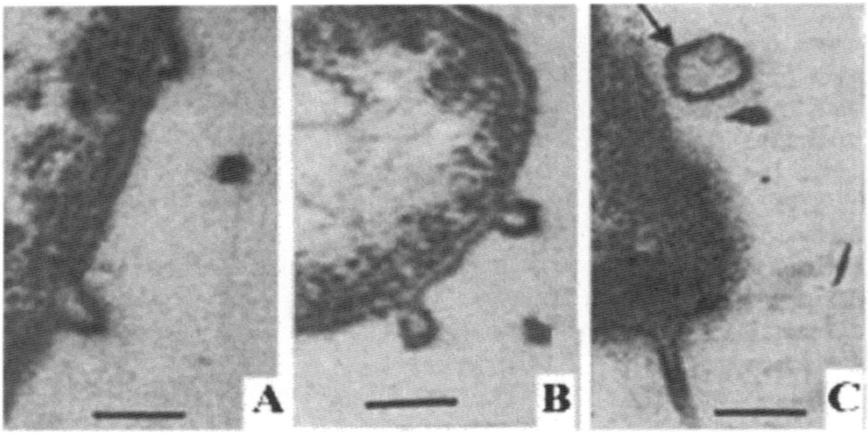

Fig. 4.3 Ultrastructural expressions (electron micrograph of ultrathin sections of actively growing cells) of the stages (A, B and C) in the formation of extracellular membrane vesicles as a novel secretory mechanism of *V. cholerae* cells. (From Chatterjee and Das 1966, 1967, Chatterjee and Chaudhuri 2003)

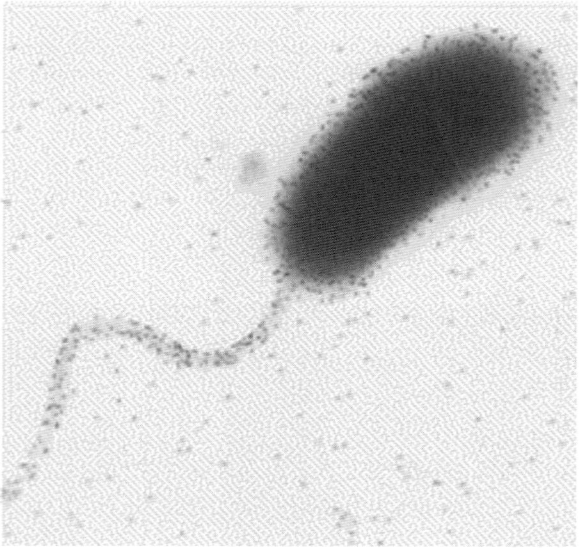

Fig. 4.4 *V. cholerae* O1, Ogawa serotype, immunolabeled with anti-B-determinant monoclonal antibody complexed with protein A and colloidal gold. The distribution of the gold particles (*black dots*) over the surface and the sheathed flagellum of the vibrio represent the distribution of B-antigen associated with the O-antigen of LPS. (Photograph obtained through the kind courtesy of U.H. Stroeher)

killing other bacteria. A detailed structural and functional study of the extracellularly released membrane vesicles of *V. cholerae* containing LPS and other materials has not been reported so far.

4.4 Chemical Composition of Lipid A

On acid hydrolysis (1% acetic acid, 100 °C, 2 h), the LPS is normally split off into lipid A and PS moieties. The lipid A fraction can be recovered as a white waxy solid that is soluble in chloroform and insoluble in water and constitutes about 27–30% by weight of *V. cholerae* LPS (Raziuddin and Kawasaki 1976, Raziuddin 1977). Like *Salmonella* and other Gram-negative bacterial species, the lipid A of *V. cholerae* was found to consist of a sugar backbone linked with fatty acids. D-Glucosamine was found to be the only amino sugar that was present in the lipid A of *V. cholerae* O1 strains irrespective of serotype (Ogawa or Inaba) or biotype (classical or El Tor) (Raziuddin 1977, Kabir 1982). Similarly, the lipid A of many non-O1 vibrios studied, including *V. cholerae* O139, and of some other vibrio species (*V. metchnikovii, V. parahaemolyticus*, etc.), contained only glucosamine as a component sugar, suggesting that the lipid A backbone is a glucosamine disaccharide, as in many Gram-negative bacterial LPS (Hisatsune et al. 1993a, Rietschel et al. 1973, Denecke and Colwell 1973). In *V. cholerae* lipid A, the ester-bound phosphate group in the D-glucosamine backbone appears to be unsubstituted, like in *Escherichia coli* (Hase and Rietschel 1976, Rosner et al. 1979), and the glycosidic phosphate group is substituted with phosphoryl ethanolamine. It differs from *Salmonella* lipid A, where the ester bound phosphate group is substituted partially by an L-4-amino-arabinosyl residue (Muhlardt et al. 1977).

The fatty acid composition of *V. cholerae* LPS appeared to be complex. Some 27 fatty acids were detected in the LPS of *V. cholerae* 569B (Inaba, O1), of which five were present in trace amounts and three had structures that were not conclusively proven (Armstrong and Redmond 1974), and this finding contrasted interestingly with the relatively simple range (eight fatty acids only) present in the LPSs of some related vibrios (Rietschel et al. 1973). Further, the most abundant hydroxylated fatty acid was 3-hydroxy lauric acid (3-OH-$C_{12:0}$), rather than the more usual 3-hydroxy myristic acid (3-OH-$C_{14:0}$) (Armstrong and Redmond 1974). Hisatsune et al. (1979) studied the fatty acid composition of LPSs of *V. cholerae* 35A3 (Inaba), NIH90 (Ogawa) and 4715 (Non-O1). The three strains presented similar fatty acid compositions, with the exceptions of (i) a minor deviation in the relative proportions of the component fatty acids, and (ii) the presence of odd-numbered normal and 3-hydroxy fatty acids, particularly in the LPS from strain 35A3 (Inaba). Although $C_{12h:0}$ was previously reported to be the most abundant 3-hydroxy fatty acid in LPS from both 569B (Inaba) and El Tor (Inaba) strains (Raziuddin and Kawasaki 1976, Armstrong and Redmond 1974), the quantity of $C_{14h:0}$ was slightly greater than that of $C_{12h:0}$ in the three strains studied by them. Raziuddin and Kawasaki (1976) studied only the Inaba-type *V. cholerae* strains and obtained identical results, with the exception of minor quantitative deviations. According to Broady et al. (1981), the LPSs from the *V. cholerae* strains 569B (Inaba), 162 (Ogawa), H-11 (NAG, non-O1) and 95R (rough) presented almost identical fatty acid compositions. The LPSs contained smaller amounts of hexadecenoic ($C_{16:1}$), stearic ($C_{18:0}$) and oleic ($C_{18:1}$) acids which were ester-bound. Further, the compositions and the distributions of the fatty acids in the lipid A of O139 LPS were very similar to those of the lipid A of O1 *V. cholerae* strain NIH 41 (Ogawa) LPS (Hisatsune et al. 1993a). It thus appeared that, barring some minor quantitative variations in the contents of individual fatty acids, the underlying general pattern in the fatty acid composition of LPS of different strains of *V. cholerae* was that (i) $C_{14:0}$ and $C_{16:0}$ were the main non-hydroxy fatty acids, and (ii) $C_{12h:0}$ and $C_{14h:0}$ were the main 3-hydroxy fatty acids present. Some branched-chain fatty acids were occasionally found only in trace quantities, and the presence of 2-hydroxy fatty acids remains to be confirmed.

Rietschel (1976) studied the absolute configuration of 3-hydroxy fatty acids present in LPSs from various bacterial groups, including *V. cholerae*, and concluded that for all LPSs, the 3-hydroxy acids, regardless of chain lengths, branching, 3-O-substitution or type of linkage, possessed the D-configuration. This was confirmed by the findings of Broady et al. (1981) on the *V. cholerae* strains belonging to different biotypes, serotypes, and serogroups. Further, since the D-3-hydroxy fatty acids were not found in other cell wall lipids, they could be considered markers of lipid A.

With respect to the nature of the linkages, Raziuddin (1977) found that approximately equal amounts of fatty acids, $C_{16:0}$, $C_{18:1}$, and 3-OH-lauric acid (3-OH $C_{12:0}$), were involved in ester linkages, while 3-OH myristic acid (3-OH-$C_{14:0}$) was the only amide-linked fatty acid. That 3-hydroxy myristic acid (3-OH-$C_{14:0}$) was the only amide-linked fatty acid detected in various strains, including the O139 of *V. cholerae* and also in some related vibrios, was confirmed by different investigators (Rietschel et al. 1973, Armstrong and Redmond 1974, Raziuddin and Kawasaki 1976, Raziuddin 1977, Hisatsune et al. 1979, Broady et al. 1981).

4.5 Chemical Structure of Lipid A

The backbone of the lipid A of *V. cholerae* O1 is composed of a (1→6)-β-D-glucosamine disaccharide with an ester-bound phosphate group at the C-4 position of the upstream glucosamine residue and a phosphate residue bound to the glycosidic hydroxyl group at the C-1 position of the downstream glucosamine one (Broady et al. 1981). The phosphate group at the C-1 position of the downstream glucosamine residue is substituted by the pyrophosphoryl ethanolamine (P-P-Etn) group. The backbone structure is thus represented by the formula (Broady et al. 1981):

P-4-G1cN(1→6)-β–D-Glcn-1-P-P-Etn

This backbone structure is well conserved between *V. cholerae* O1 and O139; the only difference lies in the fatty acid moieties attached to the backbone through ester

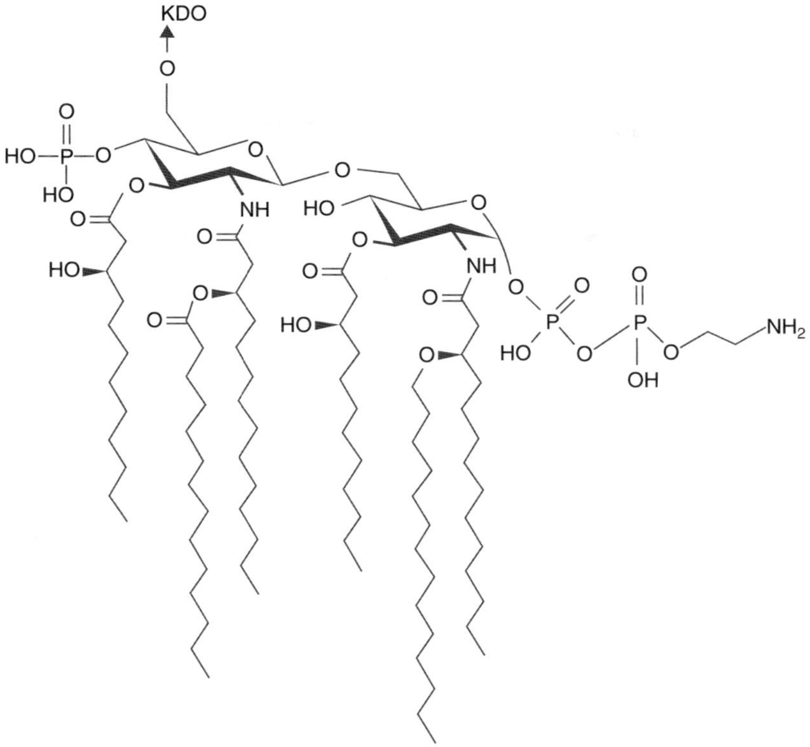

Fig. 4.5 Chemical structure of lipid A of *V. cholerae* O1 (Broady et al. 1981, Villeneuve et al. 2000). (From Chatterjee and Chaudhuri 2003)

linkage. The overall structures of the lipid A of *V. cholerae* O1 and O139 are shown in Figs. 4.5 and 4.6, respectively. Initially there was some doubt about the exact positions of the ester-linked phosphate group and the individual ester-bound fatty acid residues (Broady et al. 1981) in the lipid A of *V. cholerae* O1, but these were settled subsequently (Villeneuve et al. 2000). With respect to *V. cholerae* O139, the complete structure of lipid A was arranged by Boutonnier et al. (2001) following the works of Kabir (1982) and Wilkinson (1996).

4.6 3-Deoxy-D-Manno-Octolusonic Acid (Kdo)

Kdo is a characteristic constituent of Gram-negative bacterial LPS (Rietschel et al. 1984). Kdo connects the core-PS and lipid A (Rietschel et al. 1984). The undetectability of Kdo in LPS from the genus *Vibrio* was first pointed out by Jackson and Redmond (1971), then by Jann et al. (1973) for O1 *V. cholerae* 569B (Inaba), and then by Hisatsune et al. (1978) for non-O1 *V. cholerae* serogroup O3 and for the

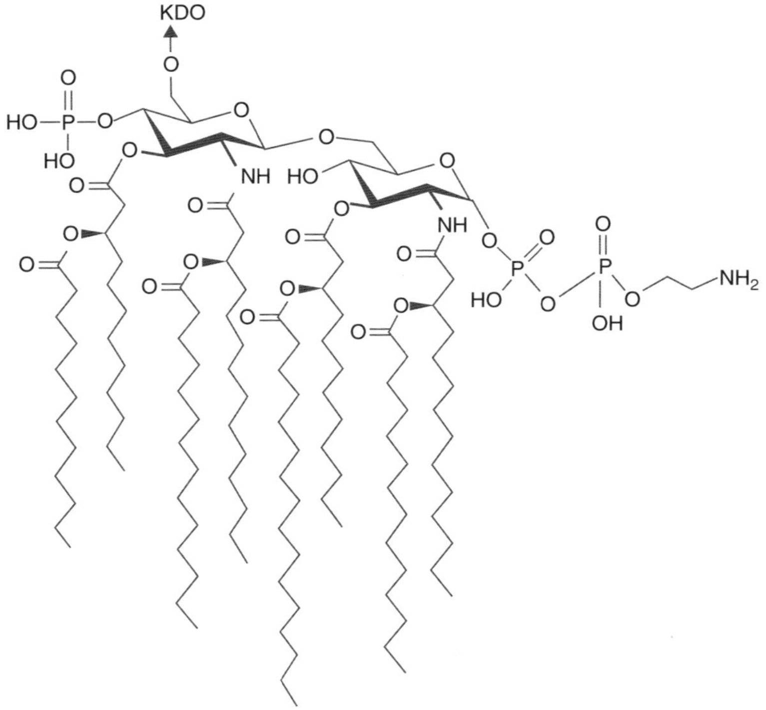

Fig. 4.6 Chemical structure of lipid A of *V. cholerae* O139 (Boutonnier et al. 2001). (From Chatterjee and Chaudhuri 2003)

whole family of *Vibrionaceae* (Hisatsune et al. 1982) by the periodate-thiobarbituric acid test (Weissbach and Hurwitz 1959) with conventional conditions of hydrolysis. The presence of Kdo-5-phosphate in the hydrolysates of LPSs of *V. cholerae* 95R and 569B, obtained by using strong acid, was first reported by Brade (1985). Caroff et al. (1987) also proposed that Kdo phosphate was present in LPSs of *V. cholerae* strains of serotypes Inaba and Ogawa (biotype El Tor) from studies of their reactivity in the periodate-thiobarbituric acid test after treatment with hydrofluoric acid. Strong acid hydrolysis released phosphorylated Kdo from the LPS of *V. cholerae*, which could be identified by gas liquid chromatography (GLC) and mass spectrometry (MS) after reduction and permethylation. It was subsequently established that LPS of *V. cholerae* contained one residue of Kdo that was phosphorylated (Brade 1985, Kondo et al. 1988, 1992) at the C-4 position and substituted with heptose from the outer core region at the C-5 position (Kondo et al. 1992). The reason for the detection of Kdo-5-P by GLC/MS (Brade 1985) could have been the migration of the P group from the C-4 to the C-5 position under the acidic conditions used during its isolation (Vinogradov et al. 1995). The undetectability of Kdo in LPS after conventional hydrolysis and the occurrences of (i) phosphorylated Kdo in strong acid hydrolysates and (ii) only one molecule of Kdo-4P in the core region are now considered taxonomic characteristics of *V. cholerae* (Kondo et al. 1992).

4.7 Core-PS: Chemical Constituents

4.7.1 O1 Vibrios

The chemical composition of the core-PS from *V. cholerae* O1 revealed that glucose, heptose, fructose, and glucosamine were invariably present (Hisatsune et al. 1989, Haishima et al. 1988, Raziuddin 1980b, Sen et al. 1979). Heptose was present mainly as L-glycero-D-manno heptose (Haishima et al. 1988, Hisatsune et al. 1989). L-Glycero-D-gluco-heptose was also initially detected as a minor component in *V. cholerae* 569B, but was not confirmed in later studies (Vinogradov et al. 1995). Studies with rough mutants of *V. cholerae* O1 (95R, a derivative of Ogawa 162) and with smooth *V. cholerae* 569B (Inaba) showed that the core PS was composed of glucose, fructose, heptose (L-glycero-D-manno heptose), glucosamine, and Kdo (Vinogradov et al. 1995).

The detection of D-fructose in LPS of *V. cholerae* makes an interesting story. Fructose occurs only rarely as an LPS component (Kenne and Lindberg 1983). Kondo et al. (1988) demonstrated that, among the vibrios studied, O1 *V. cholerae* was the only group for which all strains contained fructose in LPS. When LPS of *V. cholerae* was heated in dilute acetic acid at 100 °C, release of fructose was observed (Hisatsune et al. 1989). On heating LPS of Gram-negative enterobacteria under identical conditions, Kdo is normally released. In the LPS of most Gram-negative bacteria, Kdo is known to connect the PS moiety with lipid A (Wilkinson 1977). It was accordingly presumed that, in the LPS molecule of *V. cholerae*, fructose and not Kdo provided the connecting link between lipid A and core-PS (Jann et al. 1973). In another study (Redmond et al. 1973), it was suspected that fructose was essentially involved in the antigenicity of LPS because the release of fructose from LPS was accompanied by the loss of antigenicity. This was later ruled out because (a) fructose was found to be present in the rough strains of *V. cholerae* O1 (Kaca et al. 1986), and (b) the decrease in antigenicity did not parallel the release of fructose and required more hours of heating in dilute acetic acid (Kaca et al. 1986). Further chemical analysis revealed that D-fructose does not link the core-PS and lipid A but is present as a branch in the core region (Hisatsune et al. 1989), which was later confirmed by others (Vinogradov et al. 1995).

4.7.2 Non-O1 Vibrios

Like O1 vibrios, *V. cholerae* O139 was reported to contain glucose, heptose (L-glycero-D -manno heptose), glucosamine, fructose, and Kdo in the core-PS region (Hisatsune et al. 1993b). D-Fructose was found in a branch linked to the 6-position of a D-glucose residue (Cox et al. 1996). In *V. cholerae* O139 strain NRCC 4740, Kdo was found as Kdo-phosphate, like in O1 vibrios, and it was suggested that Kdo was phosphorylated at the C-4 position (Cox et al. 1996). Using a different strain of O139 (MO10-T4), which lacked a capsular polysaccharide (CPS) and produced a short-chain LPS, biphosphorylated Kdo residues could be detected where, in addition to the phosphate group in the usual C-4 position, the Kdo residue contained 2-aminoethyl phosphate (PEtn) at the C-7 position (Knirel et al. 1997). Although biphosphorylated Kdo residue is uncommon in *V. cholerae*, it was detected in some strains of enterobacteria (Holst and Brade 1992). The LPS of the O139 strain (MO10-T4) contained, in addition, (i) an O-acetyl group connected to a secondary position, tentatively O4 of the α-linked glucosyl group, and (ii) an additional putative fructose residue of unknown location on the LPSs of some species (Knirel et al. 1997). The significance of the differences in the core-PS structures of the two strains of O139 is not yet clear.

Very few of the other non-O1 vibrios have been extensively studied so far. Since the non-O1 vibrios are composed of a wide range of serogroups, some variations have been encountered. Basically, the core-PS of non-O1 vibrios, like O1 vibrios, contained heptose, glucose, fructose, glucosamine, glucosaminitol (in H-11), Kdo, and galactose (in H-11) (Kondo et al. 1988, Vinogradov et al. 1993). Fructose was found in some serogroups (O3, O5, O7, and O8) (Hisatsune et al. 1983), but not in O6 (Kondo et al. 1988). Moreover, in a study conducted on 44 serogroups of non-O1 vibrios, fructose was not found in as many as nine serogroups (cited in Kondo et al. 1992). A considerable amount of D-glycero-D-manno heptose was found in some strains of serogroup O3 of *V. cholerae* in addition to L-glycero-D-manno heptose (Kondo et al. 1988). D-Glycero-D-manno heptose, rarely found in Gram-negative bacterial LPS (Wilkinson 1977), was found in significant amounts in some strains of *V. cholerae* O3 (Kondo et al. 1988).

4.8 Core-PS: Chemical Structure

The chemical structures of the core-PS of *V. cholerae* of different serogroups are presented in Fig. 4.7. The chemical structures reveal the common features of the *Vibrio* cores: the single Kdo residue substituted by a phosphate residue at the 4-position (Kondo et al. 1988), and the presence of D-fructose that is linked to the 6-position of a D-glucose residue (Kondo et al. 1993b) and is located in a branch (Kaca et al. 1986) of the core-PS. Further, in common with many Gram-negative enteric bacteria (Holst and Brade 1992), the trisaccharide α-Hep*p*-(1-3)-α-Hep*p*-(1-5)-α-Kdo is present in all of these vibrio strains, although the heptose residues are not phosphorylated in the vibrios (except for the non-O1 strain H-11).

A comparison of the structures of O1 and O139 core-PS shows that the two structures are almost identical. The chemical evidence thus supports the genetic data that suggested that only the *wbe* gene cluster (the DNA region responsible for

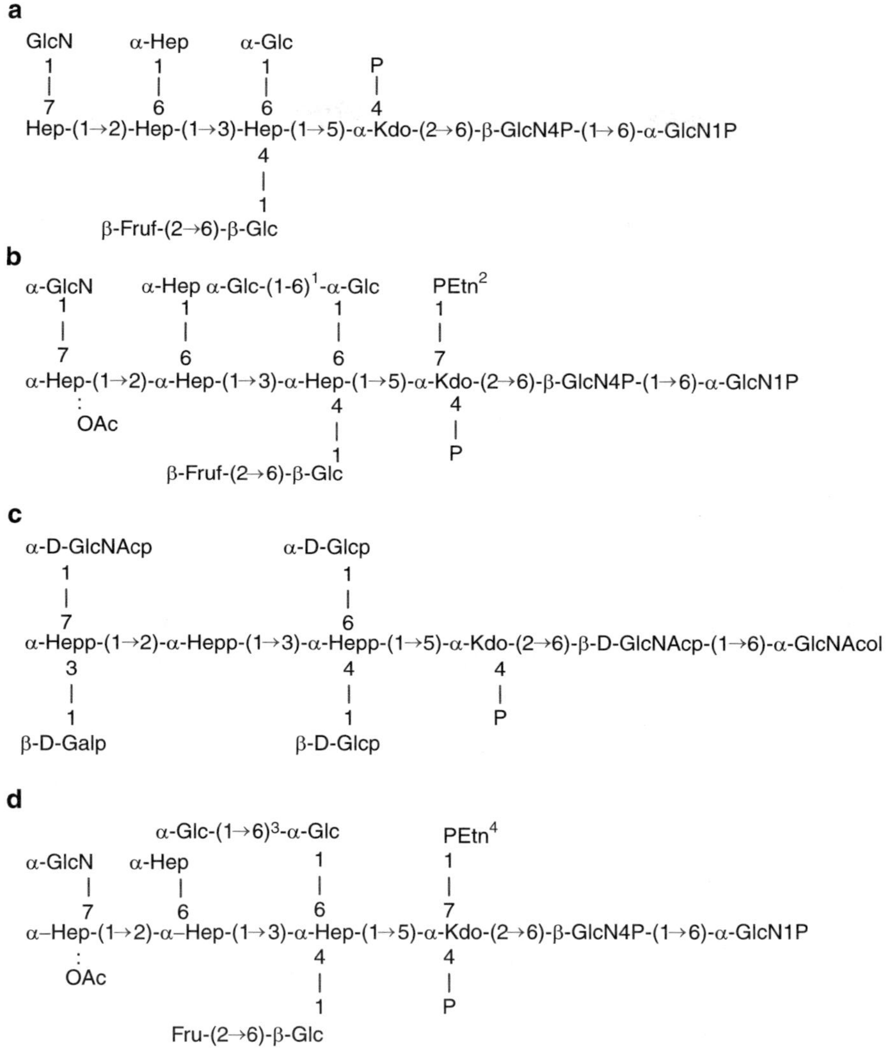

Fig. 4.7 Chemical structures of the core-PSs of *V. cholerae* of different serogroups. **a**, O1 strains 95R (Ogawa) and 569B (Inaba) (Vinogradov et al. 1995); **b**, O139 strain NRCC 4740 (Knirel et al. 1997, Ramamurthy et al. 1993); **c**, non-O1, strain H-11 (Vinogradov et al. 1993); **d**, O22 strain NRCC 4904 (Cox et al. 1997, Knirel et al. 1998); *1*, reported to be present by Cox and Perry (1996) but not reported by Cox et al. (1997); *2*, detected in strain MO10-T4 by Knirel et al. (1997); *3*, not confirmed by Knirel et al. (1998) in strain 169-68; *4*, detected in strain 169-68 by Knirel et al. (1998). (From Chatterjee and Chaudhuri 2003)

O-antigen biosynthesis) was altered in the biogenesis of O139 from O1 (Comstock et al. 1996). Figure 4.7 presents another interesting feature. The structures of the carbohydrate chains of the lipid A core region in the LPSs of *V. cholerae* rough strain 95R (Ogawa) and smooth strain 569B (Inaba) are identical. Finally, no significant difference is found in the core-PS structures of the LPSs of *V. cholerae* O22 and O139.

4.9 O-PS: Chemical Constituents

4.9.1 O1 Vibrios

The sugar composition of the O-PS of *V. cholerae* O1 was investigated, and a specific amino sugar, 4-amino-4,6-dideoxy D-mannose (perosamine), was detected as a component (Redmond 1979). The O-PS of *V. cholerae* O1 was found to be a regular $\alpha(1\rightarrow2)$-linked chain of D-perosamine (Kenne et al. 1979), the amino groups of which were acylated with 3-deoxy-L-glycero-tetronic acid (Kenne et al. 1982, Hisatsune et al. 1993a, Ito et al. 1994). No difference between Ogawa and Inaba serotypes in the sugar composition of O1 strains could be detected at that time (Kenne et al. 1979, 1982). Redmond (1978) reported the presence of 4-amino-4-deoxy-L-arabinose in the LPS of *V. cholerae* O1 serotype Ogawa and suggested that this sugar might be a determinant of the Ogawa-specific antigen. However, Kabir (1982) could not find 4-amino-4-deoxy arabinose in the LPS of *V. cholerae* serotype Ogawa and ruled out the role of this sugar as a determinant of the Ogawa-specific antigen. Hisatsune et al. (1993a) and Ito et al. (1994) reported, for the first time, that a previously undescribed sugar, 4-amino-4,6-dideoxy-2-*O*-methyl mannose (2-*O*-methyl perosamine), was present only in the O-PS of the Ogawa serotype of *V. cholerae* O1, and they proposed that this sugar might be an important constituent of the Ogawa serotype determinant. Ito et al. (1994) performed phenotypic conversion experiments in vitro and found that the sugar disappeared when the serotype conversion occurred from Ogawa to Inaba. It was argued that the 2-*O*-methyl perosamine was labile to acid and hence could not be detected earlier. It is now known that the Ogawa and Inaba serotypes differ only by a single 2-*O*-methyl group that is present in the upstream (nonreducing) terminal perosamine unit (Fig. 4.8) of the Ogawa O-PS, and that this is absent in the terminal perosamine residue of the Inaba O-PS.

4.9.2 Non-O1 Vibrios

The most prominent of the non-O1 vibrios is *V. cholerae* O139, which, unlike other non-O1 vibrios, was found to cause the disease cholera with all its clinical

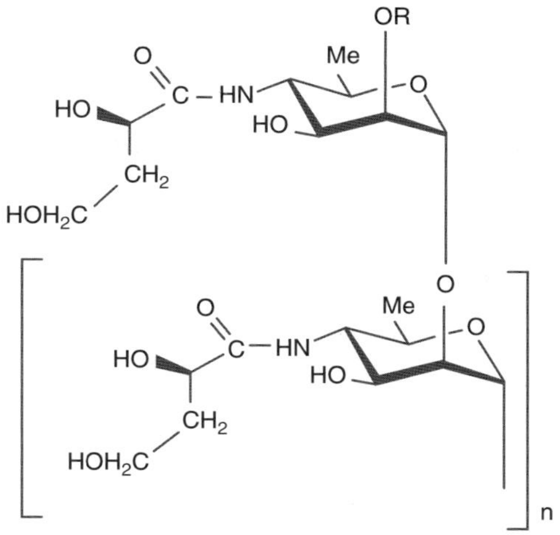

Fig. 4.8 Chemical structure of the O-PS of *V. cholerae* O1. The O-PS structures of the two serotypes, Inaba and Ogawa, are the same except at the position O-2 of the upstream, terminal perosamine group; here, R = CH_3 in the Ogawa strain and R = H only in the Inaba strain; *n* represents the number of repeating units, which may be between 12 and 18 (Hisatsune et al. 1993a, Ito et al. 1994). (From Chatterjee and Chaudhuri 2003)

manifestations. Although serologically unrelated to O1, studies (Calia et al. 1994) revealed that O139 was biotypically closely related to the *V. cholerae* O1 El Tor biotype. In terms of pathogenic characteristics, and specifically with respect to CT production (Johnson et al. 1994), the main virulence factor, *V. cholerae* O139, was indistinguishable from El Tor *V. cholerae* O1 strains. It was presumed that there were differences between the immune responses against O1 and O139 strains, which might be of considerable interest in terms of protection (Albert et al. 1994). Analysis of the sugar composition of LPS from *V. cholerae* O139 revealed a striking feature in that perosamine, a characteristic sugar component of the O-PS of *V. cholerae* O1, was absent (Hisatsune et al. 1993b). Detailed investigations revealed that the O-antigen of *V. cholerae* O139 contained only one unit (Cox et al. 1996, Cox and Perry 1996, Knirel et al. 1997), unlike most O-PSs (which usually consist of repeating units), and that it was a phosphorylated hexasaccharide consisting of colitose, galacturonic acid, quinovosamine, galactose, and glucosamine residues (Cox et al. 1996, Knirel et al. 1997). This altered surface polysaccharide epitope enabled serogroup O139 to avoid previously acquired host immunity to the serogroup O1 strain, and this contributed significantly to the early success of the organism in producing epidemic cholera (Ramamurthy et al. 1993).

Another distinguishing feature of *V. cholerae* O139 is that it produces, unlike O1, a capsular PS (CPS), which will be discussed later (Sect. 4.11). Further, the O-PS of *V. cholerae* O139 is unique in the sense that it contains colitose, which is generally not found in vibrios (Hisatsune et al. 1993b). Only the serogroup O22 has so far been found to contain colitose (Cox et al. 1997, Knirel et al. 1998). *V. cholerae* O139 strains collected from four epidemic regions of India (Chennai, Vellore, Kolkata, and Bangladesh) showed similar sugar compositions (Hisatsune et al. 1993b), indicating that they were quite homogeneous chemotaxonomically and therefore belonged to a new chemotype of non-O1 *V. cholerae* LPS (Hisatsune et al. 1993b).

Among the other non-O1 vibrios, which consist of some 200 or more serogroups formerly known as NAG or nonagglutinating vibrios, the O-PSs of a few have been studied so far. The amino sugar D-perosamine, a characteristic constituent of the O-PS of *V. cholerae* O1, is generally absent in non-O1 vibrios, while galactose is present in many of these serogroups (Sen and Mukherjee 1978, Chowdhury et al. 1991). Other sugars found in LPS of various non-O1 vibrios include L-rhamnose, heptose, fructose, glucosamine, galactosamine, and *N*-acetyl-2-amino-2-deoxy mannose (ManNAc) (Sen and Mukherjee 1978, Chowdhury et al. 1991, Knirel et al. 1997). On the other hand, O-PSs of non-O1 vibrios were found to contain several unusual sugars, e.g., ascarylose (3,6-dideoxy-L-arabino-hexose), Sug^1 or bacillosamine (2,4-diamino-2,4,6-trideoxy-D-glucose) and fucosamine (2-amino-2,6-dideoxy-L-galactose) in serogroup O3 (Chowdhury et al. 1991); Sug^2 (5-aceta midino-7-acetamido-3,5,7,9-tetradeoxy-L-glycero-β-L-manno-nonulosonic acid) in serogroup O2 (Kenne et al. 1988a); 2-acetamido-3-(*N*-formyl-L-alanyl)amino-2,3-dideoxyglucuronamide (GlcNAc3NAcylAN), 2,3-diacetamido-2,3-dideoxy mannuronamide (ManNAc3NAcAN) and 2,3-diacetamido-2,3-dideoxyguluronic acid (GulNAc3NAcA) in serogroup O8 (Kocharova et al. 2001a); 2,4-diacetamido-2,4,6-trideoxyglucose (QuiNAc4NAc) in both serogroups O8 and O5 (Kondo and Hisatsune 1989); and 2-acetamido-2,6-dideoxy-D-glucose (QuiNAc) in serogroup O22 and O139 (Knirel et al. 1998).

On the other hand, the sugar composition of the LPS from serogroup O76 (Kondo et al. 1996) has much in common with that of the serogroup O1, except that (i) perosamine in the serogroup O76 is in the L configuration in contrast to the D configuration of the perosamine in O1, (ii) a small amount of D-galactose is present in O76, and (iii) the L-perosamine is *N*-acylated with an (*S*)-(+)-2-hydroxy propionyl group in the serogroup O76. Another serogroup of *V. cholerae*, O144, was found to contain homopolymers of D-propionyl-α-L-perosamine in its O-PS (Sano et al. 1996).

4.10 O-PS: Chemical Structure

Figures 4.8 and 4.9 present the structures of the O-PS of *V. cholerae* of different serogroups and illustrate that the O-PS structure of any particular serogroup is distinct. About 12–18 monosaccharide repeating units ($n = 12$–18) are present in the O-PS of *V. cholerae* O1 (Villeneuve et al. 2000), which is in accordance with the high molar ratio of D-perosamine obtained in earlier studies (Haishima et al. 1988, Hisatsune et al. 1989). *V. cholerae* O139 lacks the "ladder pattern" characteristic of O-antigen-producing O1 strains when examined on SDS-PAGE (Manning et al. 1994).

$^{-}$O O
\\//
P
/ \\
4 6
α-Col-(1→2)-β-Gal
1
|
3
β-GlcNAc-(1→4)-α-GalA(1→3)-β-QuiNAc-(1→ **0139**
4
|
1
α-Col

→4)-β-D-QuipNAc-(1→4)-Sug^{1}p-(2→4)-β-D-Galp(1→ **02**

→2)-α-D-Hepp-(1→4)-α-L-FucpNAc-(1→3)-β-Sug^{2}p-(1→ **03**
3
|
1
α-Ascp

→4)-β-D-GlcpNAc3Acy1AN-(1→4)-β-D-ManpNAc3NacAN-(1→4)-α-L-Gulp-NAc3NAcA(1→3)-β-D-QuipNAc4Nac-(1→ **08**

→3)-α-D-ManpNAc-(1→4)-β-D-GlcpA-(1→3)-β-D-Galp-(1→3)-β-D-GlcpNAc-(1→ **010**

OAc*
|
3 4
α-Col-(1→2)-β-GalA
1
|
3
α-GlcNAc-(1→4)-α-GalA-(1→3)-β-QuiNAc-(1→ **022**
4
|
1
α-Col

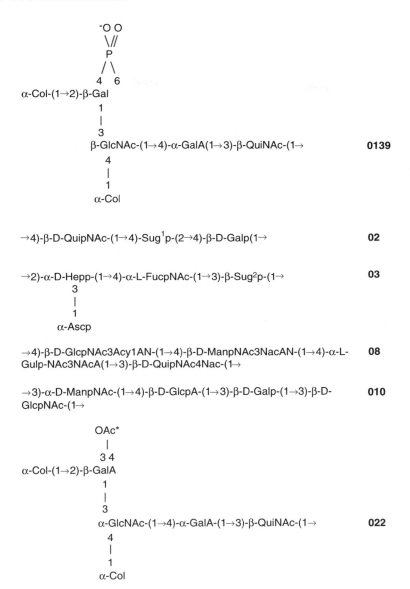

Fig. 4.9 Chemical structures of the repeat units of O-PS of *V. cholerae* of different non-O1 serogroups. Abbreviations are as given in the text. *Knirel et al. (1998) could not confirm the presence of this residue. References are: O139 (Cox et al. 1996, Cox and Perry 1996); O2 (Kenne et al. 1988a); O3 (Chowdhury et al. 1991); O8 (Kocharova et al. 2001a); O10 (Kjellberg et al. 1997); O22 (Knirel et al. 1998, Cox et al. 1997). (From Chatterjee and Chaudhuri 2003)

Figure 4.9 shows that an altered structure from perosamine homopolymer of *V. cholerae* O1 is present in the O-PSs of non-O1 vibrios, which generally consists of complex units containing tri-, tetra-, or pentasaccharide repeating units. Further, the repeating unit often contains unusual sugars. The O-PS structures of two different strains of *V. cholerae* O22 are more or less similar, except that the position of the *O*-acetyl group at β-GalA as shown by Cox et al. (1997) could not be confirmed by Knirel et al. (1998). Structural analysis of the O-PSs of LPSs of some other serogroups of *V. cholerae* and of some other vibrio species have now been completed (not shown in Fig. 4.10). Of these, it is worth mentioning here that O155 (Senchenkova et al. 1998) and O9 (Kocharova et al. 2001b) both have pentasaccharide repeating units, and O6 (Bergstrom et al. 2002) and *V. mimicus* (Landersjo et al. 1998) both have tetrasaccharide repeating units. The important point to note here is that both *V. mimicus* and O155 contain an immunodominant group, galactosyl residue substituted with a cyclic phosphate (D-galactose-4,6-cyclophosphate), in their O-PS structures, in common with the O-PS structures of LPS and CPS of *V. cholerae* O139, and that this appears to be the basis of serological cross-reactivity between O139 and O155 or *V. mimicus*.

The O-PSs of *V. cholerae* O22 and O139 are similar in chemical composition and structure and differ only in the presence of an *O*-acetylated β-GalA residue in the former instead of β-Gal-4, 6P in the latter (Knirel et al. 1997) and in the anomeric configuration of GlcNAc (Cox et al. 1997). *V. cholerae* O22, unlike O139, lacks a CPS but possesses an O-PS structure comparable to that of *V. cholerae* O139. The similarity in the O-PS structures of the serogroups O139 and O22 can be elaborated further. Both serogroups contain the same trisaccharide GlcNAc-GalA-QuiNAc, and in both cases the *N*-acetylglucosamine residue is disubstituted at the 3- and 4-positions. Further, rare terminal 3,6-dideoxy-L-xylo hexose (colitose) residues are present in both the O22 and O139 serogroups. To elaborate on the differences in the O-PS structures of these two serogroups, it may be noted that in O22, the *N*-acetylglucosamine has the α-configuration and not the β-configuration observed previously for O139. The substituent at the 3-position also differs; it is a galacturonic acid residue in O22 and a cyclic phosphorylated galactose in serogroup O139. Further, it may be noted that the phosphate group is absent in serogroup O22 and that such groups are often found to be immunodominant. Barring these small differences, the striking structural similarity between the serogroups O22 and O139 provides a chemical basis for the serological cross-reactions observed between the strains belonging to these two serogroups. The similarity further adds support to existing data suggesting that the serogroup O22 is the donor strain for the new genetic material forming the O139 serogroup *V. cholerae*. This structural evidence, when considered in conjunction with the facts that (i) serogroups O22 and O139 are the only cholera serogroups to produce colitose and (ii) a significant homology has been found between genes of the *wb** clusters of serogroups O139 and of O22 (Vimont et al. 1997), makes it very likely that the serogroup O22 is the progenitor strain. The genetic homology between O139 and O22 will be discussed further in Chap. 5 of this book.

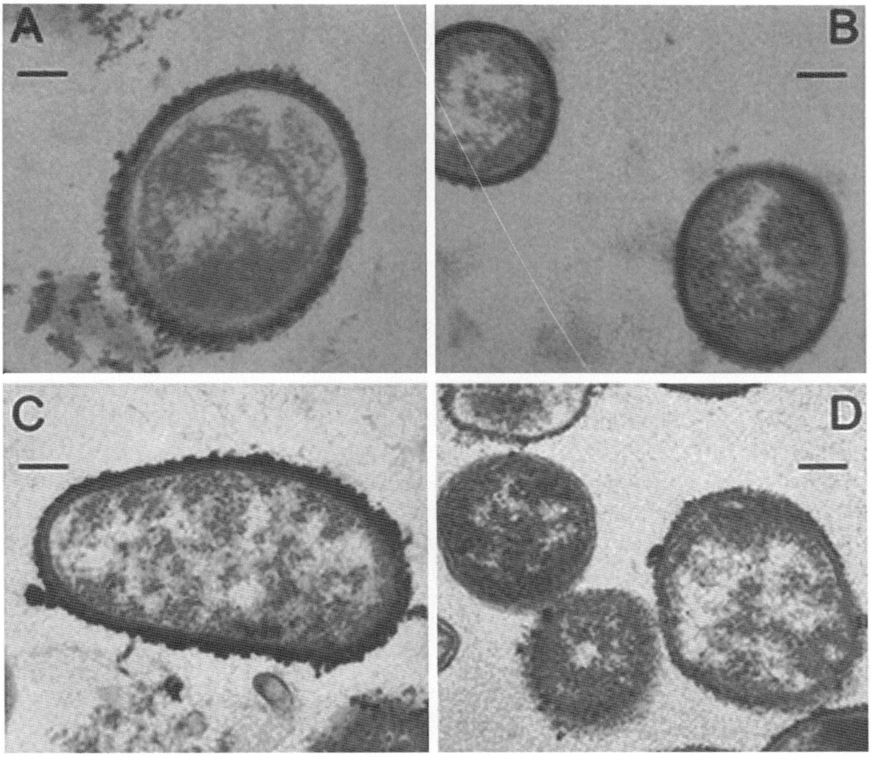

Fig. 4.10 Electron micrograph of thin sections of wild-type *V. cholerae* NRT36S (*A*) and its translucent mutants, TR3 (*B*), TR296 (*C*) and TR17 (*D*), stained with polycationic ferritin. NRT36S displayed a heavy, complete capsule surrounding the cell; TR3 did not have a complete capsule, but had some patches of capsule materials; both TR17 and TR296 had a much thinner capsule. Mutants were obtained by conjugation between wild-type *V. cholerae* strain NRT36S and a donor strain *E. coli* S17λpir/putKm-2 and the selection of colonies displaying translucent phenotype on LB agar. This phenotype suggests that genes involved in capsule biogenesis have been disrupted by the transposon. EM pictures of all the mutants were consistent with the amounts of capsule observed by the size exclusion chromatography of the capsule preparations. (Fig. 5 of Chen et al. 2007 reproduced with kind permission from O.C. Stine)

4.11 The Capsular Polysaccharide (CPS)

4.11.1 Colony Morphology and CPS

Many of the non-O1 vibrios are known to produce CPS. The O1 vibrios have so far not been shown to produce CPS. The non-O1 strains have been shown to shift between an encapsulated form with opaque colony morphology and an unencapsulated or minimally encapsulated form with translucent colony morphology. The degree of opacity correlates with the amount of capsular material that can be extracted from the cells (Johnson et al. 1992). *V. cholerae* O139, a non-O1 strain, was shown to cause large outbreaks of cholera in the Indian subcontinent in 1992 (Ramamurthy et al. 1993). Since the O139 strain was found to be toxinogenic and did not agglutinate with either polyclonal or monoclonal antisera directed against the *V. cholerae* O1 antigen, more attention was directed towards the study of this O139 strain and its various constituents. When incubated on agar plates, the *V. cholerae* O139 Bengal strain AI-1838 showed two distinct colony morphologies, translucent and opaque. Both colony types were tested for agglutination with rabbit anti-O139 antisera and showed a positive reaction. Subsequently, Johnson et al. (1994) tested eight strains, including AI-1838, belonging to the O139 serogroup and found that all eight strains had moderately opaque colony morphology on initial streaks, but that the translucent sectors and colonies appeared after subculturing. Similar changes in colony morphology were not found when more than 100 strains of O1 serogroup were examined. Comstock et al. (1995) showed by Tn*phoA* mutagenesis that the loss of capsular material was associated with the loss of opacity of the colony morphology in *V. cholerae* O139 Bengal.

4.11.2 CPS: Site of Occurrence

The presence of CPS in *V. cholerae* O139 Bengal was demonstrated by electron microscopic studies (Johnson et al. 1994, Comstock et al. 1995, Weintraub et al. 1994, Meno et al. 1998). Johnson et al. (1994) prepared two O139 strains (AI-1855 and AI-1841) for electron microscopic examination in thin section by standard methods and stained them with polycationic ferritin. The electron microscopic photographs showed that bacteria belonging to both strains were surrounded by a relatively thin, electron-dense capsule. Weintraub et al. (1994) examined another strain (AI-1838) of O139 vibrio by ultrathin section and electron microscopy and found that a coarse, electron-dense band was heterologously distributed around approximately 30% of the cells. The thickness of the band was 18–25 nm. The *V. cholerae* O1 strains (Ogawa) examined under identical conditions lacked the surrounding electron-dense band. Meno et al. (1998) examined a *V. cholerae* O139 strain by a different electron microscopy technique (freeze substitution) and found a very thin fibrous layer on the outside of the outer membrane of the cells. In contrast, the mutants of the strain O139, strain MO10-T4 (which lacked the ability to synthesize capsule) and the strain Bengal-2R1 (which failed to synthesize both the capsule and the O-antigen of LPS), were all found to have lost the surface layer. In addition, the capsule layer could not be observed on the surface of *V. cholerae* strain O1. Figure 4.10 shows electron micrographs of the capsules surrounding the wild-type *V. cholerae* strain NRT36S and its translucent mutants.

4.11.3 CPS: Sugar Composition

Kasper et al. (1983) showed (by Sephacryl S-300 chromatography and chemical analysis of fatty acids and sugars) that the phenol-water extraction of capsulated bacteria yielded a mixture of LPS and CPS. In order to purify the LPS and CPS, Weintraub et al. (1994) subjected the lyophilized aqueous phase obtained after phenol-water extraction of the O139 Bengal strain to an extraction with phenol-chloroform-petroleum (PCP) ether and obtained two fractions, a PCP-soluble fraction (containing LPS) and a PCP-insoluble fraction (containing CPS), in the ratio of 1:2. It was ascertained that the CPS contained no lipid A or fatty acid. Monosaccharide analyses showed that CPS contained glucosamine, quinovosamine, and 3,6-dideoxy-xylo-hexose. In addition, small amounts of glucose, galactose, and trace heptose were found. Gel permeation chromatography of the CPS showed that this material was of a high molecular weight, since it eluted at the void volume on Sephacryl S-300 chromatography. It was established chemically that *V. cholerae* O139 produced a CPS, distinct from the LPS. To establish the identity of the 3,6-dideoxy-xylo-hexose, it was analyzed on three different GLC columns (Weintraub et al. 1994), one polar and two nonpolar, on which it showed the same retention time as an authentic abequose derivative (3,6-dideoxy-D-xylo-hexose). However, the analysis did not discriminate between the D and L isomers, i.e., the 3,6-dideoxyhexose could well be the L-isomer (colitose). Later on, Preston et al. (1995) used two different hydrolysis conditions, 0.5 M trifluoroacetic acid at 60 °C for 3 h and 1 M trifluoroacetic acid at 100 °C for 10 h, for the CPS of O139 strain (AI-1837) and HPLC conditions suitable for neutral and acid sugars, respectively, and demonstrated the presence of 3,6-dideoxy-xylohexose, quinovosamine, glucosamine, galactose, and galacturonic acid. Subsequently, Knirel et al. (1995) produced a more definite result with respect to the sugar composition and structure of *V. cholerae* O139 CPS. These authors showed that the CPS contained D-galactose, 3,6-dideoxy-L-xylo-hexose (colitose), 2-acetamido-2-deoxy-D-glucose, 2-acetamido-2,6-dideoxy-D-glucose (*N*-acetyl-D-quinovosamine), D-galacturonic acid and phosphate.

4.11.4 CPS: Chemical Structure

Preston et al. (1995) worked out the structure of the *V. cholerae* O139 CPS by high-performance anion exchange chromatography and ^{1}H-nuclear magnetic resonance spectroscopy. The CPS was found to contain a repeating unit consisting of six sugar residues, which included one residue each of *N*-acetylglucosamine (GlcNAc), *N*-acetylquinovosamine (QuiNAc), galacturonic acid (GalA), galactose, and two residues of 3,6-dideoxy-xylo-hexose (Xylhex). However, the workers could not unambiguously determine the absolute configuration of the monosaccharides, and accordingly the residues of 3,6-dideoxy-xylo-hexose could be designated as either colitose (the L isomer) or abequose (the D isomer). Subsequently, Knirel et al. (1995) studied the structure of O139 CPS by NMR spectroscopy in combination with methylation analysis and selective degradations, including partial acid hydrolysis at pH 3.1 and dephosphorylation with aqueous 48% hydrofluoric acid, and basically confirmed the CPS structure proposed by Preston et al. (1995). In addition, Knirel et al. (1995) was able to exactly specify the absolute configurations of the constituent monosaccharides and the presence and position of the phosphate group. Very recently, the hexasaccharide repeating unit has been isolated from the *V. cholerae* O139 CPS by digestion with a polysaccharide lyase derived from a bacteriophage specific for this serogroup. It specifically cleaves at a single position on the 4-linked galacturonic acid, producing an unsaturated sugar product, the conformation of which has been studied by molecular modeling and NMR spectroscopy (Fig. 4.11) (Adeyeye et al. 2003). The structure has been found to contain a tetrasaccharide epitope homologous to the human Lewis[b] blood group antigen, and its conformation is identical or at least closely similar to that of the intact CPS studied earlier (Preston et al. 1995, Knirel et al. 1995).

V. cholerae strain NRT36S, serogroup O31, is cholera toxin negative but produces the heat-stable enterotoxin, NAG-ST, and a capsule surrounding the outer envelope. The structure of the CPS, as deduced from high-resolution NMR spectroscopy, was found to be complex and consisted of four residues (α-D-GlcA, α-D-GlcNAc, β-L-Rha, and β-D-GlcNAc) in the repeating subunit (Chen et al. 2007). Study of the capsule biogenesis indicated that the capsule gene cluster

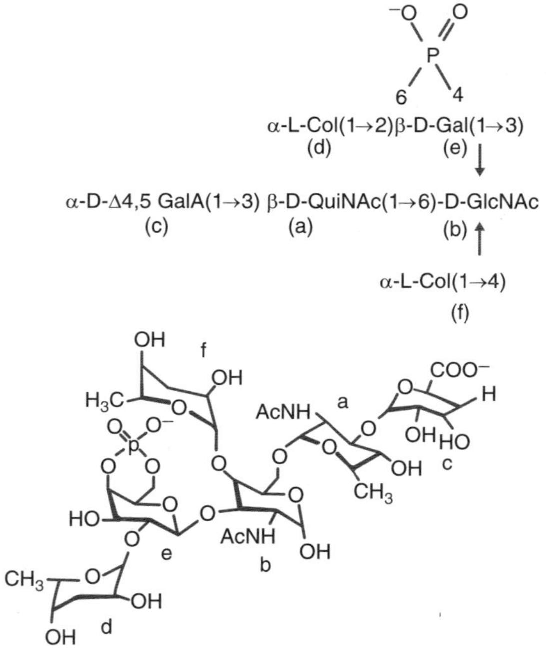

Fig. 4.11 Chemical structure (*top*) and conformation (*bottom*) of hexasaccharide obtained by digestion of the capsular polysaccharide of *V. cholerae* O139 with a lyase from a bacteriophage specific for O139. The letters *a–f* indicate the residue assignments used by the authors (Adeyeye et al. 2003). (From Chatterjee and Chaudhuri 2003)

shared the same genetic locus as that of the O-antigen of the LPS biogenesis gene cluster. This interesting co-location of the CPS and LPS biosynthesis genes was unique and would provide a mechanism for the simultaneous emergence of new O and K antigens in a single strain. The authors (Chen et al. 2007) argued that this may be a key element in the evolution of new *V. cholerae* new strains that can escape immunologic detection by host populations.

4.12 Emerging Research Trends and Future Possibilities

Some 200 or more serogroups of *V. cholerae* have already been detected. Studies of these and of any additional serogroups that may be detected in the future would be of an academic nature unless a new strain emerges that causes severe epidemic or pandemic cholera. On the other hand, research in some new directions has already been initiated, and this may prove to be of great practical use.

Recent work along these lines involve the preparation of synthetic mono- or oligosaccharide fragments that mimic the terminal mono- or oligosaccharide residues of the O-PS of *V. cholerae* LPS and studying their interactions with monoclonal anticholera antibodies with a view to developing a synthetic carbohydrate-based anticholera vaccine. Kenne et al. (1988b) reported the first synthesis of the monosaccharide repeating unit of *V. cholerae* O1, serotype Inaba. Improved syntheses, as well as the crystal structure of the methyl-α-glycoside of the monomeric repeating unit common to both Ogawa and Inaba strains, were subsequently reported by Gotoh et al. (1994). Following the discovery that the O-PSs of the two serotypes, Ogawa and Inaba, differ in that the terminal 4-*N*-tetronylated-D-perosaminyl group in the O-PS of the Ogawa strain is methylated at O-2 (Hisatsune et al. 1993a, Ito et al. 1994), the complete terminal sugar of the serotype Ogawa, methyl-α-glycoside, was synthesized and its crystal structure was determined (Lei et al. 1995a). This was followed by the synthesis of three oligosaccharides that imitate the upstream part of the O-PS of the serotype Ogawa (Lei et al. 1996) and the analogous trisaccharide related to the O-PS of the serotype Inaba (Lei et al. 1995b). As a further step in the effort to understand protective immunity to cholera, hexasaccharides representing termini of the O-PS of the two serotypes of *V. cholerae* O1 were synthesized in the form of glycosides whose aglycons allowed these substances to be linked to proteins (Ogawa et al. 1996). Further improvement in the synthesis of similar antigenic compounds has recently been reported by Ma et al. (2003).

Another recent study along these lines involved the preparation of analogs of the methyl-α-glycosides of the terminal residues of the O-PS of *V. cholerae* O1 for use as probes (Poirot et al. 2001) to study their interaction with anti-*V. cholerae* O1 antibodies (Wang et al. 1998, Liao et al. 2002). They differ from the termini of the respective O-PSs in the anomeric or absolute configuration of perosamine, the position of the *O*-methyl group in D-perosamine, and the nature of the *N*-acyl side chain. These were used to learn more about the structural requirements for binding with antibodies and also to find out the structural elements that would increase binding. Among the interesting results already achieved include the finding that some antibodies showed a remarkable tolerance to irregularities or variations in the structures of the ligands (Poirot et al. 2001). Some other studies resolved that the 2-*O*-methyl group, a small antigenic determinant, can dictate a highly specific immune response. Wang et al. (1998) carried out binding studies of anti-Ogawa Abs IgG_1s S-20-4, A-20-6, and IgA 2D6 with synthetic methyl α-glycosides of fragments up to the hexasaccharide, of the Ogawa O-PS, as well as analogs of the terminal monosaccharide, and revealed that the terminal residue accounted for approximately 90% of the maximal binding energy. They did not react with the corresponding synthetic fragments of Inaba O-PS. Further insight into the structural basis of carbohydrate recognition, and particularly *V. cholerae* serotype specificity, was obtained from the crystallographic studies of protective anticholera Abs in complex with synthetic analogs of the O-antigen (Villeneuve et al. 2000). The upstream terminal monosaccharide of the Ogawa O-PS was shown to be the primary antigenic determinant. The crystallographic study showed that the pivotal contribution was made by a structural fragment in the antigenic determinant as small as a methyl group. Chernyak et al. (2002) moved one step forward and demonstrated the induction of protective immunity by synthetic antigens that mimic the terminal hexasaccharide epitope of the O-PS of *V. cholerae* O1, serotype Ogawa, conjugated to bovine serum albumin with different carbohydrate to carrier molar ratios. The protective capacity of antiserum was evident in serum from mice immunized with all conjugates, but it was highest in the groups that received the conjugate with the lowest level of substitution. Further investigations along these lines with a view to devising synthetic analogs that will bind significantly with antibodies derived against *V. cholerae* of more than one serogroup may be very challenging but are useful for the purposes of vaccine development. The extension of such studies with the specific aim of gaining insight into molecular and atomic details relating to the structural basis of carbohydrate recognition will contribute significantly towards the development of a synthetic carbohydrate-based anticholera vaccine.

Dharmasena et al. (2007) investigated another alternative approach toward the development of a cholera vaccine that will also overcome the problems associated with the administration of LPS. These authors identified and developed peptide mimics of *V. cholerae* LPS to elicit cross-reactive immune responses against LPS. One such peptide mimic, 4P-8, was found to have a hairpin-like structure that mimicked some O-PS interactions with the monoclonal antibody (MAb) S-20-4, and also made unique ligand interactions with S-20-4. The effectiveness of the peptide 4P-8 in eliciting anti-LPS Abs was also investigated. Although the 4P-8-KLH (keyhole limpet hemocyanin) was able to elicit anti-LPS Abs in mice, the immune response was not vibriocidal or protective. The authors, however, found that boosting with 4P-8-KLH after immunizing with LPS prolonged the LPS-reactive IgG and IgM Ab responses as well as vibriocidal titers, and provided a much greater degree of protection than priming with LPS alone. The authors proposed to make further investigations into the development of peptide mimics of *V. cholerae* LPS so that a more robust immune response can be induced in animal models as a preliminary to developing a more potent cholera vaccine.

Chapter 5
Endotoxin of *Vibrio cholerae*: Genetics of Biosynthesis

Abstract The genetics of the biosynthesis of *Vibrio cholerae* lipopolysaccharide (LPS) are discussed in this chapter. While not much information is available in the literature on the genetics of the biosynthesis of lipid A of *V. cholerae*, the available information on the characteristics and proposed functions of the core polysaccharide (core-PS) biosynthetic genes is discussed. The genetic organization encoding the O-antigen polysaccharides (O-PS) of *V. cholerae* of serogroups O1 and O139, the disease-causing ones, are described, along with the putative functions of the different constituent genes. The O-PS biosynthetic genes of some non-O1 and non-O139 serogroups, particularly the serogroups O37 and O22, and their putative functions are discussed briefly. In view of the importance of the serogroup O139, the origination of the O139 strain and the possible donor of the corresponding O-PS gene cluster are analyzed with a view to providing knowledge of (i) the mode of evolution for different serogroups and (ii) the possible emergence of pathogenic strain(s) belonging to non-O1, non-O139 serogroups. Finally, the unsolved problems in this area of research and their probable impact on the production of an effective cholera vaccine are outlined.

5.1 Introduction

Lipopolysaccharide (LPS) is an important structural and functional constituent of the vibrios and of any Gram-negative bacterium in general. The physical and chemical characteristics of the *V. cholerae* LPS are presented in Chap. 4 of this book. Of the three constituents of LPS, lipid A and core-PS present more or less similar structures in the LPSs of different *V. cholerae* serogroups, whereas the O-PS of any serogroup has a distinct structure. Research on different aspects of *V. cholerae* LPS has progressed at an accelerating pace since the emergence of the non-O1 strain, *V. cholerae* O139, which caused an outbreak of the epidemic in the Indian subcontinent around 1992 (Ramamurthy et al. 1993). Gaining a knowledge of the evolution of the different serogroups of *V. cholerae* demands an understanding of the genetics of the biosynthesis of the constituent LPS. This will also help

K. Chaudhuri and S.N. Chatterjee, *Cholera Toxins*.

us to gage the possibility that one or more new pathogenic and pandemic strains of *V. cholerae* will emerge in the future. Accordingly, this chapter provides an account of our current knowledge regarding the genetics of the biosynthesis of *V. cholerae* LPS.

5.2 Lipid A

The highly conserved lipid A region is essential for outer membrane integrity in all Gram-negative bacteria (Takada and Kotani 1992). The biosynthesis of lipid A occurs at the cytoplasmic face of the inner membrane. Although a lot of knowledge has accumulated in recent years on the biosynthesis of lipid A and the transfer of Kdo in several Gram-negative bacteria (Raetz 1998), not much work is available on the genetics of the biosynthesis of lipid A, and particularly on the characteristics of the lipid A biosynthetic gene products of *V. cholerae*.

5.3 Core-PS

The core-PS of the LPS of Gram-negative bacteria, including *V. cholerae*, is also known to play a role in maintaining outer membrane stability (Heinrichs et al. 1998). Whether the core-PS also plays a part in providing virulence to the organism was not known or investigated in *V. cholerae*. Not much information was available on the genetics of the core-PS biosynthesis, while structural studies have revealed that the O1 and O139 (the disease-causing serogroups) core-PS structures are largely similar (see Chap. 3; Chatterjee and Chaudhuri 2003). Only recently, some significant information was recorded on the core-PS biosynthesis gene clusters of *V. cholerae* of different serogroups (Nesper et al. 2002a). Following the new nomenclature introduced by Reeves et al. (1996), the core-PS biosynthesis locus was designated as the *wav* gene cluster. Computer analysis showed that the *wav* gene cluster located in chromosome I comprised open reading frames (ORFs), VC0223 to VC0240, in the completely sequenced genome of the *V. cholerae* El Tor strain N16961 (Heidelberg et al. 2000). The genetic organization of the putative *wav* gene cluster as deduced from the sequence referred to above is shown in Fig. 5.1, where the Southern hybridization probes wav1 to wav4 are indicated by horizontal lines (Nesper et al. 2002a). The data summarizing the characteristics and proposed functions of the core-PS biosynthetic genes are given in Table 5.1. Investigations of 38 different *V. cholerae* strains by Southern blotting, PCR and sequence analyses revealed the presence of five different putative *wav* gene cluster types. The significant finding resulting from this was that the O1 El Tor *wav* gene cluster type (type 1) was prevalent among clinical isolates of different serogroups associated with cholera and environmental O1 strains. On the other hand, 19 unre-

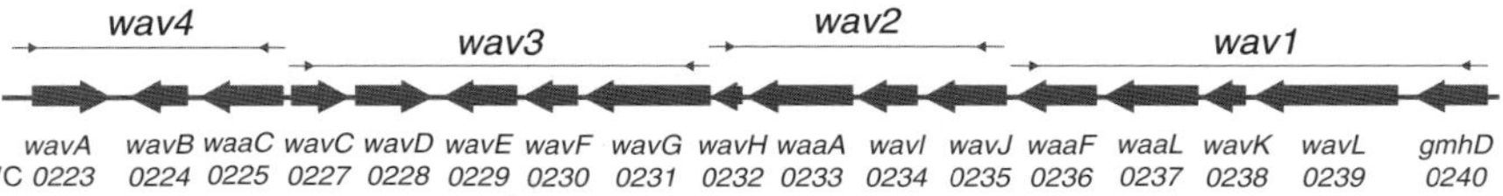

Fig. 5.1 Organization of the putative type-1 core-PS biosynthesis gene cluster (*wav*) as deduced by Nesper et al. (2002) from the sequence of *V. cholerae* El Tor strain N16961 (Heidelberg et al. 2000). *Horizontal lines with arrows at both ends* indicate the Southern hybridization probes wav1 to wav4 used for this purpose. The gene cluster is located on chromosome I of *V. cholerae* comprising ORFs VC0223 to VC0240. *Arrows* show the directions of transcription of the individual ORFs. (From Chatterjee and Chaudhuri 2004)

lated non-O1 non-O139 environmental and human isolates not associated with cholera contained four new *wav* gene cluster types that differed from each other in distinct gene loci. These data provided evidence for the horizontal transfer of *wav* genes and for limited structural diversity of the core-PS among *V. cholerae* isolates. That the type 1 *wav* gene locus was predominant in strains associated with clinical cholera suggested that a specific core-PS structure could contribute to the virulence of *V. cholerae* strains. Some other interesting observations derived from the sequence of the genome of O1 El Tor strain N16961 (website: http://www.tigr.org) included: (i) the localization of the O1-antigen biosynthesis gene cluster (*wbe*) upstream of VC0240 and the indication that in *V. cholerae* most of the LPS biosynthetic genes were clustered; (ii) that VC0237 encoded the O-antigen ligase WaaL, and; (iii) that ORF VC0233 (*waaA*) encoded the Kdo transferase that ligated the lipid A to Kdo. Table 5.1 shows the presence of ten genes (*wavA, wavB, waaC, wavC, waaA, wavI, waaF, waaL, wavL*, and *gmhD*) in all of the 38 different *V. cholerae* strains studied, indicating the presence of a common core-PS backbone structure for all the strains investigated.

Izquierdo et al. (2002) showed that the *wavB* gene of *V. cholerae* was fully able to complement the *Klebsiella pneumoniae waaE* mutants through either chemical analyses or their contributions in a biological test like resistance to nonimmune human serum. The WavB from *V. cholerae* showed identical behavior to the WaaE in the *K. pneumoniae* background in several tests. The WaaE was shown to be a β-1,4-glucosyltransferase involved in the transfer of a glucose residue to the L-glycero-D-manno-heptose I in the LPS inner core. It was concluded that the WavB protein was able to perform the same function as WaaE. It is interesting to note that although the genes encoding heptosyltransferases in *Escherichia coli* and *Salmonella enterica* were known for years, their functional characterization was difficult because of the complexity of the in vitro assay. Gronow et al. (2000) claimed to have performed the first functional characterization of these proteins of *E. coli*, heptosyltransferase I (WaaC) and II (WaaF), the glycosyltransferasees involved in the biosynthesis of the inner core of LPS. It was found that the heptosyltransferases I and II of *E. coli* were strictly monofunctional. Extensive studies are required to further characterize the products of the different genes in the core-PS genetic organization of *V. cholerae*.

Table 5.1 Characteristics and proposed functions of the core-PS biosynthetic genes of *V. cholerae* (table prepared by adding the data on classical strain O395, obtained from the NCBI genome database at http: //www.ncbi.nlm.nih.gov/genomes/lproks.cgi, to Table 2 of Chatterjee and Chaudhuri 2004 and the data of Nesper et al. 2002)

ORF[a]/Strain	New ORF designation	Predicted putative function in core-PS assembly	Accession no.[b]	Wav types[c]				
				1	2	3	4	5
VC0223/A2604	*wavA*	HepIII transferase	AAK03378	+	+	+	+	+
VC0224/A2605	*wavB*	β1,4-Glucosyl transferase	AAK03390	+	+	+	+	+
VC0225/A2606	*waaC*	HepI transferase	B64058	+	+	+	+	+
VC0227/A2607	*wavC*	Kdo kinase	BAB72027	+	+	+	+	+
VC0228/A2608	*wavD*	Unknown		+	−	−	−	−
VC0229/A2609	*wavE*	Unknown	BAB72032	+	+	+	+	−
VC0230/A2610	*wavF*	Unknown	BAB72033	+	+	+	+	−
VC0231/A2611	*wavG*	Unknown	BAB72034	+	+	+	+	−
VC0232/A2612	*wavH*	Unknown	BAB72026	+	+	+	+	−
VC0233/A2613	*waaA*	Kdo transferase	BAB72028	+	+	+	+	+
VC0234/A2614	*wavI*	Glycosyltransferase	AAK03200	+	+	+	+	+
VC0235/A2615	*wavJ*	HepIV transferase	CAA07672	+	−	−	−	−
VC0236/A2616	*waaF*	HepII transferase	AAG08397	+	+	+	+	+
VC0237/A2617	*waaL*	O-antigen ligase		+	+	+	+	+
VC0238/A2618	*wavK*	*O*-acetyl transferase	AAB99067	+	+	+	+	−
VC0239/A2619	*wavL*	Glycosyltransferase	AAL52784	+	+	+	+	+
VC0240/A2620	*gmhD*	ADP-L-glycero-D-mannoheptose-epimerase	AAC22768	+	+	+	+	+
V194	*wavM*	Glycosyltransferase (galactosyl?)	AAA60375	−	−	+	+	−
V209	*wavN*	*O*-acetyl transferase	CAC48950	−	−	−	+	−
V209	*wavO*	*O*-acetyl transferase	AAF34147	−	−	−	+	−
V192	*wavP*	Glycosyltransferase	AAK95997	−	−	−	−	+
V192	*wavQ*	Unknown		−	−	−	−	+

V192	*wavR*	Fuc4NAc pathway (TDP-4-oxo-6-deoxy-D-glucose transaminase)	AAL22774	–	–	–	–	+
V192	*wavS*	Fuc4NAc pathway	AAF33463	–	–	–	–	+
V192	*wavT*	Glycosyltransferase (galactosyl?)	AAK91721	–	–	–	–	+

[a]Either the ORF designation of the genome sequence of *V. cholerae* El Tor n16961 (VC number) and classical O395 (A numbers) or the name of the strain (V number) from which the gene has been sequenced

[b]Accession number corresponds to the protein sequence available in the database that shows a high level of similarity to the protein sequence deduced from the ORF in the same row

[c]*V. cholerae wav* gene cluster types; + shows the presence and – shows the absence of the corresponding gene (ORF) in the particular type

5.4 O-Antigen Polysaccharide (O-PS) of *V. cholerae*

5.4.1 V. cholerae *O1*

A good account of the genetics of O-antigen biosynthesis of *V. cholerae* was presented by Stroeher et al. (1998), which, along with subsequent developments in this area, form a substantial part of this chapter. The genetic regions required for the biosynthesis of the *V. cholerae* O1 O-antigen of the Inaba and Ogawa serotypes were cloned and expressed in *E. coli* K-12 by Manning et al. (1986). These genes could be localized to a region of about 16–19 kb within a 20 kb *Sac*I fragment by extensive restriction and heteroduplex analyses of independent cosmid clones. This region was shown to contain all of the information necessary for synthesis in a heterologous host (Manning et al. 1986, Ward et al. 1987) and referred to as the *wbe* region (Reeves et al. 1996). It was assumed that the enzymes required for the biosynthesis of O-antigen from common cellular intermediates were encoded in the *wbe* region. Subsequent transposon mutagenesis of this region in the *V. cholerae* O1 chromosome confirmed that this region was required for O-antigen biosynthesis (Ward and Manning 1989). Subsequently, three additional genes, designated *wbeU, wbeV*, and *wbeW*, were described and were found to be essential for O-antigen biosynthesis in *V. cholerae* O1, since mutations in these genes led to the loss of O-antigen biosynthesis (Fallarino et al. 1997). The organization of O-PS gene clusters of *V. cholerae* O1 is shown in Fig. 5.2a. These O-antigen biosynthesis genes of *V. cholerae* O1 are organized in a cluster on chromosome I of the completely sequenced strain N16961, between the open reading frames (ORFs) VC0240 (*gmhD*) and VC0264 (*rjg*) (Heidelberg et al. 2000). The gene *gmhD* encodes D-glycero-D-mannoheptose-1-phosphate guanosyltransferase, involved in LPS core biosynthesis, and is present in the left junction, while the *wbe* gene cluster is flanked by *rjg* encoding a conserved hypothetical protein with similarities to mRNA 3′-end processing factor at the right junction.

The *wbe* gene cluster of *V. cholerae* O1 consists of five major regions: (1) perosamine biosynthesis (*manC, manB, gmd*, and *wbeE*) (Stroeher et al. 1995a); (2) O-antigen transport (*wbeG, wzm*, and *wzt*) (Manning et al. 1994,1995); (3) tetronate biosynthesis (*wbeK, wbeL, wbeM, wbeN*, and *wbeO*) (Stroeher et al. 1998); (4) O-antigen modification (*wbeT*) (Stroeher et al. 1992), and; (5) additional required genes (*wbeU, wbeV*, and *wbeW*) (Fallarino et al. 1997); the genes involved in the corresponding biosynthetic pathways are cited in the neighboring parentheses. The ORF *tnp* shown in Fig. 5.2 to be present between *wbeP* and *wbeT* of the O-antigen gene cluster was later shown to be a defective insertion sequence IS1358d1. The JUMP-start sequence described by Hobbs and Reeves (1994), which contains important regulatory information (Bailey et al. 1997), lies at the start of the *wbe* operon of O1 and immediately downstream of the left junction gene, *gmhD*.

The pathway for perosamine biosynthesis involving the actions of several genes is illustrated in Fig. 5.3. It may be noted that the gene involved in the synthesis of GDP perosamine from the substrate GDP-4-keto-6-deoxymannose remains to be

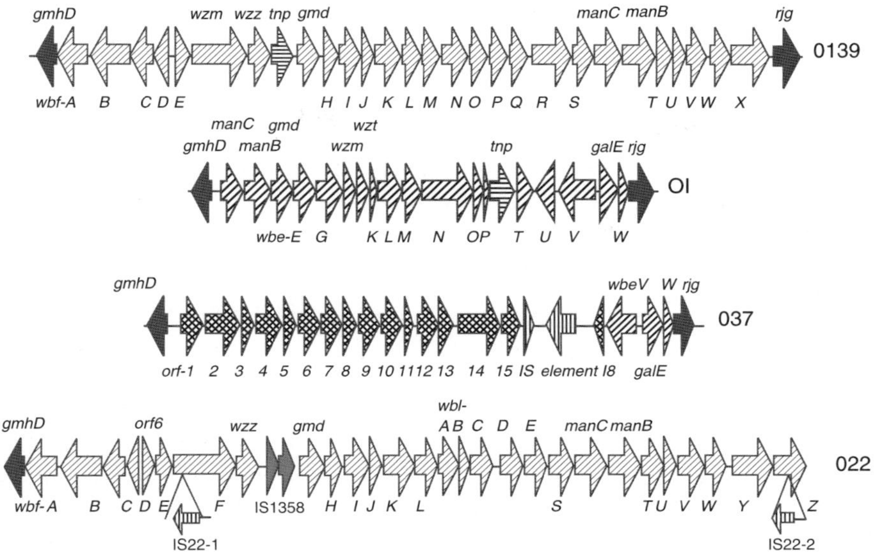

Fig. 5.2 Organization of the O-PS biosynthetic genes (*wb** region) of *V. cholerae* of four different serogroups. The directions of transcription of the different ORFs are shown by the *arrows*. The left (*gmhD*) and the right (*rjg*) junction genes are indicated by *solid arrows*. The wbe region (*manC* through *wbeW*) of serogroup O1 is 22,000 bp in length. The wbf region (*wbfA* through *wbfX*) of the serogroup O139 is 35,807 bp in length. It may be noted, however, that the ORF corresponding to *wbfX* was designated *wbfY* and another previously unidentified ORF, *wbfZ*, was identified at the 3-end of *wbfY* by Yamasaki et al. (1999). The ORF termed *trp* was later designated as IS1358 for 0139 and IS1358d1 for O1. The *wb** region (ORF-1 through *wbeW*) of serogroup O37 is 27,552 bp in length. The left junction is at *gmhD* gene, but the right junctions of O37 and O139 are different. On the other hand, the O1 right junction genes, *wbeV* through *rjg*, are conserved in the O37 *wb** region. The *wb** region of serogroup O22 is 35,900 bp in length. The *arrows below wbfF and wbfZ* show genes found in insertion sequences, encoding proteins homologous to transposases. All genes within the O22 O-PS gene cluster, with the exception of *wbfA* and *wblA* to *wblE*, have very high homology to corresponding genes found in the O139 serogroup. (From Chatterjee and Chaudhuri 2004)

identified. It may be *wbeE* or the immediately adjacent gene *wbeG*. The gene *wbeG* encodes a glycosyl transferase (Heidelberg et al. 2000) and its function has not yet been shown, although it is speculated to encode O-antigen polymerase or perosamine transferase (Stroeher et al. 1998). The genes *wzm* and *wzt* appear to be involved in polysaccharide export. The Wzm is predicted to be an integral membrane protein containing six α-helical transmembrane spanning domains and is thought to act as a carrier or pore. Wzt is homologous to export proteins and contains ATP-binding domains. It is likely that Wzm and Wzt are present as homodimers in the export complex (Manning et al. 1995).

The tetronate biosynthesis pathway is illustrated in Fig. 5.4. Here, the identities and actions of different enzymes have been drawn mostly from homology studies. The similarity of WbeN to both LuxC and LuxE (*Vibrio harveyi*) suggests that it

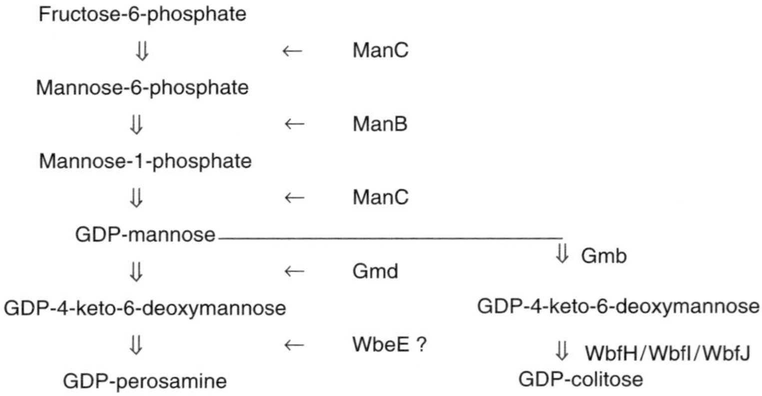

Fig. 5.3 Pathway for the biosynthesis of perosamine in *V. cholerae* O1 and colitose in *V. cholerae* O139, as proposed by Stroeher et al. (1998). The proteins acting at different stages in the biosynthesis of perosamine are ManC, ManB, ManC, Gmd, and WbeE. The action of WbeE (which shows homology to pyridoxal-binding proteins from several organisms, and many of which are involved in the synthesis of sugars similar to perosamine) as a perosamine synthetase remains, however, to be confirmed. In the O139 serogroup, GDP-mannose is converted to GDP-4-keto-6-deoxymannose and then to colitose, most likely by the actions of proteins encoded by the region containing *wbfH, wbfI,* and *wbfJ* ORFs. These ORFs show good homology to three genes in the *E. coli* O111 *wb** region, to which no defined function has been ascribed, but *E. coli* O111 also contains colitose in its LPS (Stroeher et al. 1998). Gmb is actually the Gmd homolog of *V. cholerae* O1 and plays a role in the conversion of GDP-mannose to GDP-4-keto-6-deoxymannose, which is required for the synthesis of colitose in O139 (Stroeher et al. 1995b, 1997). (From Chatterjee and Chaudhuri 2004)

has both enzymatic activities, i.e., fatty acid reductase and acyl protein synthetase. WbeM appears to be an iron-containing alcohol dehydrogenase, WbeL an acetyl-CoA synthetase, and WbeO an acetyl-CoA transferase. Among the three additional genes identified as being associated with O-antigen biosynthesis of *V. cholerae* O1, the putative functions of WbeU and WbeW are mannosyl transferase and galactosyl transferase, respectively. WbeV is, however, a conserved hypothetical protein. The ORF between *wbeV* and *wbeW*, originally designated ORF 35.7 (Fallarino et al. 1997), is homologous to *galE* of *E. coli* coding for UDP-glucose-4-epimerase (Heidelberg et al. 2000). Interestingly, the genes *wbeU* and *wbeV* are transcribed in the opposite direction to the rest of the *wbe* operon and appear to be translationally controlled (Fallarino et al. 1997). On the other hand, *galE* and *wbeW* are transcribed in the same direction as that of the *wbe* gene cluster. Chromosomal insertion mutation in *wbeV, wbeU*, or *wbeW* caused the loss of O-antigen (as stated earlier) and also resistance to the vibriophage CP-T1, suggesting that these genes were essential for O-antigen biosynthesis. Mutational studies further showed that *galE* was not required for O-antigen biosynthesis (Fallarino et al. 1997). These genes would not have been expressed in any of the constructs in which *V. cholerae* serotype-specific O-antigen was being made in *E. coli* K-12 or other heterologous hosts such as *Salmonella* (Manning et al. 1986). It was possible that these hosts might have

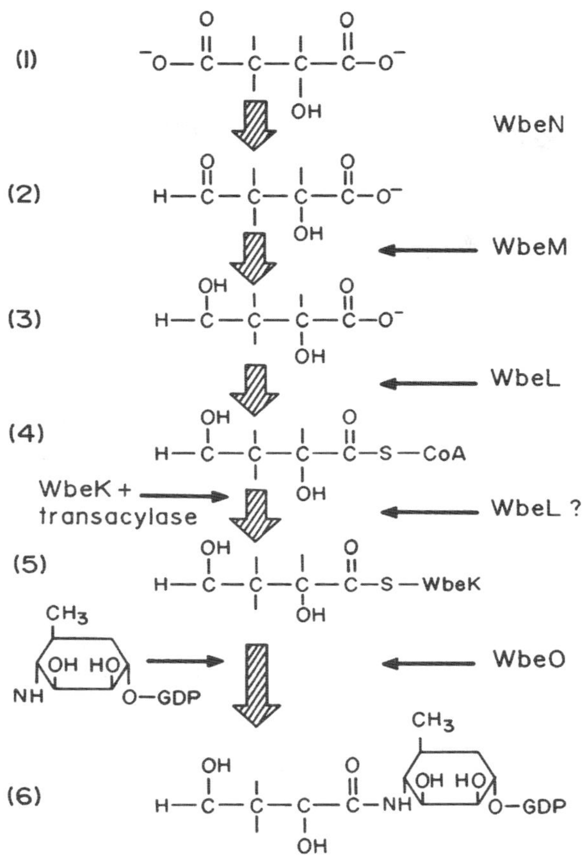

Fig. 5.4 Pathway for the biosynthesis of tetronate as proposed by Stroeher et al. (1998). The proposed substrates for the various enzymes at the different stages of biosynthesis are malate (1), aldehyde (2), dihydroxy carboxylic acid (3), a coenzyme-A (4), and an activated acyl carrier protein (ACP) (5). This ACP activated precursor can be condensed with a molecule of perosamine to give rise to a complete O-antigen subunit (6). (From Chatterjee and Chaudhuri 2004)

possessed *wbeU, wbeV*, and *wbeW* homologs, or that the expression might be at a suboptimal level (Fallarino et al. 1997).

V. cholerae of serogroup O1 is subdivided into three serotypes, Inaba, Ogawa, and Hikojima (see Chap. 3 and also Chatterjee and Chaudhuri 2003). Structural studies discovered that the O-PSs of the two serotypes, Ogawa and Inaba, differed in that the terminal 4-*N*-tetronylated-D-perosaminyl group in the O-PS of the Ogawa strain is methylated at O-2. Since the *wbe* gene cluster of *V. cholerae* O1 encodes the O-antigen biosynthesis of both Inaba and Ogawa serotypes, it is logical to assume that the gene(s) responsible for serotype specificity should lie within this region. Complete sequencing of the *wbe* operons from the Inaba and Ogawa strains revealed a single base pair deletion in the *wbeT* gene. Further analysis of *wbeT* genes from a number of Inaba and Ogawa strains indicated that Inaba strains all appeared to have a mutated/truncated *wbeT* gene. Also, the introduction of *wbeT* alone from an Ogawa into an Inaba strain allowed serotype conversion to Ogawa, and the construction of defined mutations in the *wbeT* gene of an Ogawa strain resulted in the Inaba serotype (Stroeher et al. 1992). This explained why the conversion from Ogawa to Inaba occurred at a much higher frequency than the converse, since this only required a mutation in the *wbeT* gene whereas the converse required a specific base correction. Although *wbeT* was involved in modifying the O-antigen, it was not vital for O-antigen synthesis in *V. cholerae* and existed as a separate transcriptional unit. It is, however, important to note that *wbeT* is preceded by a defective insertion element IS1358d1, which defines the end of the actual *wbe* operon. The *wbeT* and the additional *wbe* genes described by Fallarino et al. (1997) were quite separate from the rest of the *wbe* operon. A summary of the possible functions of *V. cholerae* O1 *wbe* genes is presented in Table 5.2.

5.4.2 V. cholerae *O139*

The O139 *wbf* region is about 35 kb in length and was collectively sequenced by several groups of workers. The 14.363 kb sequence of the left part of the *wbf* region, *gmhD* to *gmd*, was reported by Bik et al. (1996), the 12.938 kb right part of the *wbf* region, *wbfQ* to *wbfX*, was reported by Comstock et al. (1996), and the intermediate region was reported by Stroeher et al. (1997). *V. cholerae* O139 was a consequence of the loss of the original *wbe* genes from O1 and the acquisition of a new set. Specifically, it was demonstrated that the serogroup O139 resulted from a 22 kb deletion of the *wbe* region of O1 and replacement with a 35 kb *wbf* region encoding the O139 O-antigen (Comstock et al. 1996). The sequenced O1 and O139 *wb** regions contained only a 30-bp overlap at the right junction (Bik et al. 1996, Fallarino et al. 1997, Stroeher et al. 1998). The gene, *gmhD*, involved in core biosynthesis forms the left junction, *rjg* the right junction, and the insertion sequence IS1358 (to replace *tnp* shown in Fig. 5.2) is situated at an intermediate position in the *wbf* region (Manning et al. 1994, Stroeher et al. 1995b, c, 1998, Bik et al. 1996, Fallarino et al. 1997). The region involved in the biosynthesis of the *V. cholerae* O139 surface polysaccharide is complex and composed of several independent transcriptional units (Stroeher et al. 1998). Mutational analysis had shown that there were some genes involved in both capsule and O-antigen biosynthesis as well as genes specific for either capsule or O-antigen (Waldor et al. 1994, Comstock et al. 1995, Stroeher et al. 1995a, Bik et al. 1996).

The two open reading frames (ORFs) (to the left of IS1358) were designated (Bik et al. 1995) *wzm* and *wzz*, and their putative functions were capsule transport and O-antigen chain length regulation, as derived from homology studies with genes of some other bacteria. Mutations in both *wzm* and *wzz* appeared to affect capsule but not O-antigen biosynthesis (Bik et al. 1996). The remaining region between IS1358 and *gmhD* was reported to contain *wbfE, wbfD, wbfC, wbfB*, and

Table 5.2 *V. cholerae* O1 *wbe* genes and their possible functions (amended version of Table 3 from Chatterjee and Chaudhuri 2004)

Genes	G + C content (%)	Length of protein (No. of aa)	Putative function[a]		Locus no.[b]	
			Stroeher et al. (1998)	Heidelberg et al. (2000)	*V. cholerae* El Tor N16961	*V. cholerae* classical O395
1	2	3	4	5	6	7
gmhD	46.1	314	Core biosynthesis	ADP-L-glycero-D-mannoheptose-6-epimerase	VC0240	A2620
manC	42.1	465	PMI/GMP activity	Mannose-1-phosphate guanylyltransferase	VC0241	A2621
manB	38.6	463	Phosphomannomutase	Phosphomannomutase	VC0242	A2622
gmd	39.6	373	Oxidoreductase	GDP-mannose 4,6-dehydratase	VC0243	A2623
wbeE	40	367	Pyridoxal binding protein	Perosamine synthase	VC0244	A2624
wbeG	35.1	463	Unknown	Unknown	VC0245	A2625
wzm	35.1	257	O-antigen export (channel?)	LPS/O-antigen transport protein	VC0246	A2626
wzt	36.5	250	O-antigen export (energizer?)	LPS/O-antigen transport protein	VC0247	A2627
wbeK	33.3	80	Acyl carrier protein	Acyl carrier protein, putative	VC0248	A2628
wbeL	37.7	471	Fatty acid ligase	WbeL protein	VC0249	A2629
wbeM	42.5	374	Alcohol dehydrogenase	Iron containing alcohol dehydrogenase family protein	VC0250	A2630
wbeN	42.2	825	Fatty acid reductase	Acyl protein synthase/ acyl-CoA reductase	VC0251	A2631
wbeO	39	188	Acetyl-CoA transferase	Acetyltransferase WbeO, CysE /LacA/LpxA/ NodL family	VC0252	A2632

(Continued)

Table 5.2 (continued)

Genes	G + C content (%)	Length of protein (No. of aa)	Putative function[a]		Locus no.[b]	
			Stroeher et al. (1998)	Heidelberg et al. (2000)	*V. cholerae* El Tor N16961	*V. cholerae* classical O395
1	2	3	4	5	6	7
wbeP	43.1	73	Unknown			
IS135 8d1	42.7	65,34,68	Insertion sequence			
wbeT	31.1[c]	287[c]	Ogawa determination	WbeT protein	VC0258	A2636
wbeU	35.1[c]	370[c]	Mannosyl transferase	Mannosyl transferase	VC0260	A2638
wbeV	43.5[c]	621[c]	LPS biosynthesis	LPS biosynthesis protein	VC0259	A2637
ORF 35.7/*galE*	45.2	323	dTDP-glucose dehydratase	UDP-galactose 4-epimerase	VC0262	A2639
wbeW	44.1	184	Galactosyl transferase	Galactosyl transferase, putative	VC0263	A2640

Locus numbers of the strain classical O395 have been added in column 7

[a] Putative functions of any gene as presented by two different groups of workers have been reported here without modification

[b] Locus numbers refer to the strains *V. cholerae* El Tor n16961 and classical O395, available as completed sequences. The data on classical strain O395 were obtained from the NCBI genome database at http://www.ncbi.nlm.nih.gov/genomes/lproks.cgi

[c] Corresponding values given by Heidelberg et al. (2000) are somewhat different and are not mentioned here

wbfA, none of which showed any homology to polysaccharide biosynthesis genes, and their roles were unknown (Bik et al. 1996). It was further reported that an ORF downstream from IS1358 showed a high degree of homology to Gmd, the oxidoreductase from *V. cholerae* O1, indicating the presence of polysaccharide genes on both sides of IS1358 in *V. cholerae* O139. Comstock et al. (1996) identified a number of other ORFs that showed homology to polysaccharide biosynthetic proteins, and defined the other end-point of the novel O139 *V. cholerae* DNA.

Another point of interest to be noted here is that, unlike the *wbe* region for *V. cholerae* O1, that for O139 does not appear to be made up of clearly defined regions corresponding to the specific biosynthetic pathways. The O139 polysaccharide biosynthetic cluster has a number of redundancies. One example is that there are at least two possible galactosyl transferases and a number of UDP-galactose-4-epimerases involved in the synthesis of UDPgalactose. Figure 5.2 and Table 5.3 present the genetic organization of the *wbf* region of *V. cholerae* O139 and the possible functions of the O139 *wbf*/capsular genes, respectively. *V. cholerae* O139 was shown to have a *wzy* homolog (WbfQ/Wzy). The possible function of this homolog was speculated on, particularly since Wzy protein was known to be involved in O-antigen polymerization, and only a single O-antigen subunit sugar was substituted on the core in *V. cholerae* O139. Subsequent mutation analysis of the *wbf* /capsule gene cluster of *V. cholerae* O139 clearly showed that the Wzy homolog was solely involved in capsule biosynthesis (Stroeher et al. 1998). Another point of interest is that the *wbf* region of *V. cholerae* O139 contains only one partner of the two-component polysaccharide export system. The O139 contains only a gene (*wzt*) encoding the transmembrane protein but no energizing partner (Bik et al. 1995). Further, *wzt* was shown to be an essential gene in *V. cholerae* O139 capsule biosynthesis (Bik et al. 1996).

Yamasaki et al. (1999) independently reported the sequence of the O139 O-PS gene cluster, which in general confirmed the works published previously by others (Bik et al. 1996, Comstock et al. 1996, Mooi and Bik 1997, Stroeher et al. 1997). However, the data of Yamasaki et al. (1999) showed some differences. The size of the genes and homologous proteins encoded by the genes were different in the cases of *wbfK, wbfM, wbfQ,wbfS, manC, wbfX*, and *wbfY* when compared with earlier reports (Comstock et al. 1996, Stroeher et al. 1997). According to Comstock et al. (1996) the ORFs found next to ORF9 (*wbfW*) were ORF10 and ORF11, which were designated *wbfX* and *wbfY*, respectively. Yamasaki et al. (1999) found, on the other hand, a large ORF in the region corresponding to the *wbfX* and *wbfY* of Comstock et al. (1996), and designated this large ORF *wbfY*. Yamasaki et al. (1999) observed that the separated ORFs (*wbfX* and *wbfY*) found previously may have resulted from mutations. There was a potential gene between *wbfD* and the repeat unit and JUMPstart sequence upstream of *wbfE* which was not previously identified, but the authors were not very definite about its identity, as the ORF was too short. *wbfZ*, which was not previously documented, was identified at the 3′-end of *wbfY*. The sequence recorded by Yamasaki et al. (1999) also contained a 7.8 kb segment upstream of the O139-specific sequence and a 2.2 kb segment downstream of it.

Table 5.3 *V. cholerae* O139 *wbf* genes and their possible functions (reproduced version of Table 4 from Chatterjee and Chaudhuri 2004)

Genes	G + C content (%)	Length of protein (no. of aa)	Putative function[a]	
			Stroeher et al. (1998)	Yamasaki et al. (1999)
gmhD	46	314	Core biosynthesis	ADP-L-glycero-D-mannoheptose-6-epimerase
wbfA	34.5	505	Unknown	Unknown
wbfB	48.8	730	Unknown	Unknown
wbfC	49.1	288	Unknown	Unknown
wbfD	45.7	229	Unknown	Unknown
wbfE	44.7	196	Unknown	Unknown
wbfF (wzm)	47.3	911	O-antigen export	Putative capsule transport protein/precursor
wzz	45.9	335	Regulation of O-antigen chain length	Regulation of O-antigen chain length
IS1358	42.9	375	Insertion sequence	DNA transposition
gmd	46.6	372	Oxidoreductase;	GDP-D-mannosedehydratase oxidoreductase; involved in the biosynthesis of GDP-fucose from GDP-mannose
wbfH	43.6	308	Colitose biosynthesis	Probable UDP-glucose-4-epimerase
wbfI	41.4	390	Colitose biosynthesis	Probable CDP-4-keto-6-deoxy-D-glucose-3-dehydrogenase
wbfJ	35.2	161	Colitose biosynthesis	Unknown
wbfK	34.4	478	Unknown	Unknown
wbfL	30.2	338	Unknown	Probable fucosyl transferase
wbfM	29.3	337	Unknown	Probable bacteriophage integrase
wbfN	30.2	485	Unknown	Probable K5 antigen synthesis
wbfO	29.7	337	Glycosyl transferase	Glycosyl transferase
wbfP	33.9	310	Galactosyl transferase	Probable glycosyl transferase
wbfQ (wzy)	28	354	Rfc-like O-antigen polymerase	Unknown
wbfR	43.1	636	Asparagine synthetase	Asparagine synthetase
wbfS	41.3	377	Galactosyl transferase	Sugar transferase
manC	46.8	460	GDP-mannose pyrophosphorylase and phosphomannose isomerase	GDP-mannose pyrophosphorylase and phosphomannose isomerase
manB	46.2	483	Phosphomannosemutase	Phosphomannosemutase
wbfT	41.6	328	UDP-galactose-4-epimerase	UDP-galactose-4-epimerase
wbfU	42.7	186	Galactosyl transferase	Galactosyl transferase
wbfV	45.5	388	Nucleotide sugar dehydrogenase	Nucleotide sugar dehydrogenase
wbfW	44.4	334	Nucleotide sugar epimerase	Nucleotide sugar epimerase

(*Continued*)

Table 5.3 (continued)

Genes	G + C content (%)	Length of protein (no. of aa)	Putative function[a]	
			Stroeher et al. (1998)	Yamasaki et al. (1999)
wbfX	44.6	227	Unknown	Not reported
wbfY	46.6	378	Unknown	Unknown (prot length 646)
wbfZ	54.1	446	Not reported	Unknown

[a]Putative functions of any gene as given by two different groups of workers have been reproduced here without modification

5.4.3 V. cholerae *of Serogroups Non-O1 Non-O139*

The O-PS genetic organization of *V. cholerae* of serogroups non-O1 non-O139 has not been adequately studied. Some relevant information on the serogroups O37 and O22 are available and discussed below. Some other members of the non-O1, non-O139 serogroups have been investigated to elucidate the homologies between their O-PS biosynthetic genes and those of O139, and a brief account of this is presented below.

5.4.3.1 *V. cholerae* O37

Among the various non-O1 non-O139 serogroups, the serogroup O37 deserves attention since the strains of this serogroup have been implicated in localized cholera outbreaks in the past (Aldova et al. 1968, Kamal 1971) and thus have clinical relevance. The O37 O-PS genetic region, shown in Fig. 5.2, was sequenced, and the entire O37 *wb** genetic region was found 27,552 bp long, while the O37-specific sequence was 23,388 bp long (Li et al. 2002b). The O37 *wb** cluster is unique, since no significant DNA homology between the O37 specific sequences and the previously published *V. cholerae wb** cluster sequences was found, except at the right junction. Twenty-three ORFs were identified in the O37 *wb** region; many of the ORFs (ORF1 to ORF13 and ORF18) encoded enzymes involved in polysaccharide biosynthesis, ORF14 and ORF15 encoded hypothetical proteins of unknown functions, and ORF14 had very weak homology to *yhf0* (which encoded a hypothetical protein) of *Bacillus subtilis*. A 1549-bp promoter region separates *gmhD* and ORF1, and this region contains a putative promoter and *ops* elements found in known polysaccharide biosynthesis regions (Hobbs and Reeves 1994, Bailey et al. 1997). There is an IS element in the interval between the region that is unique to O37 and the right junction. The region downstream of the IS element has three ORFs almost identical to the O1 *wbe* cluster: *wbeV, galE*, and *wbeW*, followed by *rjg*. The homology breakpoints (i.e., the DNA sequences where the common backbone sequence ends and the serogroup-specific sequence starts) in O37 strains are different from those of the O1 and O139 junctions. In the O37 serogroup, the left and right junctions are at *gmhD* and *wbeV*, respectively, instead of at *gmhD* and *rjg* as they are in O139.

5.4.3.2 *V. cholerae* O22

The O-antigen cluster of the O22 serogroup, shown in Fig. 5.2, is 35.9 kb long (Aldova et al. 1968). Insertion sequences designated IS22-1 and IS22-2, which were flanked by repeat sequences, were found in the putative region of *wbfF* and *wbfZ*, respectively. If these ISs are deleted, genes which correspond to *wbfF and wbfZ* will remain intact. Therefore, the IS sequence interrupts the translation of the *wbfF* and *wbfZ*. The 35.9 kb region was found to contain 30 ORFs and the hypothetical *wbfF*, and the *wbfZ* found in the O139 serogroup was also present at the 3′-end of *wbfY*, as shown in Fig. 5.2 and Table 5.4. The transcriptional directions of *gmhD to wbfD* are all from right to left, while those of *wbfE* to *wbfZ* are all from left to right in accordance with those of genes in the O139 serogroup. The products IS22-1 and IS22-2 have 100% homology to each other (Table 5.4). This IS encodes a protein homologous to TnpA of *B. thuringiensis* (Baum 1994) or hypothetical protein 2131 of spinach (Zhou et al. 1988), as shown in Table 5.4, suggesting that transposons might be involved in horizontal gene transfer of the O-antigen biosynthesis gene of the O22 serogroup.

The JUMPstart sequence, which is a conserved 39 bp sequence usually located in the noncoding upstream region of several bacterial gene clusters (Hobbs and Reeves 1994), was also found in the region between *wbfD* and ORF6. In addition, just upstream of the JUMPstart sequence, seven tandemly repeated 7 bp units [G(G/A)(A/T)(C/T)CTA] were observed, the function of which was not clear. Analysis of GC content in individual genes in the O22 serogroup revealed that although the GC contents of most of the genes investigated were in the range 41–54%, the GC contents of *wbfA* and *wbfJ* to *wblE* were 28–35%, which is lower than that of the *V. cholerae* genome, indicating that *wbfA* and *wbfJ* to *wbfQ* may have originated from an organism other than *V. cholerae*.

The accession numbers that provide entry to the recorded sequences of (i) the core-PS *wav* genes or gene clusters and (ii) the O-PS gene clusters of O1 and several non-O1 serogroups as well as the complete genome of *V. cholerae* along with the corresponding references are given in Tables 5.5 and 5.6, respectively, for the convenience of general readers.

5.4.3.3 Other Non-O1 Non-O139 *V. cholerae*

Many of the other non-O1 non-O139 strains were investigated to determine the extent of homology of their O-PS genes with those of O139, but a detailed genetic analysis is not available. It was suggested that the evolution of the *wbf* region of *V. cholerae* O139 might have occurred via the acquisition of genes from a number of different ancestors. Several groups had shown that at least some of the DNA associated with O139 surface polysaccharide biosynthesis was also found in non-O1 *V. cholerae* serogroups (Bik et al. 1995, 1996, Mooi and Bik 1997). Two groups of workers (Bik et al. 1996, Stroeher et al. 1997) showed that an ORF in the region downstream of IS1358 of the O139 strain was approximately 80% identical to the

Table 5.4 Identities of potential ORFs found in the O-antigen biosynthesis region of *V. cholerae* O22 to reported homologs (data taken from Yamasaki et al. 1999; reproduced version of Table 5 from Chatterjee and Chaudhuri 2004)

ORF	Size (bp)	G + C (%)	No. of amino acids	Homologous protein encoded by	Identity (%)/(aa overlap)	Possible function
gmhD	942	45.81	314	O139 *gmhD*	97.5/314	LPS core synthesis
wbfA	1521	38.38	507	O139 *wbfA*	37.2/522	Unknown
wbfB	2190	48.60	730	O139 *wbfB*	98.9/730	Unknown
wbfC	864	49.01	288	O139 *wbfC*	99.7/288	Unknown
wbfD	687	45.55	229	O139 *wbfD*	98.7/229	Unknown
ORF6	231	49.57	77	O139 orf6	97.4/77	Unknown
wbfE	588	44.66	196	O139 *wbfE*	100.0	Unknown
wbfF[a]	2736	47.14	912	O139 *wbfF(wzm)*	99.6/911	Transport of capsular precursors
Wzz	828	46.32	276	O139 *wzz*	98.9/265	Regulation of O-antigen length
IS1358	300	40.92	100	O139 IS1358	98.0/99	Insertion sequence-like element
IS1358	552	42.75	183	O139 IS1358	97.7/176	Insertion sequence-like element
Gmd	1116	46.81	372	O139 *gmd*	99.5/372	O-antigen synthesis oxidoreductase
wbfH	924	43.79	308	O139 *wbfH*	97.1/308	Unknown
wbfI	1170	42.44	390	O139 *wbfI*	98.5/390	CDP-4-keto-6-deoxy-D-glucose-3-dehydrogenase
wbfJ	483	34.76	161	O139 *wbfJ*	98.1/161	Unknown
wbfK	1437	34.02	479	O139 *wbfK*	92.7/479	Unknown
wbfL	1038	35.15	346	O139 *wbfL*	60.9/343	Fucosyl transferase
wblA	843	35.46	281	Fuc-TIv of human	29.7/111	Alpha 1,3-fucosyl transferase
				Alpha(1,2)FT of human	29.2/185	Beta-D-galactoside 2-alpha-L-fucosyltransferase
wblB	483	32.70	161	*lacA* of *E. coli*	39.6/91	Galactoside-*O*-acetyltransferase
wblC	1020	33.13	340	O139 *wbfO*	43.8/333	Unknown
				Igt D homolog of *H. influenzae*	31.8/170	Glycosyl transferase

(*Continued*)

Table 5.4 (continued)

ORF	Size (bp)	G + C (%)	No. of amino acids	Homologous protein encoded by	Identity (%)/(aa overlap)	Possible function
wblD	1194	27.56	398	NADH dehydrogenase of *Trypanosoma brucei*	18.9/375	NADH dehydrogenase
wblE	1077	30.36	359	rfaG of *E. coli*	24.2/157	Glucosyl transferase
wbfS	1074	41.03	358	O139 *wbfS*	93.3/357	Sugar transferase
manC	1395	46.49	465	O139 *manC*	99.6/465	GDP-mannose pyrophosphorylase and phosphomannose isomerase
manB	1449	46.00	483	O139 *manB*	99.0/483	Phosphomannomutase
wbfT	963	40.77	321	O139 *wbfT*	99.7/321	UDP-galactose-4-epimerase
wbfU	549	40.75	183	O139 *wbfU*	98.9/183	Galactosyl transferase
wbfV	1164	45.75	388	O139 *wbfV*	97.9/388	Nucleotide sugar dehydrogenase
wbfW	1002	44.17	334	O139 *wbfW*	99.1/334	Nucleotide sugar epimerase
wbfY	1938	46.26	646	O139 *wbfY*	98.0/646	Unknown
wbfZ[a]	1323	54.26	441	O139 *wbfZ*	97.3/446	Unknown
IS22-1	1596	46.23	372	IS22-2	100	Unknown
IS22-2	1596	46.07	372	IS22-1	100	Unknown
				tnpA of *Bacillus thuringiensis*	25.0/36	Transposase
				Hypothetical protein 2131 of spinach	35.3/17	Transposon-mediated origin

[a]Calculated nucleotides of *wbfF* and *wbfZ* were excluded from the nucleotide sequences of IS22-1 and IS22-2

Table 5.5 GenBank accession numbers of the recorded sequences of some Core-PS biosynthetic genes or gene clusters of different *V. cholerae* strains (reproduced version of Table 6 from Chatterjee and Chaudhuri 2004)

Gene or gene cluster[a]	Strain	Serogroup, biotype or serotype	*Wav* cluster type	Accession number
waaL	P27459	O1, El Tor	1	AF443420
waaL	O395	O1, classical, Ogawa	1	AF443421
waaL	V243	O1	1	AF443422
waaL	M010	O139	1	AF443423
waaL	V244	O37	1	AF443424
waaL	V207	Non-O1, non-O139	1	AF443425
wavC–wavE	V215	Non-O1, non-O139	2	AF443426
wavI–wavK	V215	Non-O1, non-O139	2	AF443845
wavC–wavE	V194	Non-O1, non-O139	3	AF44793
wavI–wavK	V194	Non-O1, non-O139	3	AF44794
wavC–wavE	V209	O141	4	AF443847
wavI–wavK	V209	O141	4	AF444792
wavI–wavK	V208	Non-O1, non-O139	4	AF443846
wavC–waaF	V192	O6	5	AF449195
waaF–gmhD	V192	O6	5	AF449194

Data taken from Nesper et al. (2002)
[a]Cluster is represented by the names of the first and the last genes separated by a hyphen

Table 5.6 GenBank accession numbers of the recorded sequences of the complete genome and O-PS gene clusters of different *V. cholerae* strains (reproduced version of Table 7 from Chatterjee and Chaudhuri 2004)

Gene cluster	*V. cholerae* strain[a]	Accession no.[b]	References[c]
Complete genome			
Chromosome 1	El Tor N16961	AE 003852	Heidelberg et al. (2000)
Chromosome 2	El Tor N16961	AE 003853	Heidelberg et al. (2000)
Chromosome 1	Classical O395	CP000626	Heidelberg (unpublished)
Chromosome 2	Classical O395	CP000627	Heidelberg (unpublished)
O-PS gene clusters of serogroups			
O1	569B	Y07788	Fallarino et al. (1997)
	017	X59553	Stroeher et al. (1992)
			Manning et al. (1995)
O139	MO45	AB012956	Yamasaki et al. (1999)
	MO45	X90547	Bik et al. (1996)
	AI1837	U47057	Comstock et al. (1996)
	MO45	U24571	Dumontier (unpublished)[d]
	AI1837	AF090685	Sozhamannan et al. (1999)
O37	1322–69	AF390573	Li et al. (2002)
O22		AB012957	Yamasaki et al. (1999)

[a]The particular strain sequenced
[b]Accession number which will permit entry into the data bank
[c]Publication in which the sequencing is reported
[d]Obtained from the database of accession number U24571

gmd gene of *V. cholerae* O1. Further, some of the regions downstream of IS1358 were found in non-O1 and non-cholera vibrios (Comstock et al. 1995). These authors showed in addition that homologous DNA was present in not only a wide number of non-O1 *V. cholerae* strains but also in a variety of non-cholera vibrios. These studies thus indicated that a number of the *V. cholerae* O139 surface polysaccharide genes had their origins within the *Vibrionaceae*. However, the exact mechanism by which these genes were transferred remains unknown.

When DNA fragments derived from the O139 O-antigen biosynthesis gene region were used as probes, it was possible to divide the entire O139 O-antigen biosynthesis gene region into five classes, designated I–V, based on the reactivity pattern of the probes against reference strains of *V. cholerae* representing serogroups O1–O193 (Yamasaki et al. 1999). The genes in Class I included eight probes; the range of hybridization of these probes with other non-O1 non-O139 serogroups, as determined by the colony hybridization method, varied from 29 to 150 (Yamasaki et al. 1999). Genes in class II reacted with O22, O45, O53, O57, O61, O62, O75, O92, O105, O107, O112, O125, O132, O139, O141, O151, and O152 serogroups. Genes in class III reacted with only O22 and O139 serogroups. Genes in class IV reacted with only the O139 serogroup. Class V included three probes; the range of hybridization of these probes with other non-O1 non-O139 serogroups varied from 21 to 92. This study thus gave only an idea of the similarity or dissimilarity of the O-PS genetic organizations of the 154 non-O1 non-O139 serogroups with that of O139. In another interesting study covering O139 and several non-O1 non-O139 serogroups of *V. cholerae*, Sozhamannan et al. (1999) sequenced the genes downstream of the O139 *wbfX* gene and analyzed the genes flanking the *wb** gene cluster in other serogroups, including 43 non-O1 non-O139 strains. The principal results obtained included: (i) the *gmhD* and *rjg* genes were present in all of the *V. cholerae* strains examined and the IS1358 element in 61% of the strains; (ii) the *gmhD* and *rjg* genes were probably arranged similarly, flanking the O-antigen gene cluster, and; (iii) *rjg* was not an essential gene for cell viability but it was a hot spot for genetic rearrangements. These results indicated a cassette-like organization of the *wb** region, with the conserved genes (*gmhD* and *rjg*) flanking the divergent, serogroup-specific *wb** genes and IS1358. It was suggested that homologous recombination via the junction genes could be the mechanism for the emergence of new pathogens.

The possibility of homologous recombination-mediated exchange of O-antigen biosynthetic gene clusters (*wb**) was tested by suitably designed experiments (DNA dot blotting, IS1004 finger-printing and restriction fragment length polymorphism analysis) involving 300 *V. cholerae* strains of non-O1 non-O139 serogroups (Li et al. 2002b). Four non-O1 strains (serogroups O27, O37, O53, and O65) were found to have an O1 genetic backbone, suggesting an exchange of *wb** clusters. These strains were presumably derived from an epidemic strain by *wb** cluster exchange and subsequently diverged. DNA sequence analysis of the *wb** regions of one serogroup (O37) supported this conclusion. Although the virulence regions of these four strains were quite heterogeneous, they contained the entire VPI and a pre-CTXφ (CTXφ without the *ctx* genes) or a CTXφ wild-type. This study thus identified four non-O1 non-O139 serogroups that acquired pathogenic potential, the implication of which is discussed later.

5.5 Progenitor of *V. cholerae* O139

Several studies have shown that the *V. cholerae* O139 strain is phylogenetically and phenotypically very similar to O1 El Tor strains. Like El Tor strains, O139 Bengal strains have tandemly repeated chromosomal cholera toxin genes (Das et al. 1993, Johnson et al. 1994). Bengal strains also have zonula occludens toxin (Zot), accessory cholera toxin (Ace) (Johnson et al. 1994), and they have genes for and express the TcpA pilus (Hall et al. 1993, Morris 1994). Chromosomal DNAs from Bengal and El Tor strains give similar banding patterns by pulsed field gel electrophoresis and by Southern blot analysis with labelled rRNA (ribotyping) (Morris 1994, Berche et al. 1994, Popovic et al. 1995). They are also identical by multilocus enzyme electrophoresis analysis (Johnson et al. 1994). However, at least two important differences exist between O1 El Tor and Bengal strains. Manning et al. (1994) have demonstrated that O139 possesses a truncated LPS O-side chain which is unreactive with O1-specific antiserum, and that a large portion of the DNA corresponding to the *wbe* region of O1 strains is missing from the O139 strains. In addition, O139 expresses a polysaccharide capsule and the associated novel sugars resulting in the O139 serogroup (Johnson et al. 1994, Manning et al. 1994, Waldor et al. 1994, Weintraub et al. 1994, Faruque et al. 2003) (for details, see Chap. 4 and Chatterjee and Chaudhuri 2003).

The mechanisms underlying the origination of O139 and the possible future emergence of other pathogenic *V. cholerae* strains have obviously drawn the attentions of recent investigators (Faruque et al. 2003), and at least two probable hypotheses have been put forward. The first one proposed that a transposition event mediated by the element IS1358 resulted in the replacement of the O1 *wbe* genes with the O139 *wbf* genes (Stroeher et al. 1995a, 1997, Stroeher and Manning 1997, Mooi and Bik 1997). It was found that approximately 40% of all strains belonging to serogroups O1–O155 contained this or a very closely related element (Stroeher et al. 1998). *V. anguillarum*, a closely related species, also contained this element. Many of the *V. cholerae* and *V. anguillarum* serogroups that contained this element had multiple copies, indicating the mobile nature of the elements. Among other important features of this element, IS1358, it was interesting to note its homology to the H-repeat of the Rhs element in *E. coli* (Zhao et al. 1993, Hill et al. 1994, Stroeher et al. 1995a). The H-repeat present in the Rhs element of *E. coli* had been proposed to play a role in large chromosomal rearrangement (Zhao et al. 1993, Hill et al. 1994). Similarly, an H-repeat element found to be associated with the *S. enterica wb** region had also been proposed to play a role in recombination between different *wb** operons (Xiang et al. 1994). It was proposed that *V. cholerae* and *V. anguillarum* strains might exchange *wb** genes, since the homologs of *bplA* and *vipA* of *V. anguillarum* O2 were found in a number of non-O1 *V. cholerae* serogroups (Stroeher et al. 1998). The presence of IS1358 in many serogroups of both *V. anguillarum* and *V. cholerae* (Jedani et al. 1998) is suggestive of such exchanges. However, the exact role of IS1358 in the evolution and rearrangements of *wb**/capsule regions was unknown, and although IS1358 of *V. cholerae* O139 and a number of the *V. anguilarum* serogroups appeared to have all the features of a functional IS element, it was not possible to demonstrate transposition. The second hypothesis involved the role of junction genes in a homologous recombination event resulting in the replacement of the O1 *wbe* region by the O139 *wbf* region (Comstock et al. 1996, Mooi and Bik 1997, Stroeher et al. 1998). With respect to the *wb** regions of O1 and O139 strains, the genes downstream of *wbfX* were not known, and it was presumed that *wbfX* was the last gene of the *wbf* cluster. Sozhamannan et al. (1999) determined the sequences of 4 kb further downstream of the *wbfX* gene and, in addition to what was stated earlier, their study revealed that the *rjg* gene downstream of *wbfX* was 1341 bp long, and that the first 17 bp of this were different in O1 and O139 strains. Their results (as stated earlier) supported the second hypothesis and further revealed that the preferential linkage of IS1358 to the *wb** regions of most *V. cholerae* strains analyzed supported the idea that IS1358 was acquired by homologous recombination, rather than by transposition, because a random transposition event was likely to deliver the IS element to any part of the chromosome (Stroeher et al. 1997).

While O139 strains have been suggested to acquire a unique DNA region (Waldor and Mekalanos 1994, Bik et al. 1995, 1996) by homologous recombination via the junction genes, the mechanism of how the genes in this DNA region were acquired from a donor strain remained largely unknown and gave rise to many speculations. One such speculation was that the DNA was transferred via conjugation, since many of the *V. cholerae* O139 Bengal isolates carried a large conjugation plasmid (Stroeher et al. 1998). Alternatively, a phage-mediated transduction and incorporation of the relevant DNA could be another possibility. Thus, the possibility of generating new serogroups through the reassortment of genes involved in the synthesis of surface polysaccharides assumed great practical importance. Subject to the inadequate state of knowledge about the actual mode of gene transfer, it appeared reasonable to believe that an O1 El Tor strain was the progenitor of the O139 serogroup. Further, it was important to identify the donor of the O139 O-PS specific (*wbf*) gene cluster.

5.6 Genesis of O139 O-PS Gene Cluster

Several earlier studies had already suggested possible nonpathogenic serogroups of *V. cholerae* as a donor for the *wbf* cluster and the involvement of generalized transducing phages or conjugative plasmids as vectors for the large DNA region in the recombination event (Mooi and Bik 1997, Sozhamannan et al. 1999). The idea that *V. cholerae* O22 might be a putative source of exogenous DNA resulting in the emergence of the *V. cholerae* O139 had been suggested by different workers on the basis of (i) structural and chemical analysis of their LPS (Chap. 4; Chatterjee and Chaudhuri 2003), (ii) the O-antigenic relationship (Isshiki et al. 1996), (iii) cloning analysis and hybridization tests (Dumontier and Berche 1998, Yamasaki et al. 1999), (iv) the identical JUMPstart sequences found in strains of the two serogroups (Yamasaki et al. 1999), (v) the presence of seven tandemly repeated 7-bp units, G(G/A)(A/T)(C/T)CTA, in the two serogroups (Bik et al. 1996), and (vi) the sequence analysis of a 4.5 kb fragment containing IS1358 and its adjacent genes in the two serogroups (Dumontier and Berche 1998). Sequencing of the entire *wb** regions of the strains of these two serogroups was done and the similarities and differences between these two gene clusters have already been shown (Fig. 5.2).

V. cholerae strains from serogroups O141 and O69 possessed some of these O139 antigen- and capsule-biosynthesis genes upstream of the IS1358 element (Bik et al. 1996). However, O69 and O141 strains could not be exogenous DNA donors, as the *wbfF* and *wzz* genes in these strains differed substantially from those of *V. cholerae* O139 (Bik et al. 1996). Further, the O139 antibodies did not cross-react with *V. cholerae* O69 and O141 antigens and so the O139, O69 and O141 did not produce cell wall polysaccharides with common antigenic determinants (Bik et al. 1996). The O139 *wbf* region comprising *wbfA-wzz* genes is found in part in *V. cholerae* O69 and O141. Sequencing of these regions of O69 and O141 revealed that the O139 DNA had not been directly acquired from these serogroups (Bik et al. 1996). In contrast, it was reported that the *V. cholerae* strains O22 and O155 possessed antigen factors in common with *V. cholerae* O139 (Bik et al. 1996, Isshiki et al. 1996). However, *V. cholerae* O155, in which several copies of IS1358 were found, did not possess any of the flanking genes (Dumontier and Berche 1998).

All of these findings taken together suggested that strains of *V. cholerae* O22 from the environment might have been the source or donor of the exogenous DNA resulting in the emergence of the new epidemic strain O139 (Dumontier and Berche 1998, Faruque et al. 2003) from a progenitor El Tor strain of *V. cholerae* O1, as presented schematically in Fig. 5.5.

5.7 Concluding Remarks

Many new serogroups of *V. cholerae* have been and are still being discovered. There were around 155 known serogroups in 1998 (Stroeher et al. 1998), and in 2002 this number had risen to some 200 or more (Li et al. 2002b). The structure of the O-PS of *V. cholerae* of any particular serogroup has been found to be unique (Chatterjee and Chaudhuri 2003). The genetic organization encoding the O-PS biosynthesis is thus quite susceptible to change, but the factors responsible for effecting such changes are still largely unknown. On the other hand, several non-O1 and non-O139 serogroups have been found to acquire pathogenic potential in epidemic

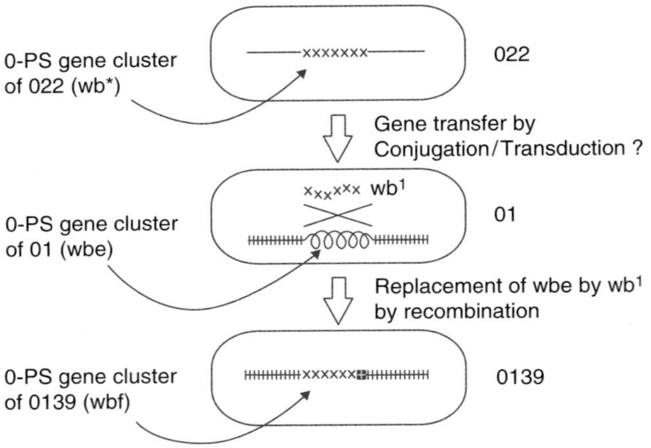

Fig. 5.5 Origination of the *V. cholerae* O139 strain as conceptualized from the data published by several authors. A major fraction (wb^1) of the O-PS gene cluster (wb^*) of a strain of O22 is donated to an ancestral strain of O1 by a phage-mediated transduction or conjugation process. The gene cluster wb^1 replaces the O-PS gene cluster (*wbe*) of O1 and gets integrated into the O1 chromosome by a recombination mechanism (×), thereby converting the ancestral O1 strain into a new O139 strain of O-PS gene cluster, *wbf*. The details, including the quantitative aspects of this gene transfer and exchange mechanism, remain to be confirmed by further experimentation. Although the O-PS gene clusters of O22 (wb^*) and O139 (*wbf*) are almost the same size, each one contains some genes which differentiate the two serogroups. This diagram is not to scale. (From Chatterjee and Chaudhuri 2004)

genetic backgrounds (Li et al. 2002b) and may cause future epidemics. The existing *V. cholerae* vaccines would provide little or no protection against such newly identified pathogenic strains. This situation thus demands that new vaccine development strategies should be evolved rapidly.

It is expected that, like the *V. cholerae* O1 strains, not all of the strains belonging to the O139 serogroup will be pathogenic. Molecular analysis of several O139 strains isolated in the recent past revealed the existence of genetically diverse strains and included both toxinogenic and nontoxinogenic variants. The new variants of *V. cholerae* O139 included strains with new ribotypes, CTX genotypes and altered antimicrobial resistance (Mitra et al. 1996, Faruque et al. 1997). The nontoxinogenic O139 strains included *N*-[1-(thienyl)cyclohexyl] piperidine (TCP)-positive variants that were susceptible to toxinogenic conversion by CTX phage (Waldor and Mekalanos 1996, Faruque et al. 2000), as well as TCP-negative strains. The *gmhD* locus (assumed to have served as a site of homologous recombination in the gene transfer event that led to the origination of O-PS gene cluster of *V. cholerae* O139), after amplification from all toxinogenic O139, El Tor and CTXφ-susceptible nontoxinogenic O139 strains, was similar in size and also similar to those produced by non-O1 non-O139 strains isolated from the environment (Faruque et al. 2000). It was presumed that these CT-negative O139 strains lacked the entire TCP pathogenicity island. From these studies, Faruque et al. (2003) concluded that these CT-negative O139 strains possibly originated from non-O1 progenitors, that the O139 antigen was present in different lineages, and that the O139 serogroup comprised both epidemic and nonepidemic strains derived separately from different progenitors. The identification of these progenitors and the mechanisms of gene transfer in such cases are, however, the subject of further investigations.

Various changes in genetic and phenotypic properties of O139 strains have now been documented, including (i) amplification or rearrangement of the CTXφ prophage, (ii) acquisition of new CTXφ prophage, (iii) restriction fragment length polymorphisms in conserved rRNA genes (ribotypes), (iv) diversity in the multilocus enzyme electrophoresis patterns, (v) change in the antimicrobial resistance pattern (antibiogram) (Mukhopadhyay et al. 1998), etc. Factors which determine the emergence and domination of particular clones of toxinogenic *V. cholerae* and the displacement of existing ones have remained largely unknown. Molecular studies revealed that *V. cholerae* strains possessed a distinctive class of integrons, which are gene expression elements that may capture ORFs and convert them to functional genes (Mazel et al. 1998). Until recently, all known integrons were associated only with genes conferring antibiotic resistance (Recchia and Hall 1995). It was suggested that integrons in *V. cholerae* might also play a role in the acquisition of pathogenic genes as well as genes for different biological functions (Mazel et al. 1998). Further work on the role of integrons in the *V. cholerae* system is expected to provide a better insight into the mechanism of acquisition of genes by this organism. It is believed that the continual emergence of new toxinogenic strains is an essential component of the natural ecosystem, and that this is likely to further complicate the development of an effective cholera vaccine (Li et al. 2002a). An effective—and possibly hypothetical at this moment—cholera vaccine should remain unaffected by such spontaneous changes occurring in nature.

Chapter 6
Endotoxin of *Vibrio cholerae*: Biological Functions

Abstract The salient features of the biological functions of the lipopolysaccharide (LPS) are presented in this chapter, including (i) its endotoxic activities, (ii) its antigenic properties, (iii) immunological responses to it, and (iv) its phage receptor activities. The biological functions of the capsular polysaccharides (CPS) of *Vibrio cholerae* are also discussed briefly as a relevant topic. The roles of LPS and other extracellular polysaccharides in the (i) intestinal adherence and virulence of the vibrios and (ii) biofilm formation by the organism are analyzed on the basis of the available data. Every effort has been made to highlight, where applicable, the gaps in our knowledge. The need for continuous serogroup surveillance and monitoring of environmental waters as well as the role of LPS in the design of newer cholera vaccines are discussed briefly in conclusion.

6.1 Introduction

The previous chapters have presented physical and chemical characterizations of the LPS and an in-depth study of the genetics of its biosynthesis. The O-antigen polysaccharide (O-PS) has been shown to be an important constituent of the LPS. These analyses have highlighted the significant role of LPS in the causation and spread of cholera. As a result, the biological functions of LPS and its different constituents, particularly in relation to the causation and spread of cholera epidemics, have naturally become very important targets for study. While lipid A is the key constituent that exhibits endotoxic properties, the O-PS of LPS is mainly responsible for its immunogenicity and the production of vibriocidal antibodies in the host, and has led to the design of different types of conjugated or nonconjugated cholera vaccines. An account of these functional aspects of LPS is presented in this chapter. Further, this chapter takes the opportunity to discuss the role of LPS (i) in the formation of biofilm that enables the organism to survive in hostile natural environments, (ii) in the intestinal adherence and colonization of the organisms—an important step in the causation of the disease in the infected host, and (iii) as the receptor for cholera phages and thus in bacterial population control in aquatic environments. The role of capsular polysaccharide (CPS) associated with *V. cholerae* of several serogroups in the expression of the virulence of the organisms is also briefly discussed.

6.2 Endotoxic Activities

The LPSs of many Gram-negative bacteria are known to exhibit a wide spectrum of endotoxic activities (Luderitz et al. 1971). In accord with this, the LPS of *V. cholerae* was shown to exhibit several endotoxic activities, e.g., pyrogenicity, lethality to mice, local Shwartzmann reaction, and limulus lysate gelation (Raziuddin 1978, Kabir and Mann 1980). *V. cholerae* LPS was also shown to exhibit mitogenic effects and to possess adjuvant properties (Kabir and Mann 1980). It induced in vitro proliferation of murine spleen lymphocytes and also murine intestinal lymphocytes, as measured by the uptake of ^{3}H-thymidine (Kabir and Mann 1980).

6.2.1 Role of Lipid A

Studies on different Gram-negative bacteria have shown that, among all the chemical constituents of LPS, lipid A is most responsible for its endotoxic activities (Galanos et al. 1972, Rietschel et al. 1975). Chemical isolation of lipid A confirmed that it was the active domain responsible for the induction of all known pathophysiological LPS effects (Westphal and Luderitz 1954, Galanos et al. 1985). Since lipid A of *V. cholerae* LPS possesses a general structural similarity to the lipid A of many other Gram-negative bacteria (Broady et al. 1981), it is expected that the *V. cholerae* lipid A would behave in a similar way to them. Lipid A obtained from five different strains of *V. cholerae* and complexes with bovine serum albumin did in fact exhibit the following endotoxic activities: (i) bone marrow reaction, (ii) limulus lysate gelation, (iii) pyrogenicity, (iv) mouse/chick embryo lethality, (v) tumor hemorrhage, and (vi) complement inactivation (Raziuddin 1979). Experiments conducted in parallel with LPS and lipid A of these strains revealed that lipid A was the major factor contributing to the endotoxic properties of LPS, and that lipid A–BSA complexes were slightly more active than the parental LPS in many of the tests carried out (Raziuddin 1979).

6.2.2 Roles of Particular Constituent Chemical Groups

Specific chemical modifications of endotoxins have proven to be a very efficient tool for recognizing the participation of particular groups in their toxic activities (McIntire et al. 1976, Chedid et al. 1975). In one set of experiments (Raziuddin 1980a), LPS of *V. cholerae* was treated separately with succinic anhydride, phthalic anhydride (both of which produce significant decreases in fatty acids, particularly the ester-linked fatty acids), and dinitrophenyl ethylene diamine (which produces an increase in total fatty acid content), and the resulting changes, if any, in the endotoxic activities (e.g., lethal activities in chick embryos and mice and local Shwartzmann reaction in rabbits) were noted. These studies indicated that the ester-linked fatty acids and in particular the 3-hydroxy lauric acid in LPS of *V. cholerae* play a crucial role in eliciting some of its toxic effects. This is in agreement with the finding that the ester-linked fatty acids were generally found to be the important factors determining the toxicity of LPS in Gram-negative bacteria (Luderitz et al. 1971, Rietschel et al. 1971, 1975, Takada and Kotani 1992). The succinyl and phthalyl derivatives of LPSs from *Salmonella typhimurium* and *S. minnesota* were less toxic in mice than the original LPS (Chedid et al. 1975). In *Escherichia coli*, succinylation and phthalylation of LPSs reduced the toxicity and pyrogenicity without any considerable loss in ester-linked fatty acids (McIntire et al. 1976). Rietschel et al. (1971), however, showed that succinylation of glycolipids from *S. minnesota* R595 had no effect on pyrogenicity as well as lethality in animals. Alkali-digested LPSs from *V. cholerae* O1 that had lost ester-linked fatty acids were nontoxic to rabbits and mice (Raziuddin 1978). In a different set of experiments, mutants of *V. cholerae* 569B resistant to common antibiotics and neutral and anionic detergents were isolated (Paul et al. 1990). The outer membranes of these strains showed a significant deficiency in the acylation of lipid A. The contents of amide- and ester-linked fatty acids in the lipid A of these strains were reduced to 50–56% and 29–37%, respectively. This defect was specific for lipid A, as there was no change in the acylation of phospholipids. The reduction in the fatty acid content of lipid A was reflected in the endotoxic properties of LPSs of these mutant strains. LPS from both mutant strains exhibited markedly low endotoxicity in the localized Shwartzmann reaction, limulus gelation assay, and in the complement fixation assay. All of these studies indicated the role of fatty acids in the endotoxicity of LPS and ruled out the possibility that any other chemical constituent of LPS is involved in all of its endotoxic activities.

6.2.3 Effect on Cell Morphology

LPSs from many different enteric bacteria were found to alter human endothelial cell morphology in vitro in a species-dependent and dose-dependent manner (Seifert et al. 1991). The LPSs acted directly on endothelial cells as well as synergistically with interferon-gamma (IFN-γ) to alter cell morphology, and were therefore likely to contribute to the vascular pathology of Gram-negative infections. Seifert et al. (1991) observed that LPSs derived from cells produced no alteration of human endothelial cell morphology. However, Islam et al. (2002) directly added endotoxins from different enteric pathogens (including *V. cholerae* Inaba 569B) to the neutrophils in suspension, and observed that each one of the endotoxins tested changed the shape of the neutrophils. The relevance of these findings to human infection is not immediately obvious.

6.2.4 *Effect on Neutrophil Chemotaxis*

While the endotoxins themselves were not found to be chemotactic (Proctor 1985), they were found to induce chemotactic activity in plasma through the activation of complement (Sveen 1978). Bignold et al. (1991) showed that endotoxins from various bacteria, including *V. cholerae*, inhibited chemotaxis of neutrophils to IL-8. No endotoxin affected chemotaxis to formyl peptide or was itself chemotactic for neutrophils. It was suggested that chemotaxis to IL-8 might be mediated by cellular mechanisms different from those involved in chemotaxis to formyl peptide. Islam et al. (2002) directly added endotoxins from different enteric pathogens (including *V. cholerae* Inaba 569B) to the neutrophils in suspension, and found that each one of them stimulated the neutrophils to acquire locomotor morphology. Based on earlier observations on the responses of neutrophils (Shields and Haston 1985) and monocytes (Islam and Wilkinson 1988) to chemotactic factors, the authors concluded that the endotoxins acted as chemotactic factors.

6.2.5 *Effect on Hemagglutinating Activity of Bacterial Cells*

In several human pathogens, the hemagglutinating ability of bacterial cells was closely correlated with the ability of the bacteria to adhere to the host intestine (Nagayama et al. 1994, Alam et al. 1996). Alam et al. (1997) found that hemagglutination was a common function of the polysaccharide moiety of LPSs from important human enteropathogenic bacteria, including O139 Bengal. O139 LPS showed the highest hemagglutinating activity. The authors argued that since cell-mediated hemagglutination was correlated with bacterial adherence, hemagglutination induced by the polysaccharide moiety of LPS indicated that LPS was a potential adhesin.

6.3 Antigenic Properties

The LPS molecule contains three distinct regions: the lipid A region, which is hydrophobic and forms part of the lipid bilayer of the outer membrane; the core oligosaccharide; and the O-antigen (O-Ag). The outermost region, the O-Ag, provides the major antigenic variability of the cell surface. There are several classification systems for the O-Ag of *V. cholerae* (Kaper et al. 1995). The typing scheme of Sakazaki and Shimada (1977) is the most widely used system and uses sera against heat-killed organisms. This scheme involved 138 different serogroups. The serogroup O139 and others beyond 139 were added to it subsequently.

V. cholerae O1 has been divided into two biotypes—classical and El Tor, which are further subdivided into three serotypes: Inaba, Ogawa, and Hikojima. The three serotypes have been distinguished on the basis of three antigenic determinants or epitopes—A, B, and C—associated with the O-Ag of the LPS and defined using cross-absorbed antisera (Burrows et al. 1946, Sakazaki and Tamura 1971, Redmond et al. 1973). These epitopes are absent in rough mutants lacking the O-Ag (Hisatsune and Kondo 1980, Gustafsson and Holme 1985, Ward and Manning 1989). All three serotypes share a common epitope, A. Inaba strains express A and C epitopes, whereas the Ogawa and Hikojima serotypes express A, B, and a lesser amount of the C epitope (Sakazaki and Tamura 1971, Redmond 1979). The Hikojima subtype is rare and unstable, is not recognized by many investigators, and appears to be a variant of the Ogawa serotype (Chatterjee and Chaudhuri 2003, Stroeher et al. 1992, Hisatsune et al. 1993b). Analysis of the chemical structure of the O-Ag has shown it to be composed of a homopolymer containing the amino sugar D-perosamine substituted with 3-deoxy-L-glycerotetronic acid, which may be the A epitope (Kaper et al. 1995, Stroeher et al. 1992) present in both Inaba and Ogawa serotypes. Gustafsson and Holme (1985) carried out some immunochemical studies and gel permeation chromatography of the polysaccharide fractions extracted from the Ogawa, Inaba, and Hikojima serotypes of *V. cholerae* O1 strains. The authors showed that the A epitope was present as multiple determinants, supporting the view that the perosamine polymer was the structural basis for this epitope, and that the B and C epitopes were present as single determinants on each polysaccharide chain. Wang et al. (1998) carried out binding studies of the anti-Ogawa antibodies IgG1 S-20-6 and IgG1 S-20-4 with synthetic methyl a-glycosides of fragments (up to the hexasaccharide) of the Ogawa O-PS, as well as with analogs of the terminal monosaccharide, and revealed that the terminal residue accounted for approximately 90% of the maximal binding energy. Their study showed that the terminal monosaccharide of the O-Ag of Ogawa LPS, bearing the 2-*O*-methyl group, was probably the serotype-specific determinant for the Ogawa strain, i.e., the B epitope. Further, binding studies with monoclonal antibody (MAb) that recognized an epitope common to both Ogawa and Inaba serotypes suggested that it partially recognized the core and the O-PS of the LPS (Wang et al. 1998). The epitope that is common to the Ogawa and Inaba serotypes was further characterized by purifying the core and the O-PS linked to the core of O1 LPS by preparative electrophoresis (Villeneuve et al. 1999). The O-PS was then subjected to periodate oxidation to destroy sugars. ELISA studies of the binding of these purified saccharide fragments to a MAb, IgG3, that recognized both the Ogawa and Inaba serotypes showed that both the core and the O-PS were involved in this common epitope, which was supposed to be the C epitope. In order to understand the structural basis of carbohydrate recognition and the *V. cholerae* serotype specificity, Villeneuve et al. (2000) undertook crystallographic studies of protective anticholera antibodies in complex with synthetic analogs of the O-Ag. The crystal structure of the murine Fab S-20-4 from a protective antibody (Ab) specific for the LPS Ag of the Ogawa serotype was determined in its unliganded form and in complex with synthetic fragments of the Ogawa O-PS. The authors presented a model of the Ab–O-PS complex based on the previously determined structure of the Fab–disaccharide complex. This model (Fig. 6.1) showed that the upstream terminal perosamine is bound inside the Ab binding cavity, with the 2-*O*-methyl group lying

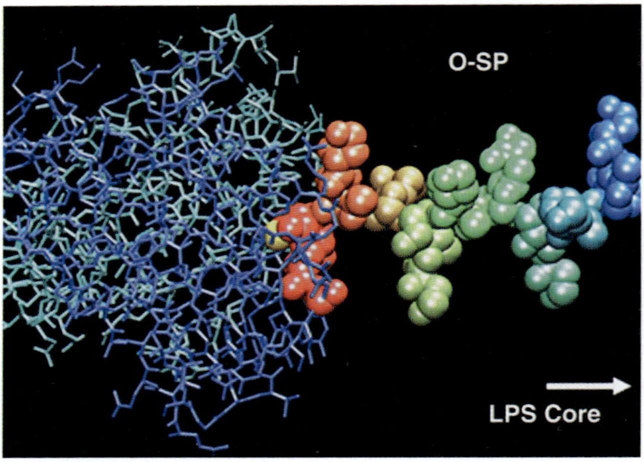

Fig. 6.1 Methylated α-glycoside of the disaccharide of *V. cholerae* O1 O-PS, serotype Ogawa, was synthesized, complexed with the Fab fragment of monoclonal mouse Ab S-20-4 specific for the LPS Ag of the Ogawa serotype, and crystallized. The figure shows the model of the Ab–O-PS complex derived from an X-ray crystallographic study of the Fab–disaccharide complex. Individual perosamine residues are shown in different colors; the light and heavy chains of the Ab variable domains are shown in *light and dark blue*, respectively. The upstream terminal perosamine is bound inside the Ab-binding cavity (the 2-*O*-methyl group at the center of the interface is shown in *yellow*). (Figure 5 of Villeneuve et al. (2000), reproduced with the kind permission of the *Proceedings of the National Academy of Sciences USA* and P.M. Alzari)

at the center of the interface. The second perosamine residue is positioned at the exterior of the binding site and makes fewer contacts with the Ab residues, whereas subsequent sugar residues are not involved in the interaction. This study thus confirmed that the upstream terminal monosaccharide of the O-PS was the primary antigenic determinant or epitope, and explained the serotype specificity of anti-Ogawa Abs by showing the pivotal contribution made by a tiny structural fragment in the antigenic determinant: a methyl group.

6.4 Immunological Responses

6.4.1 Vibriocidal Antibody Level and Immunity

LPSs of enteric pathogens are immunogenic and appear to be the primary protective antigen (Levine et al. 1979, Chitnis et al. 1982a, Manning et al. 1986, Apter et al. 1993). *V. cholerae* O1 LPS was shown to induce protective immune responses in humans and animals (Neoh and Rowley 1970, Freter and Jones 1976, Winner et al. 1991, Gupta et al. 1992, Apter et al. 1993, Jonson et al. 1996, Losonsky et al. 1997, Qadri et al. 1999), and thus its use as a protective immunogen for cholera vaccine development is widely accepted (Mosley 1969, Svennerholm and Holmgren 1976, Winner et al. 1991). The gene clusters that determined the biosynthesis of the LPS O-Ag of O1, of both the Inaba and Ogawa serotypes, were cloned and expressed in *E. coli* K-12 (Manning et al. 1986). The O-Ags expressed by *E. coli* K-12 had the specificity of Abs raised against *E. coli* K-12 that harbored one of these clones, and were as highly protective in the infant mouse model system as the Abs. Abs to these Ags are generally measured using a vibriocidal assay (Anonymous 1979, CDC 1994), which records the killing of cells in the presence of immune sera and compliment. Mosley et al. (1969) provided evidence for the involvement of systemic Abs in protecting against cholera and demonstrated a correlation between serum vibriocidal Ab level and protection against cholera. Elevated vibriocidal Ab levels were correlated with protection against both O1 colonization and disease (Glass et al. 1985). The titer of vibriocidal Abs increased after natural infection with O1 or O139 (Qadri et al. 1995a, 1997) and after oral or parenteral vaccination (Jetborn et al. 1986, Svennerholm et al. 1984). The field trials of vaccines showed that the level of vibriocidal antibodies in serum was the best measure of induced immunity, since it correlated with the elicitation of a protective intestinal immune response against cholera (Levine et al. 1988a, Wassermann et al. 1994). The majority of vibriocidal Abs were absorbed with LPS. It was, however, suggested by others that vibriocidal Abs were simply markers of protection and that other Abs, probably of local origin, provided protection (Levine and Pierce 1992). However, similar to protection against experimental cholera caused by O1 bacteria (Svennerholm and Holmgren 1976), a strong synergistic protective effect was achieved when anti-O139 bacterial Abs and anti-CTB (cholera toxin subunit B) Abs were combined, even though the antitoxin immunity by itself was only marginally effective (Jonson et al. 1996). The mechanism of such synergistic action is, however, not known.

6.4.2 Monoclonal Antibodies

Monoclonal Abs directed against *V. cholerae* O1 were found to recognize the polysaccharide moiety of the corresponding LPS (Bougoudogo et al. 1995), to be vibriocidal, and to agglutinate O1 strains. When subcutaneously injected into neonatal mice, this Ab protected mice against an oral challenge with O1 (Bougoudogo et al. 1995). Protection was serotype dependent. LPS is a type 1 T cell-independent antigen. In some other pathogenic organisms, it was shown that, at high levels, purified LPSs were mitogenic for mouse B cells, resulting in panIg production. At low levels, LPS could induce LPS-specific IgM and IgG Abs in humans and mice when the appropriate conditions were present (Dreisbach et al. 2000, Cryz et al. 1984). In the case of cells, LPS could induce isotype switching to IgG and IgA Abs that were protective (Qadri et al. 1999, Winner et al. 1991, Svennerholm and Holmgren 1976, Svennerholm 1975). The exact conditions needed for anti O-PS Ab responses to form in humans with cholera have not been defined. Anti-LPS IgM was thought to be operative in vitro vibriocidal assays in which a positive response correlated with clinical protection in humans (Losonsky et al. 1997). This Ab promoted lysis of vibrio cells in vitro in the presence of guinea pig complement (Glass et al. 1985, Clemens et al. 1991, Wassermann et al. 1994, Losonsky et al. 1996). Studies with human volunteers showed that primary vibriocidal responses seemed to correlate better with IgM titers than with IgG ones (Losonsky et al. 1996). It was, however, believed that IgA and IgG, rather than IgM, played a role in protection against cholera (Svennerholm 1975, Svennerholm and Holmgren 1976, Winner et al. 1991, Levine and Pierce 1992, Qadri et al. 1999). Colonization of the intestine by *V. cholerae* evoked a mucosal immune response in the host, including secretion of IgA Abs (secretory IgA or sIgA) that were thought to be involved in limiting the duration of the primary infection and in imparting resistance to subsequent oral challenge (Levine et al. 1979, Svennerholm et al. 1984, Jetborn et al. 1986). The response included polyclonal sIgA Abs directed against both CT and LPS (Levine et al. 1979, Jetborn et al. 1986). Apter et al (1993) demonstrated that anti-LPS sIgA was much more effective than anti-CT sIgA at preventing *V. cholerae*-induced diarrheal disease in suckling mice. Although IgA or IgG could be induced by *V. cholerae* LPS, the difficulty involved in generating these generally T cell-dependent Abs might contribute to cholera vaccine failure rates. Exposure to *V. cholerae* in the form of either infection or vaccination with intact bacteria could induce protective Ig characteristic of both T cell-independent (IgM) (Kossaczka et al. 2000) and T cell-dependent responses (IgG and IgA) (Winner et al. 1991, Qadri et al. 1999, Kossaczka et al. 2000). Experimental manipulation of the anti-LPS immune responses so that secretory IgA or IgG specific for *V. cholerae* O-PS are optimally induced will facilitate the development of an effective cholera vaccine.

Gupta et al. (1992) showed that conjugates prepared by the binding of hydrazine-treated LPS from *V. cholerae* O1, serotype Inaba, to CT were safer and induced both serum IgM and IgG Abs with vibriocidal activity and IgG anti-CT. These authors suggested how serum vibriocidal Abs might prevent cholera. Serum Abs, especially those of the IgG class, penetrate into the lumen of the intestine; it is likely that complement proteins are also present. Contact with the walls of the intestine then occurs due to peristalsis. The inoculums of the vibrios that survive the gastric acid are probably low and the organisms have short polysaccharides on their LPS; this trait is associated with a high susceptibility to the complement-dependent action of serum Abs; the ingested *V. cholerae* are lysed on the intestinal mucosal surface (Gupta et al. 1992).

The role of complements has, however, been debated for years. It has been found that while O1 is readily and reliably lysed by complement in the presence of specific Ab, this is not the case with O139 strains because of the presence of capsular layers. Attridge et al. (2000) devised a modified assay system and showed that O139 strains were lysed by the Ab and complement. They observed that the earlier assay method and not the capsule production provided the major impediment to lysis by Ab and complement. Other workers (Steele et al. 1974, 1975, Bellamy et al. 1975) questioned the role of complement in the protective effects of specific Ab, at least in the infant mouse cholera model (IMCM). They showed that the enzymic fragment $F(ab')_2$ of the IgG molecules retained full protective activity despite losses in complement fixation. They proposed that the protection offered by the Ab was achieved by the crosslinking of bacteria and thereby reducing the number of organisms adsorbed to the intestinal wall. Subsequently, Attridge et al. (2004) showed that MAbs prepared against toxin-coregulated pili (TCP) isolated from O1 El Tor were able to provide biotype-specific protection against experimental cholera in infant mice, although the MAbs were not lytic in the presence of complement. They suggested that Abs to TCP protected by directly blocking colonization of the mucosal surface rather than any complement-dependent lysis. However, it was observed that long-term protection against cholera could be accomplished in the absence of a detectable anti-TCP immune response (Kaper et al. 1995). In the context of all of these observations, the exact role of complements in the protection against cholera remains unclear.

6.4.3 Immunoglobulin Subclasses

The immunoglobulins (Igs), IgA and IgG, have subclasses that are known to exhibit different functions. Knowledge about the subclass distribution of specific Abs in infection caused by O1 or O139 is limited. Jertborn et al. (1988) showed that CT induced responses of the four IgG subclasses (IgG1, IgG2, IgG3, and IgG4) and the IgA1 subclass in sera of the cholera vaccinees and patients. A study of North American volunteers showed that secondary challenge with O1 resulted in LPS-specific responses of the IgG1 and IgG3 subclasses (Attridge et al. 2004), whereas after primary exposure, the major response to LPS was of IgG4 Abs. The LPS-specific IgG1 and IgG3 responses in the North American volunteers were highly associated with the vibriocidal activity in the IgG fraction, suggesting that these subclasses might also contribute to vibriocidal Abs. Qadri et al. (1999) made a comparative study of the subclass distribution of the mucosal and systematic Ab responses in patients infected with O1 or O139 to two Ags, LPS and CT. They assessed the Ab-secreting cells (ASC) in the circulation, which served as a proxy indicator of the mucosal immune response. LPS-specific ASCs of both IgA1 and IgA2 subclasses were found, with the IgA1 ASC response predominating in both O1- and O139-infected patients. Both groups of cholera patients showed significant increases in LPS-specific IgG1, IgG2, and IgG3 Abs in plasma. Again, both groups of patients showed CT-specific ASC responses of the different IgG and IgA subclasses in the circulation. The authors showed that, despite possessing a capsule and an LPS that is structurally different from that of O1, O139 induced Ab subclasses similar to those seen in O1 cholera. Further investigations are required to decide whether the LPS-specific responses of the different subclasses can be used as an alternative marker of immunity, and whether the vaccines against O1 and O139 cholera can be developed to stimulate Ab subclasses that are likely to offer protection.

6.4.4 Antibody Assay for Encapsulated Cells

Recent studies evaluating the usefulness of the vibriocidal assay for O139 infections have produced conflicting results, and strain-to-strain variability in the sensitivity of the vibriocidal assay to fully encapsulated O139 strain has been reported (Morris et al. 1995, Tacket et al. 1995, Losonsky et al. 1997). The possibility that the CPS might interfere with complement-mediated killing of the organisms prompted some workers to develop modified vibriocidal assay methods. Losonsky et al. (1997) established a modified vibriocidal assay using another O139 target strain; strain 2L, an unencapsulated insertion mutant of parent strain AI-1837, which retained the truncated O side chain. However, the modified vibriocidal assay for fully encapsulated O139 strain AI-1837 and for the unencapsulated insertion mutant strain 2L produced a very modest vibriocidal response in volunteers challenged with O139 that is not specific. Boutonnier et al. (2001), on the other hand, prepared a conjugate of the polysaccharide moiety (O-specific polysaccharide plus core) of the LPS of O139 (pmLPS) and tetanus toxoid (TT), and tested its immunological properties using BALB/c mice. The conjugate (pmLPS-TT) elicited high levels of IgG Abs, peaking three months after the first immunization and declining slowly during the following five months. Antibodies elicited by the conjugate that recognized both CPS and LPS from O139 were vibriocidal and were protective in a neonatal mouse model of cholera infection. The authors claimed that conjugation of the O139 pmLPS enhanced its immunogenicity and conferred T-dependent properties on this polysaccharide. Boutonnier et al. (2003) further developed a new method (microtiter plate assay) for determining the vibriocidal Ab titer, which was considered equally convenient and efficient for both O1 and O139 serogroups. Their method was also found to be convenient for measuring the activity of animal sera and mouse MAbs.

A new assay using blocking of the limulus amebocyte lysate (LAL) reaction in a microtiter plate was developed (Chang and Sack 2001) in order to detect Abs to O139 LPS that would be less susceptible to the confounding effects of the capsule. It was shown that Abs to *V. cholerae* blocked the LAL reaction and that LAL titers were comparable to the vibriocidal titers. Also, blocking of the gel reaction was serotype specific. However, practical use of this method in field studies required further investigations. Nandy et al. (1996), in a different approach, raised antisera to the truncated form of O-polysaccharides (TFOP) linked to the core of O139 LPS, and found that anti-TFOP Abs and their Fab (IgG) fragments induced passive protection against challenge with colonial variants of encapsulated O139 strains in the suckling mouse model of experimental cholera. The authors found that such protection was mediated by inhibition of intestinal colonization.

6.4.5 Synthetic Oligosaccharides

The use of a synthetic, O-PS-based immunogen was found to eliminate the toxicity problems associated with native LPS. Chernyak et al. (2002) prepared immunogens

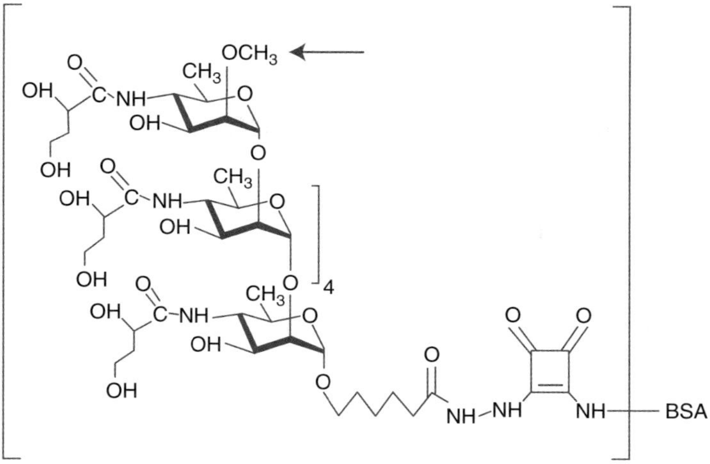

Fig. 6.2 Chemical structure of the neoglycoconjugate immunogen (CHO [Ogawa terminal hexasaccharide]–BSA) complex, as obtained from the works of Chernyak et al. (2001, 2002). BSA was linked to the chemically synthesized, linker-equipped hexasaccharide fragment of the O-PS of *V. cholerae* O1 serotype Ogawa. *Arrow* shows the 2-*O*-methyl group in the terminal sugar of the Ogawa serotype, which is replaced by the 2-OH group in the Inaba serotype. Three different immunogens (A, B, and C) based on the synthetic Ogawa epitope that varied in terms of the number of hexasaccharide residues and were covalently coupled to BSA were used to test immune responses in female BALB/c mice. The CHO to BSA molar ratios in the three immunogens were: A, 15.5:1; B, 9.2:1; and C, 4.6:1. (From Chatterjee and Chaudhuri 2006)

by linking bovine serum albumin (BSA) to the chemically synthesized, linker-equipped hexasaccharide fragment of the O-PS of O1, serotype Ogawa, by appropriate chemical methods (Chernyak et al. 2001, Saksena et al. 2003) (Fig. 6.2). Conjugates with different carbohydrate (CHO)-to-carrier (BSA) molar ratios were tested for immunogenicity and efficacy in mice. All of the conjugates tested were found to be immunogenic, and a correlation was found between vibriocidal activity and protection. The protective capacity of antiserum was evident in serum from mice immunized with all conjugates, but it was highest in the groups that received the conjugate with the lowest level of substitution (conjugate C). The corresponding mice received fewer immunizations with conjugate C. The level of substitution and the number of immunizations affected the repertoire profile of the anti-Ogawa epitope response, but the reasons for this differential protection were not known and required further investigations. Subsequently, a series of conjugates made from Inaba di-, tetra-, and hexasaccharide and BSA were prepared (Ma et al. 2003, Meeks et al. 2004) and found to be immunogenic in mice, inducing IgM and the T-dependent IgG1 subclasses. However, the Inaba-specific Abs, IgM and IgG1, were neither vibriocidal nor protective in the infant mouse cholera model (Meeks et al. 2004). Again, the exact reason for the functional differences between the anti-Inaba and anti-Ogawa Abs remained unexplained. In contrast to the anti-Inaba CHO-BSA sera, the secondary, anti-whole LPS sera were vibriocidal. The authors thus suggested that the Abs induced by the Inaba CHO-BSA conjugates did not bind with enough affinity or specificity to native LPS when expressed on the bacterial surface.

The authors have undertaken a program of developing neoglycoconjugates (NGC) as boosters for cholera vaccine responses that have waned (Wade 2006). They believe that the NGCs are an attractive alternative because they can be delivered parenterally without the attending inflammation of LPS, and they have a carrier component to enhance B cell memory and antibody-isotype switching. The efficacy of Ogawa hexasaccharide NGCs of different linker lengths was investigated. The authors observed that priming B cells with suboptimal doses of native LPS, followed by the NGCs of various saccharide lengths, led to very efficient production of vibriocidal antibodies. Mice immunized 28 days apart with immunogens containing the shortest or medium-length linkers, but not the longest, produced vibriocidal and protective antibodies (Saksena et al. 2006).

6.5 Role in the Intestinal Adhesion and Virulence of the Vibrios

Several studies (Neoh and Rowley 1970, Manning et al. 1986, Gupta et al. 1992, Manning et al. 1994) implied that the O-Ag represented a protective Ag and was involved in the adherence and colonization of *V. cholerae*. Fuerst and Perry (1988) demonstrated LPS on sheathed flagella of *V. cholerae* O1 by protein A–gold immunoelectron microscopy. The flagellum on *V. cholerae* cells was found to be essential for in vitro attachment and enhanced initial colonization of the host intestinal surface in the infant mouse cholera model (Attridge and Rowley 1983). The LPS on the flagellum was thus found to function as a carrier of adhesions (Chitnis et al. 1982b, Booth et al. 1985). On the other hand, Mukhopadhyay et al. (2000) observed that, in the mouse model, anti-LPS Abs induced passive protection through microagglutination and/or immobilization of vibrios, which did not allow the vibrios to adhere to and colonize the intestine. The different animal models or cultured cells may not, however, be completely suitable for evaluating the factors that are essential for the colonization of the human gut. To get a more realistic picture, Benitez et al. (1997) studied the interactions between O1 and O139 with the highly differentiated mucin-secreting cells HT29-18N2, which were derived from the human colonic adenocarcinoma HT29 cell line. Choleragenic vibrios were shown to adhere to and multiply on monolayers of these cells. Their adherence was partially inhibited by LPS. The authors further showed that the flagella, the active *toxR* gene, and the virulence cassette were not essential for binding. The authors emphasized that the interactions studied mimicked important events accompanying intestinal colonization, and as such provided a new approach for studying factors involved in the intestinal colonization of the vibrios.

Several studies correlated LPS mutations with colonization defects in *V. cholerae* (Waldor et al. 1994, Chiang and Mekalanos 1998) and other organisms

(Bilge et al. 1996, Licht et al. 1996, Zhang et al. 1997), but the mechanism by which LPS mutations decreased colonization remained unclear. Iredell et al. (1998) studied some *wbe*::Tn mutants (which were resistant to phages known to use the O-antigen as their receptor), and tried to explain the role of LPS in virulence of *V. cholerae* O1. The authors found that the mutants were unable to assemble TCP on their surface, but that the major subunit TcpA could be found as an intracellular pool. These mutants could be complemented back to wild-type using the cloned *wbe* region, implying that the functional TCP assembly was dependent upon an intact LPS. This was significant in the background to the finding that TCP is a protective Ag in animal models (Taylor et al. 1987, Sharma et al. 1989, Voss et al. 1996), and it has been shown to be an essential colonization factor for both human and infant mouse, and for both classical and El Tor strains (Taylor et al. 1987, Sambrook et al. 1989, Voss et al. 1996). While searching for genes required for colonization, Chiang and Mekalanos (1999) identified *manB* and *wbeL* mutants as colonization-defective strains. They found no defect in the TCP production of *wbe* mutants and could not detect any reduction in the TCP expression in a *gmd*::Tn5*lac* mutant of *V. cholerae* O395 strain. The authors accordingly concluded that the colonization defect associated with *wbe* mutations was unrelated to defects in TCP assembly, but on the other hand, they suggested that LPS itself was important for colonization. They argued that although the exact role of LPS in colonization was unclear, the possibility that LPS defects might render bacteria more susceptible to gut-associated bactericidal substances appeared sensible, particularly since LPS was known to be involved in resistance to antibiotics and complement-mediated killing (Roantree 1957, Raetz 1996).

Nesper et al. (2001) isolated a bacteriophage K139-resistant mutant of *V. cholerae* O1 El Tor with intact O-Ag but altered core oligosaccharide and also a mutation in the *galU* gene. They further isolated another *gal* mutant (inactivated *galE*), which was defective in the catabolism of exogenous galactose but synthesized an apparently normal LPS. They found that the *galU* and a rough LPS mutant (R-LPS), but not the *galE* mutant, were defective in colonization; a phenotype also associated with O-Ag negative mutants. Their study further showed that *galU* and R-LPS mutants were more sensitive to short-chain organic acids, cationic antimicrobial peptides, the complement system, the bile salts, and other hydrophobic agents, indicating that the outer membrane of these organisms could not provide an effective barrier function. The O-antigen negative strains were also found to be sensitive to complement and cationic peptides, but displayed significant resistance to bile salts and short-chain organic acids. This study indicated the involvement of *galU* in *V. cholerae* virulence, correlated with the observed change in LPS structure, and a role for both *galU* and *galE* in the environmental survival of *V. cholerae*. In a more recent study, the authors (Nesper et al. 2002b) further investigated the role of the LPS O-side chain and CPS of O139 in intestinal colonization using genetically engineered mutants. Their results showed that the loss of the LPS O-side chain or CPS resulted in an approximately 30-fold reduction in the colonization of the infant mouse small intestine. Their study further indicated that, as far as *V. cholerae* O139 strain was concerned, the presence of both the LPS O-side chain and CPS was important during the colonization process. On the other hand, Attridge et al. (2001) obtained bacteriophage JA-1 (which uses the capsule as the receptor) mutants of several phenotypes, exhibiting loss of capsule and/or O-Ag from the cell surface, whereupon they studied their residual complement resistances and infant mouse colonization potentials and showed that the production of O-Ag was of much greater significance than the presence of capsular material. During the courses of these studies, additional factors (other than the known colonization factors) involved in the colonization and acid tolerance of *V. cholerae* were subsequently identified (Merrell et al. 2002). Several genes were identified whose activities during colonization were not previously appreciated (Merrell et al. 2002). The functions of these genes included the production of factors involved in metabolic activities, regulation of cellular processes, transport, adaptation to stress, and some unknown functions. These authors identified nine new factors as being crucial to the *V. cholerae* acid tolerance response (previously identified to be important for the epidemic spread of cholera), and showed that mutations in the genes *gshB, hepA*, and *recO* resulted in a 1000-fold reduction in colonization (Merrell et al. 2002).

6.6 LPS as Phage Receptor

V. cholerae LPS, like the LPSs of many Gram-negative bacteria, was found to act as the receptor of several cholera phages (Maiti and Chatterjee 1971, Maiti et al. 1977, Guidolin and Manning 1985). Mukerjee's Group IV cholera phage differentiates the classical and El Tor biotypes of *V. cholerae*; the classical ones being sensitive to and the El Tor ones resistant to these phages (Mukerjee 1963b). The basis of this differentiation was traced to the LPSs of the two biotypes. The phage φ149 was inactivated by the classical LPS but was resistant to the El Tor LPS (Maiti and Chatterjee 1971, Maiti et al. 1977). Adsorption of cholera phage φ149 to isolated classical LPS followed first-order reaction kinetics, the 50% phage inactivating concentration of LPS (IC50) being 7 μg/ml. After treatment of LPS by 0.5% (w/v) sodium deoxycholate at 37 °C, the LPS largely lost its phage-inactivating capacity and the IC50 value rose to 3.6 mg/ml (Maiti and Chatterjee 1971, Maiti et al. 1977). This was in accord with the fact that sodium deoxycholate dissociated LPS of Gram-negative bacteria into very small units with subsequent loss of biological activity (Ribi et al. 1966, Lindberg 1967). When cholate was removed by extensive dialysis, the phage-inactivating capacity of LPS was restored significantly, the IC50 value being 570 μg/ml of LPS. The cholate alone could not inactivate the phages by any significant degree (Maiti and Chatterjee 1971). LPS isolated from Inaba or Ogawa serotypes and classical or El Tor biotypes of *V. cholerae* showed identical phage-inactivating capacities for the phage CP-T1 (Guidolin and Manning 1985). On the other hand, LPS from a CP-T1-resistant mutant exhibited no phage-inactivating capacity. The mutant was shown to lack the O-Ag by bactericidal assays utilizing a MAb directed against the O-Ag side chain of *V. cholerae* LPS. The absence of O-antigen in the phage-resistant strain was further confirmed by an

SDS-PAGE study of ^{32}P-labeled LPS (Guidolin and Manning 1985). Similarly, another phage, VCII, specific to O1 classical strains was found to have receptors in the O-Ag of LPS, and the VCII-resistant mutants lacked the O-Ag (Ward and Manning 1989).

Bacteriophage K139 was originally isolated from a *V. cholerae* O139 strain and was found to belong to the Kappa phage family (Reidl and Mekalanos 1995). Further analysis revealed that this phage was widely distributed among clinical El Tor strains and was also found as a defective prophage in classical O1 strains (Reidl and Mekalanos 1995, Nesper et al. 1999). K139 was perhaps the first vibriophage for which the entire genome was sequenced (Kapfhammer et al. 2002). The tail fibers were thought to be involved in receptor binding. The presumed tail fiber genes of the phage K139 were sequenced and analyzed, and two conserved and two variable regions were identified. Three different tail fiber types were discovered, depending on the different combinations of the variable regions. Since the C-terminal part of the tail fiber was believed to be involved in receptor binding (Kapfhammer et al. 2002), it was speculated that the variable regions of the K139 phages determined their binding abilities to different O-Ag receptors. Phage binding studies with purified LPSs of different O1 serotypes and biotypes revealed that the O1 O-Ag served as the phage receptor. Analysis of the LPSs of spontaneous phage-resistant mutants revealed that most of them synthesized incomplete LPS molecules composed of either defective O1 O-Ag or core oligosaccharide (Nesper et al. 1999). Upon applying hypervirulent phage K139 cm9 to O1 El Tor strains, different phage-resistant mutants were isolated, and these were found to express different LPS mutations. Interestingly, several mutants were found to be linked not with the O1 O-Ag but with the core structure. Such mutants indirectly implicated the core region of the LPS in secondary phage infection steps (Nesper et al. 1999). Among the O-Ag defective mutants, one mutant was characterized for the loss of O-Ag due to the transposition of IS1004 into the *wbeW* gene encoding a putative glycosyltransferase. In a later study (Nesper et al. 2000), one *wbeW*: IS1004 serum-sensitive mutant was treated with normal human serum, and several survivors showing precise excision of IS1004, restoring O-Ag biosynthesis and serum resistance, were detected. Further, by screening for phage resistance among clinical isolates and performing LPS analysis of nonlysogenic strains, one strain was identified with decreased O-Ag presentation and a significant reduction in the ability to colonize the mouse small intestine. Several other cholera phages were found to have receptors that were not in the cell wall LPS but in other structures associated with the organism (Albert et al. 1996, Jouravleva et al. 1998, Waldor and Mekalanos 1996). Albert et al. (1996) reported a phage JA-1 that infected *V. cholerae* O139 and used capsular polysaccharide as its receptor. Several JA-1 phage-resistant mutants were isolated that showed a range of phenotypes with loss of capsule or O-antigen from the cell surface (Attridge et al. 2001). A lysogenic cholera phage 493 was isolated from a *V. cholerae* O139 strain which utilized the MSHA type IV pilus as a receptor (Jouravleva et al. 1998). The filamentous cholera phage CTXφ was known to utilize the TcpA pilus as the receptor (Waldor and Mekalanos 1996).

6.7 Biofilm Formation and the Structure of LPS

Biofilm formation by bacteria is of great importance in respect to their survival in natural environments and the causation of epidemic outbursts of the disease. Biofilm can develop on abiotic surfaces and generally consists of bacterial cells entwined in a protective matrix of extracellular polysaccharides. *V. cholerae* is a natural inhabitant of aquatic ecosystems and is known to attach to different environmental surfaces. Adhikari and Chatterjee (1969) reported the formation of thick pellicles on the surfaces of static liquid cultures of several mannose-sensitive hemagglutinating strains of *V. cholerae* El Tor, and found a direct correlation between the formation of a special type of pili on the bacterial surface and pellicle formation. Tweedy et al. (1968) confirmed the presence of pili on the surface and produced evidence that the *Vibrio* strains that exhibited a weaker hemagglutination reaction were comparatively poor at pili formation. Recently, *V. cholerae* El Tor has been reported to form three-dimensional biofilm on abiotic surfaces (Watnick et al. 1999, 2001, Watnick and Kolter 1999, Nesper et al. 2001, Chiavelli et al. 2001) and on simple static liquid cultures (Yildiz and Schoolnik 1999), in agreement with the earlier observations of Adhikari and Chatterjee (1969). O1 El Tor N16961 required the MSHA, a type IV pilus, and the flagellum to associate with abiotic surfaces (Watnick et al. 1999, 2001, Chiavelli et al. 2001) in LB broth, whereas O139 strain MO10 depended only on the flagellum for surface association (Watnick and Kolter 1999). For subsequent development of a three-dimensional biofilm, both of the strains required the presence of the *vps* genes, which are responsible for the synthesis of an exopolysaccharide-based adhesive extracellular matrix (Watnick et al. 1999, Watnick and Kolter 1999, Yildiz and Schoolnik 1999). Watnick and Kolter (1999) further reported, using transposon mutagenesis, that the genes involved in biofilm formation included those encoding (i) the biosynthesis and secretion of the type IV pilus (MSHA), (ii) the synthesis of exopolysaccharide, and (iii) flagellar motility. Accordingly, they suggested that the three steps in the process of biofilm formation were: (i) type IV pilus and the flagellum accelerate attachment to the abiotic surface, (ii) flagellum mediates the spread along the abiotic surface, and (iii) exopolysaccharide forms the three-dimensional biofilm architecture. The exopolysaccharide initially forms the so-called slime layer on the surface of the bacteria. The biofilm formation is normally associated with the change from a normal, smooth colony morphology to the rugose one of the bacteria (Mizunoe et al. 1999). The rugose colony morphology was the result of increased synthesis of the VPS exopolysaccharide (Yildiz and Schoolnik 1999, Wai et al. 1998), and the transcriptional regulation of the *vps* genes, which are required for the synthesis of the VPS exopolysaccharide, was altered in these strains (Yildiz et al. 2001). Thus, these variants rapidly formed biofilm in LB broth that were much thicker than those formed by smooth-colony variants of *V. cholerae*. Electron microscopic examination of the rugose-form *V. cholerae* El Tor strain TSI-4 revealed thick, electron-dense exopolysaccharide materials surrounding polycationic ferritin-stained cells, while the ferritin-stained material was absent around the translucent form of the strain TSI-4. Scanning electron microscopic examinations further revealed that the surface of the biofilm was colonized by actively dividing rod-shaped cells. The exopolysaccharide materials allowed the rugose strains to acquire resistance to osmotic and oxidative stress, such that they were capable of causing human disease (Morris et al. 1996). *V. cholerae* O139 strain MO10 was also shown to produce exopolysaccharide leading to biofilm formation in response to nutrient starvation, with a concomitant change from a normal smooth colony morphology to a rugose one (Mizunoe et al. 1999). It was further demonstrated by immunoelectron microscopy that there was an epitope that was common to the exopolysaccharide Ag of *V. cholerae* O1 strain TSI-4 (rugose form) and that of O139 strain MO10 (Mizunoe et al. 1999).

Since the entire *V. cholerae* O1 genome sequence was available (Heidelberg et al. 2000), a method was developed for the whole-genome characterization of the biofilm phenotype through the use of microarray-based expression profiling (Schoolnik et al. 2001). The important objectives of this study were the differential expression pattern between the sessile and planktonic populations of the same culture, the identification of genes selectively expressed during different stages of biofilm development, and the identification of genes differentially expressed during the adaptation of a mature biofilm to various changes in the fluid phase, etc. Hango and Watnick (2002) subsequently identified a transcriptional repressor in *V. cholerae* that inhibited exopolysaccharide synthesis and biofilm development. It was shown that the repressor was the *V. cholerae* homolog of *E. coli* CytR, a protein that represses nucleoside uptake and catabolism when nucleosides are scarce.

The influence of biofilm formation on the structure of LPS or vice versa among *V. cholerae* cells is likely to form another important field of study. In *Pseudomonas aeruginosa*, studies had indicated that changes in LPS phenotype affected adherence properties and influenced biofilm formation (Rocchetta et al. 1999). Nesper et al. (2001) studied several aspects, including the resistance to phage K139.cm9 of and biofilm formation by the different *galU* and *galE* mutants of El Tor. Among the spontaneous phage K139.cm9-resistant strains, they found strains with a rugose colony morphology that constitutively synthesize an exopolysaccharide and produce biofilm on abiotic surfaces. They introduced *galU* and *galE* mutations into the rugose variant P27459res105 and found that both mutations yielded smooth colony forms, suggesting that *galU* and *galE* mutants were unable to synthesize the exopolysaccharide and could not form the biofilm. The activated carbohydrate moieties, like UDP-glucose and UDP-galactose, were often involved in the synthesis of different surface structures of bacteria (Nesper et al. 2001). Enzymes for the biosynthesis of UDP-glucose and UDP-galactose are UDP-glucose-pyrophosphorylase, encoded by *galU*, and UDP-glucose-4-epimerase, encoded by *galE* (Lin 1996). The fact that *galU* and *galE* were found to be essential for the formation of a biofilm by the phage-resistant rugose variant suggested that the synthesis of UDP-galactose via UDP-glucose was necessary for the biosynthesis of exopolysaccharide. Kierek and Watnick (2003a) recently reported the formation of *vps*-independent biofilm of *V. cholerae* in model seawater. Although Ca^{2+} was shown to be required for the formation of *vps*-independent biofilm (Kierek and Watnick 2003b), the exact mechanism underlying the Ca^{2+} dependence of *vps*-independent biofilm formation has not yet been established. It was, however, shown that (i) both MSHA and flagellum were required for the formation of *vps*-independent biofilm, (ii) both the O-Ag and the capsule of *V. cholerae* O139 promoted this biofilm formation, (iii) spontaneous unencapsulated variants of O139 also exhibited markedly increased surface association, (iv) Ca^{2+} was an integral component of the *vps*-independent extracellular biofilm matrix, and (v) the biofilm formed in true seawater exhibited O-Ag polysaccharide dependence and disintegrated upon exposure to true freshwater (Fig. 6.3). LPS was thus found to play a significant role in biofilm formation.

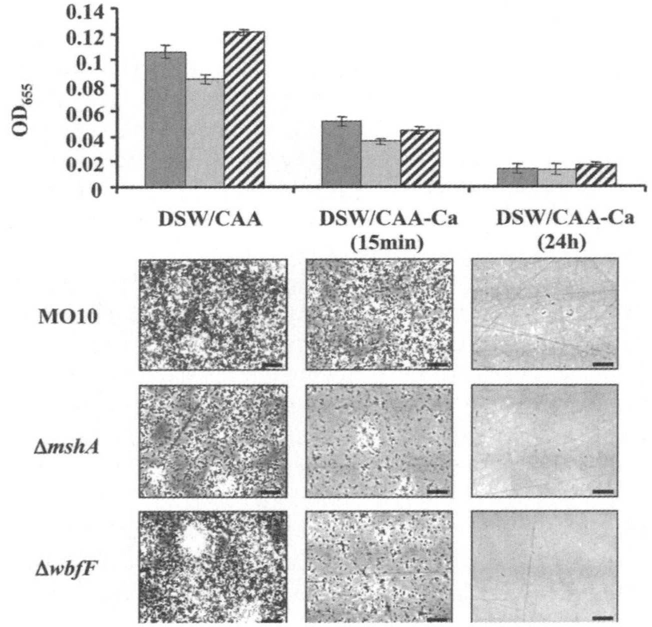

Fig. 6.3 Quantification (*upper*) and phase contrast microscopy (*lower*) of wild-type *V. cholerae* (MO10; *black bars), ΔmshA* mutant (*gray bars*) and *ΔwbfF* mutant (*striped bars*) biofilms after incubation in DSW medium (including casamino acids, CAA) for 24 h and then 15 min and 24 h after the replacement of the bathing medium with DSW medium lacking Ca^{2+} (–Ca) (Kierek and Watnick 2003a, b). DSW medium is a defined salty medium based on the composition of artificial seawater (Kierek and Watnick 2003b). The *ΔwbfF* mutant has transposon insertion in the gene responsible for the export of the capsule precursor, and the *ΔmshA* mutant is the one harboring a deletion in the gene *mshA*. The figure illustrates how the biofilms formed by the wild type and the mutants of *V. cholerae* disintegrated rapidly after the removal of Ca^{2+} from the medium. (Reproduced from the paper of Kierek and Watnick (2003b) with the kind permission of the *Proceedings of the National Academy of Sciences USA* and P.I. Watnick)

6.8 Capsular Polysaccharide (CPS)

Both LPS and CPS of the strain O139 were found to be immunogenic. They reacted in an enzyme immunoassay with rabbit Abs generated against heat-killed bacteria (Johnson et al. 1994). Waldor et al. (1994) carried out immunoblot analyses of either whole cell lysates or LPS preparations and obtained three electrophoretic forms of the O139 Ag, i.e., two slowly migrating forms and one rapidly migrating one that appeared identical to O139 LPS. All three forms of the antigen shared an epitope defined by an O139-specific MAb. A serum-sensitive nonencapsulated mutant was isolated that lacked only the slow-migrating forms. The slow-migrating forms did not stain with silver, whereas the rapidly migrating form did, indicating that the former might constitute highly polymerized O-Ag side-chain molecules that were not covalently bound to core-PS and lipid A, i.e., the O-Ag capsule. This is in agreement with the observations of other workers (Johnson et al. 1994, Weintraub et al. 1994), that the *V. cholerae* O139 serogroup Ag includes both the LPS and the CPS.

The presence of capsule on the *V. cholerae* O139 strain contributed to its virulence in several ways. The CPS made the strain more resistant to killing by normal human serum, and the loss of capsule was associated with loss of resistance (Comstock et al. 1995). Johnson et al. (1992) derived an unencapsulated mutant of the O139 strain by transposon mutagenesis and showed that it was readily killed by serum, while the encapsulated one was protected significantly. The unencapsulated mutant was less virulent in a mouse model than the encapsulated parent. Another factor contributing to the virulence of the non-O1 strains of *V. cholerae* in general was septicemia (Johnson et al. 1992). Non-O1 *V. cholerae* strain NRT36S produced a polysaccharide capsule that determined colony morphology, serum resistance, and virulence in mice. The causation of such extraintestinal disease has not been found with *V. cholerae* O1 strains. A third reason for the increased virulence of the encapsulated strains of *V. cholerae* compared to the nonencapsulated ones was that both the LPS and CPS were important for their colonization of the small intestine of the newborn mouse (Weintraub et al. 1994). There was also evidence that the CPS mediated adherence to epithelial cells (Waldor et al. 1994). Using the intestinal epithelial cell line Caco-2, a clear correlation between the amount of capsular material expressed and the avidity of binding to Caco-2 cells was found. In view of all of these findings, Johnson et al. (1994) observed that the presence of capsule on O139 strains had profound implications for vaccine development. There was already a report of septicemia caused by an O139 strain (Jesudason et al. 1993). Further, in keeping with observations of other non-O1 isolates (Safrin et al. 1987), sepsis was found to occur in a patient with underlying liver disease. Since O139 strains might follow the pattern of other non-O1 strains, the risk of dissemination would be greatest in persons with chronic underlying illness (Safrin et al. 1987, Panigrahi et al. 1992). Thus, whether it is advisable to administer oral attenuated vaccines that still carry the capsule to persons who may have underlying illness is questionable. Further works addressing these problems will be of practical importance.

6.9 Concluding Remarks

6.9.1 Recognition of LPS and Activation of the Innate Immunity of the Host

The molecular mechanisms involved in the recognition of LPS of Gram-negative organisms and the initiation of host response have been reviewed by various authors (Raetz and Whitfield 2002, Caroff et al. 2002, Diks et al. 2004). At the extracellular stage, LPS has to be bound to a transport molecule, a lipid-binding protein (LBP), which facilitates its binding to a surface protein, CD14. CD14 then brings LPS into the proximity of the cell membrane. The LPS-binding protein, MD-2, then opsonizes LPS so that it can be recognized by another protein, TLR4, to initiate signal transduction. LPS is then briefly released into the lipid bilayer, where it interacts with a complex of receptors, e.g., heat shock proteins (HSPs) and others, depending upon the cell type. TLR-mediated signaling activates signal transduction pathways (such as NFκβ, JNK/p38, NF/IL6, and IRF) that induce the transcription of cytokines (such as TNF-α and the type 1 interferons), and these in turn stimulate immune function and control the expression of a variety of inducible immune response genes. A recent study has shown that LPS acts through the TLR4-MyD88-dependent signaling pathway and induces INF-α, IL-1β, and MIP-3α and significantly smaller amounts of IFN-β, nitric oxide, and IP-10 in macrophages (Zughaier et al. 2005). Further studies on LPS are required to at least gain a better knowledge of its interaction with the B cells involving TLRs.

6.9.2 Serogroup Surveillance and Monitoring

The structure of the O-PS of any serogroup has been found to be unique (Chap. 4; Chatterjee and Chaudhuri 2003). The genetic organization encoding O-PS biosynthesis is quite susceptible to change, but the factors responsible for effecting such changes are still largely unknown (Chatterjee and Chaudhuri 2004). Thus, a new serogroup or any of the known serogroups may acquire pathogenic potential in an epidemic genetic background and may cause future epidemics. This situation demands continuous and strict surveillance and requires monitoring for the emergence of either a new serogroup or any of the known serogroups with pathogenic potential, so that appropriate vaccines can be devised promptly. Blokesch and Schoolnik (2007) have recently provided an interesting example of serogroup conversion of *V. cholerae* in natural aquatic reservoirs. It was found that the growth of *V. cholerae* on a chitin surface induces competence for natural transformation, a mechanism for intra-species gene exchange. Their study showed that the acquisition of the O139 gene cluster by an O1 El Tor strain could be mediated by a natural transformation occurring within a community of bacteria living on a chitin surface. The O139 derivatives of this transformation event were not killed by bacteriophages that attack O1 strains and produced an O139-specific LPS, capsular layer, and antigenic determinant. The authors put forward a mechanism of gene transfer and serogroup conversion, from O1 to O139, on a chitin surface immersed in natural seawater, and hypothesized that this ecological niche and mechanism for gene transfer might provide a system for serogroup switching in this species. This finding further emphasizes the need for continual serogroup surveillance of this species in natural aquatic environments.

Phages are known to play a role in the emergence of pathogenic clones, and may also be involved in territorialism between different strains. For example, CT genes were shown to be transferred to nontoxinogenic strains through a lysogenic filamentous phage (Waldor and Mekalanos 1996), and the emergence and dominance of O139 in Bangladesh and India during 1992–1993 may have involved phages as both a means of horizontal gene transfer and a bacteriocidal selective mechanism. Faruque et al. (2005a) monitored environmental water samples from Bangladesh for nearly three years and found that significantly more of these samples contained either a phage targeting LPS as its receptor or a phage-susceptible strain than both. Intraepidemic periods were characterized by water samples containing cholera phages but no viable bacteria. Faruque et al. (2005b) further observed that host-mediated phage amplification during the cholera epidemic likely contributed to increased environmental phage abundance, decreased load of environmental *V. cholerae*, and, hence, the collapse of the epidemic. The authors thus put forward the important suggestion that environmental surveillance for vibriophages could be useful in tracking outbreaks, predicting epidemics, and anticipating emergence of new serogroups. Further, vibriophages might also be employed as biological control agents in areas that suffer cholera epidemics (Faruque et al. 2005a, b).

However, any effective surveillance and monitoring of the aqueous environments demands the availability of rapid diagnostic tests for cholera. Several such tests were available for O1 or O139 using LPS Ag (Colwell et al. 1992, Carillo et al. 1994, Agarwal et al. 1995, Hasan et al. 1995). A multistep colloidal gold-based colorimetric immunoassay known as SMART was also developed for the direct detection of O1 or O139 in stool samples (Hasan et al. 1994, Qadri et al. 1995b). Nato et al. (2003) described the development of a diagnostic test, the one-step immunochromatographic dipstick test, for the rapid detection of O1 or O139. This test was based on LPS detection using colloidal gold particles and immunochromatography, and was claimed to show very high specificity and sensitivity; it could thus provide a simple tool for epidemiological surveys. Similar simple and rapid tests should also be developed for the different serogroups of *V. cholerae*. Robert-Pillot et al. (2002) devised a method for the improved and specific detection of *V. cholerae* in environmental samples involving the culture of a selective medium and a colony hybridization assay with an oligonucleotide probe. The rapid detection of pathogenic vibrios using biochemical and immunological markers, PCR and DNA microarray techniques will be a very challenging task. This will require the use of DNA sequences specific for virulence genes or vibrio species to build and optimize a DNA microarray chip that specifically identifies pathogenic isolates of various vibrio species (Anonymous 2003). The successful development of such a technique will provide a less time-consuming diagnostic strategy for use in the surveillance and monitoring of the estuarine or environmental water samples.

6.9.3 LPS and Cholera Vaccine

Since first isolation of *V. cholerae* by Koch in 1883, several cholera vaccines have been developed and evaluated in clinical trials (Levine and Pierce 1992, Levine and Kaper 1995, Mekalanos et al. 1995). The involvement of LPS O-Ag in the design and preparation of cholera vaccine using recombinant DNA technology and synthetic carbohydrate chemistry has been a rather recent and alternative approach. An attempt to construct a cholera vaccine using recombinant DNA technology has used the attenuated *Salmonella typhi* vaccine strain Ty21a containing cloned *V. cholerae* genes expressing the O-Ag (Forrest et al. 1989, Tacket et al. 1990, Attridge et al. 1991). Yet another novel idea for the design of a carbohydrate-based cholera vaccine originated from the work of Villeneuve et al. (2000), and was discussed earlier. Abs specific for the terminal perosamine could selectively protect against the Ogawa serotype but would fail to recognize the Inaba serotype. Therefore, protective antibodies against both serotypes could, the authors argued, bind to the inner part of the O-PS and/or sugar residues defining the core of the LPS molecule. This idea appears promising but is still to be implemented in practice. It was, however, proposed by another group of workers (Robbins et al. 1995) that serum IgG Abs conferred protection against enteric diseases by inactivating the inoculum on the mucosal surfaces. At the level of laboratory animals, systemic administration of IgG Abs specific for the O-PS of *V. cholerae* O1 was found to protect neonatal mice against loss of weight and death following intragastric challenge with *V. cholerae* O1 (Bougoudogo et al. 1995). It may be pertinent to note here the fundamental limitation that natural infection does not occur in animals, although a few of the animal models have yielded useful information relevant to human disease. Further, the infant mice or the infant rabbits may be susceptible to such infection, but only for a relatively short time after birth. It is therefore generally recognized that it is the volunteer challenge studies that can give the most useful information about the human disease. A recent review (Provenzano et al. 2006) of LPS-based immunogens acknowledged that LPS is the only immunogen proven to induce protective antibody in humans, discussed the role of anti-LPS antibodies in protection from cholera, the importance and the potential role of B cell subsets in protection based on their anatomical location, and introduced the antigen-receptor specificity of various subsets. It was observed that a universal one-dose cholera vaccine, as sought by many investigators, may not be realistic because of variables intrinsic to the host and the immune response to different LPS serotypes.

Cellular cholera vaccines are poor immunogens and have T cell-independent properties (Mosley 1969, Mosley et al. 1969, Sack et al. 1991). Besides, the use of LPS as a vaccine or in cellular vaccine often results in adverse reactions due to its endotoxic properties. The recipient of the cellular vaccine usually has high levels of IgM anti-LPS Ab for about six months. The rapid decline in this IgM vibriocidal activity explains the short-lived protection conferred by cellular vaccines (Neoh and Rowley 1970, Clemens et al. 1991, Szu et al. 1994). With a view to eliminating these undesirable properties of the cellular vaccines, two groups of workers have produced conjugate vaccines by coupling "detoxified" LPS to protein carriers (Gupta et al. 1992, Chang and Sack 2001). Gupta et al. (1992) produced deacylated LPS (DeALPS) by treating LPS with hydrazine, thereby reducing the endotoxic properties of LPS to clinically accepted levels. Conjugate vaccines were prepared by binding DeALPS from *V. cholerae* O1, serotype Inaba, to CT variants CT-1 and CT-2 (used mainly as carrier proteins which are also immunogenic) with a spacer, and they were evaluated (Phase I trial) in healthy volunteers (Gupta et al. 1998). The conjugates elicited the highest levels of IgG anti-LPS vibriocidal Abs, which persisted longer than those elicited by the whole-cell vaccine. The authors, however, expected to further improve on the level of IgG anti-LPS achieved with their conjugates so far. Although the pmLPS conjugate vaccine prepared by Boutonnier et al. (2001), already discussed in Sect. 6.4.4 of this chapter, was found to be protective in the neonatal mouse model of cholera infection, it is also yet to be evaluated clinically. An earlier study (Johnson et al. 1995) showed that an O139 CPS-TT conjugate vaccine induced protection in a rabbit ileal loop model of experimental cholera. Recently, O139 CPS conjugated with a recombinant mutant diphtheria toxin was shown to elicit high levels of serum anti-CPS IgG in mice with vibriocidal activity (Kossaczka et al. 2000). The use of a chemically synthesized, linker-equipped hexasaccharide fragment of the O-PS conjugated to BSA as an immunogen, as discussed earlier, has shown very promising results. However, the use of these synthetic conjugates as vaccine candidates requires further investigation and clinical evaluation.

While studying the B cell responses to LPS epitopes, Wade (2006) found that the magnitude of serum anti-LPS antibody titers and the capacity to induce vibriocidal antibodies (IgM) are influenced by the initial immunizing serotype of LPS, the structure of the LPS immunogen (native LPS versus NGC), and the order of the serotype immunization in a prime boost immunization strategy. It was observed that the host species can affect the immunization response to LPS immunogens. It was suggested that the host's B cell repertoire can influence the immunization efficacy, which should be taken into account when developing new generations of cholera vaccine. In another study (Wade 2006) involving the recombinant exotoxin from *P. aeruginosa* conjugated to O1 serotype-specific polysaccharides (mono-, di-, and hexasaccharide), it was observed that the functional B cell epitopes on the LPS differ from those of the neoglycoconjugates (NGC), and that the order of immunization and the serotype of the boosting conjugate can influence the epitope specificity and function of the antisera.

The goal of having a long-standing vaccine effective against more than one serogroup and/or serotype is yet to be achieved. Although the task ahead may be challenging, it certainly demands greater attention of researchers in the context of the long-standing problem of global defense against recurring cholera epidemics.

Chapter 7
Cholera Toxin (CT): Structure

Abstract This chapter presents our current knowledge of the primary, secondary, tertiary, and quaternary structures of the cholera toxin (CT) molecule, as derived using various physicochemical techniques and particularly X-ray crystallography. The holotoxin molecule contains six subunits: one A and five B subunits. The A subunit can be separated into two fragments—A1, containing 192 (positions 1–192 in the sequence) amino acids; and A2, containing 48 (positions 193–240 in the sequence) amino acids—by proteolytic cleavage at the site between residues Arg (192) and Ser (193) and reduction of the disulfide bond between the residues Cys (187) and Cys (199). Holotoxin contains a symmetrical pentamer of five identical B subunits surrounding a cylindrical central pore. While the A1 fragment of the A subunit sits on the top of the B pentamer as a wedge-shaped structure, the elongated A2 fragment consists of an alpha helix plus a "tail" that extends through the pore formed by the B pentamer, and it is mainly responsible for the interaction between the A and B subunits. Structural studies have beautifully revealed the nature of binding between the B subunits and the GM1 pentasaccharides of the epithelial cells as a "two-fingered grip," and how the A1 fragment in the cytosol of the epithelial cell is activated by binding with the ARF-GTP complex. Structural studies further provide valuable insights into possible approaches to the design of drugs aimed at inhibiting the action of CT.

7.1 Introduction

As stated in Chap. 3 of this book, *Vibrio cholerae* produces several extracellular products that exhibit toxic effects of some sort in human cells. Among these, the one known as cholera enterotoxin or cholera toxin (CT)—or choleragen in the early days—produces dehydrating diarrhea (clinically characteristic of cholera) in humans. After *V. cholerae* culture supernatant had been passed through membrane filters (blocking all particles of bacterial size or larger), a massive accumulation of fluid in the ligated rabbit ileal loop resulted when the filtrate was injected directly into its lumen (De 1959), and the disease of cholera was produced in the

K. Chaudhuri and S.N. Chatterjee, *Cholera Toxins*.

infant rabbits when administered orally (Dutta et al. 1959). These experiments established for the first time that the cholera toxin is an exotoxin. The cholera toxin was subsequently purified (Finkelstein and LoSpalluto 1969, 1970), and the purified toxin as well as the toxoid (subsequently known as the B subunit of CT) were made commercially available. These developments enabled investigators to study different properties of these molecules (the CT holotoxin and its subunits, A1, A2, and B), including their structures (primary, secondary, tertiary, and quaternary) and their interactions with other agents, including the membranes of the intestinal epithelial cells. The structure of the cholera toxin (CT) molecule, as elucidated by X-ray crystallography, and some of its other relevant properties are presented and discussed in this chapter.

7.2 Isolation and Purification

The hypertoxigenic strain 569B is generally used for the production and purification of cholera enterotoxin or cholera toxin (CT). On purification, two factors with differing biological activities were identified (Finkelstein and LoSpalluto 1969, 1970, Spangler 1992). One of the factors contained the full toxicity—the active principle—and was termed choleragen, which subsequently became known as the cholera toxin, CT. The second factor was termed choleragenoid (the B subunit of CT), was noncytotoxic, and was subsequently found to act as a toxoid. The crude CT is recovered from the culture supernatant by coprecipitation with hexametaphosphates. Following dialysis against phosphate and centrifugation to remove insoluble materials, the choleragen (CT) and choleragenoid (CTB or toxoid) are purified in a single step by ion exchange chromatography on Whatman PII phosphocellulose. If toxin substantially free from choleragenoid is required, the protein may be dialyzed and rechromatographed. Subsequently better methods of purification were available.

The oligomeric structure of the holotoxin, CT, was determined by an elegant series of experiments involving gel electrophoresis and other techniques (Gill 1976, 1978, van Heyningen 1976, Kunkel and Robertson 1979a, b, Gill et al. 1981). It was found that CT is composed of five B subunits and one A subunit consisting of two polypeptides linked by a single disulfide bond. Gill (1976) suggested that the B subunits were arranged in a ring with A on the axis, that A2 was required for reassembly, and that the B oligomer was required for the entry of subunit A into the cytosol.

The isoelectric point (pI) of the purified choleragenoid (B oligomer) is 7.8, and that of the holotoxin (CT) is around 6.7. The molar absorptivity (ε_{280}) of CT is 9.6×10^4, and $(A^{1\%})_{1\,cm} = 11.41$ at 280 nm. The molar absorptivity of choleragenoid is 1.43×10^4, and $(A^{1\%})_{1\,cm} = 9.56$ with the substances in 200 mM Tris–HCl buffer, pH 7.5 (Finkelstein and LoSpalluto 1970, Moss et al. 1976, Spangler and Westbrook 1989, Spangler 1992). Lyophilized preparations of CT are stable during long-term storage at 4 °C, but long-term storage in solution at 4 °C or room temperature storage results in extensive isoelectric heterogeneity (Spangler and Westbrook 1989,

Spangler 1992), and this results in six isoelectric species, AB_5, AB_4B', AB_3B_2', AB_2B_3', ABB_4', and AB_5'. These charge variants can be visualized by the isoelectric focusing of intact CT, which showed six bands, the primary species having a pI of 6.9. The CT holotoxin is completely disaggregated at pH 3.2 at room temperature and in the absence of any denaturing agent (Mekalanos et al. 1983). It can also be completely dissociated into monomeric subunits by heating in 0.1% sodium dodecyl sulfate (SDS) (van Heyningen 1976). On the other hand, addition of SDS without heating, followed by immediate electrophoresis in a denaturing (SDS-containing) gel, results in the dissociation of subunit A from intact pentameric B (Gill 1976, Spangler and Westbrook 1989). The intact B pentamer is also dissociated from the A subunit by heating to 65 °C for 5 min in the absence of denaturing agent. Dissociation by treatment with low pH or heat in the absence of denaturing agents results in the rapid precipitation of the A subunit, leaving the B pentamer in solution (van Heyningen 1976). Crystals of choleragen or cholera toxin (CT) were grown successfully from batches of freshly isolated, isoelectrically pure cholera toxin (Sigler et al. 1977, Spangler and Westbrook 1989), and the three-dimensional structure of CT at 2.5 Å resolution was determined by X-ray crystallography (Zhang et al. 1995a. The isolated and purified CTB sub-unit was also crystallized and its structure determined independently by X-ray crystallography at 2.4 Å resolution (Zhang et al. 1995b).

7.3 Primary Structures of CT and LT

Cholera toxin (CT) and *Escherichia coli* heat-labile enterotoxin (LT) are structurally and functionally similar AB_5 (one A and five B subunits) toxins with over 80% sequence identity. The amino acid sequences of the A (CTA) and B (CTB) subunits of the holotoxin CT are shown in Figs. 7.1 and 7.2, respectively. The amino acid sequences of the *E. coli* LTA and LTB subunits are also shown in the same figures. While the A subunit contains 240 amino acids, the individual monomers of the B pentamer contain only 103 amino acids. These figures, however, do not include the amino acids present in the signal peptides, which initially remain attached to the A and B subunits of CT. The amino acid sequences of CTA from the strains 2125 and 569B were found to be identical. Similarly, the amino acid sequences of CTB from strains 569B and 62746 were found to be identical. There are, however, minor strain-to-strain variations in terms of the amino acid sequences of CTA and CTB. The gene of the *E. coli* toxin, LT, shares considerable structural homology with that of the *V. cholerae* toxin, CT. At the nucleotide level, the A and B cistrons of the LT and CT are 75 and 77% homologous, respectively. Overall, the amino acid sequences are largely conserved between the two (with more than 80% sequence homology), with the differences scattered throughout the sequence except for the region around the cleavage site between the subunits A_1 and A_2 (residues 192 and 193), where the homology drops to about 33% between amino acids 190 and 212.

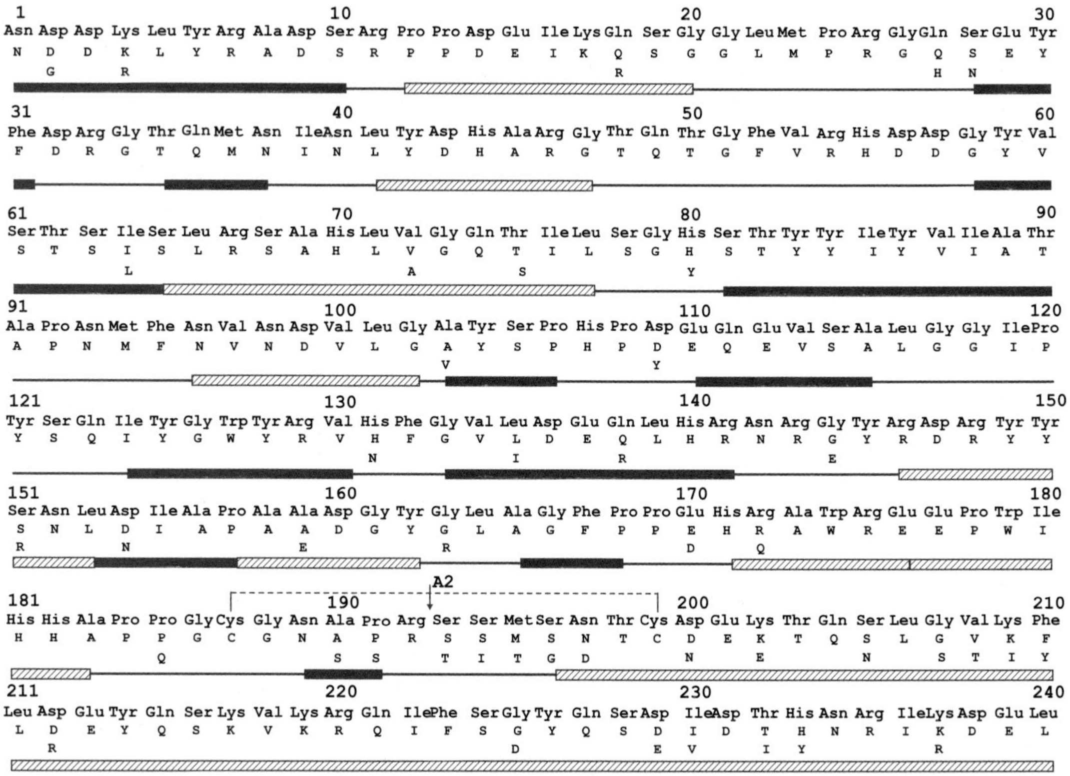

Fig. 7.1 Amino acid sequence and secondary structure of the cholera toxin A subunit. Both the *three-letter* and the *one-letter symbols* of the amino acids and their respective positions (*numbers*) in the sequence are given. Each fourth row shows the sequence homology between CT and LT of *E. coli*. *One-letter symbols in each fourth row* show the positions at which the sequences differ. *Arrow* shows the site between residues 192 (arginine) and 193 (serine) where cleavage takes place to separate the A1 and A2 fragments. Secondary structures are shown in each fifth row: *striated bars*, α-helix; *solid black bars*, β-chain; *continuous line*, coil

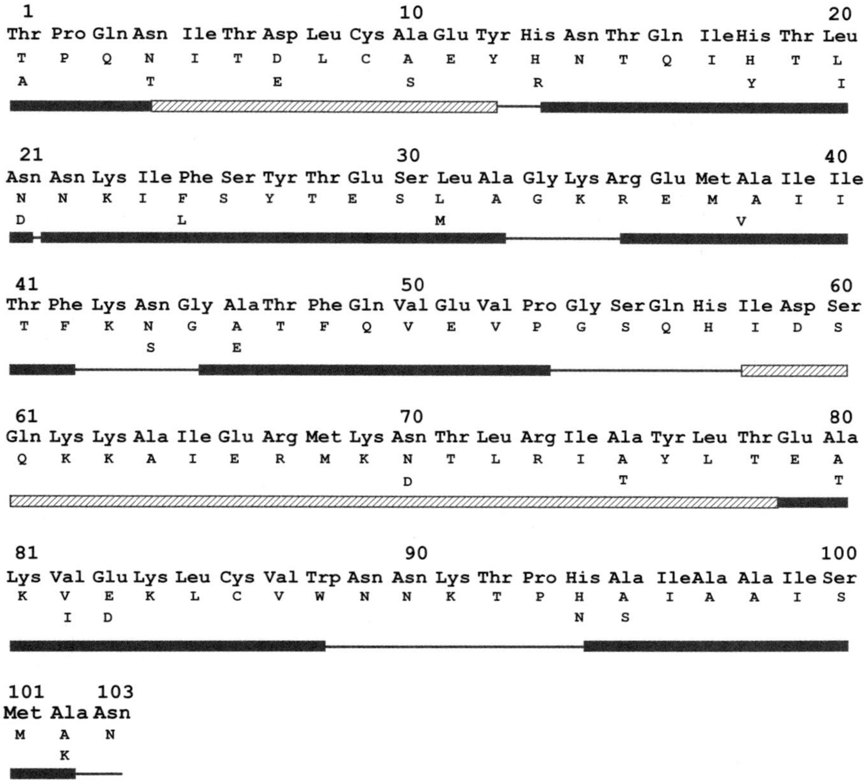

Fig. 7.2 Amino acid sequence and secondary structure of the cholera toxin B subunit containing 103 amino acids in relation to *E. coli* LT. *Bars and lines* have the same meaning as in Fig. 7.1

The A subunit of CT (28 kDa) is proteolytically cleaved by *V. cholerae* between Arg (192) and Ser (193). The resulting A1 (residues 1–192; 23 kDa) and A2 (residues 193–240; 5 kDa) subunits remain linked together by a disulfide bridge (Fig. 7.3). As will be discussed in detail later, the A subunit, CTA, is partially reduced to form CTA1 and CTA2 subunits after its arrival in the endoplasmic reticulum of the host cell. The A_1 fragment (residues 1–192) displays ADP-ribosyl transferase activity, and an A_2 fragment (residues 193–240) mediates interaction with the B subunit pentamer, which serves to bind the holotoxin to the eukaryotic cell receptor. The mature B subunit contains 103 amino acids with a subunit weight of 11.6 kDa.

7.4 Structures of CT and LT from X-Ray Crystallography

The first structure elucidated for an AB_5 toxin was that of the LT holotoxin (of *E. coli*) in 1991 (Sixma et al. 1991). This was followed closely in 1992 by an examination of the structure of LT bound to lactose (Sixma et al. 1992), and by a

Cholera toxin A subunit (uncleaved)

```
1          187        192           199            240
Asn-Asp-...-Cys-Gly-...-Arg-Ser-...-Thr-Cys-...-Lys-Asp-Glu-Leu
            |                           |
            S---------------------------S
```

Cholera toxin A subunit (proteolytically cleaved)

```
1          187        192 ↓ 193         199            240
Asn-Asp-...-Cys-Gly-...-Arg    Ser-...-Thr-Cys-...-Lys-Asp-Glu-Leu
            |                              |
            S------------------------------S
```

Cholera toxin A1- subunit **Cholera toxin A2-subunit**

```
1          187        192      193      199             240
Asn-Asp-...-Cys-Gly-...-Arg    Ser-...-Thr-Cys-...-Lys-Asp-Glu-Leu
            |                              |
            SH                             SH
```

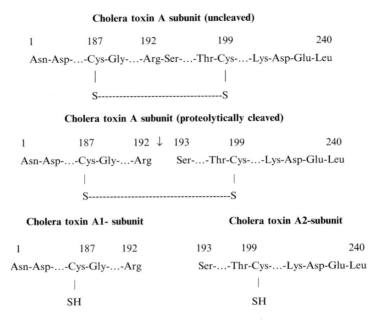

Fig. 7.3 Primary structure of the cholera toxin A subunit and its cleavage products CTA1 and CTA2. The cleavage steps involve (i) proteolytic cleavage at the site between residues 192 and 193, and (ii) reduction of the disulfide bond with the formation of the SH groups at the positions shown in the figure

higher-resolution refinement (1.95 Å) of the structure of LT in 1993 (Sixma et al. 1993). Then the crystal structure of the cholera toxin B pentamer bound to receptor GM1 pentasaccharide (Fig. 7.4) was determined in 1994 (Merritt et al. 1994a, b). The structure of the purified cholera toxin (CT) was separately solved by X-ray crystallography (Fig. 7.5) and refined at 2.5 Å resolution (Zhang et al. 1995a). Further, the crystal structure of the CTB subunit (choleragenoid) was independently solved and refined at 2.4 Å resolution (Zhang et al. 1995b). In all of these structure determinations, the structures of the B subunit of LT or CT (either as a part of the holotoxin or isolated and separated from the A subunit) resembled each other very closely. Similarly, the structures of the A subunit resembled each other closely, except for some differences in the structures of the A2 subunit of LT and CT. The *E. coli* toxin LT and the cholera toxin CT, which show 80% homology in their amino acid sequences, share the same receptor specificity, catalytic activity, and immunological properties. The two toxins are thus expected to have identical conformations, and this was confirmed by X-ray crystallography. The LT structure has thus served as a model for CT or for AB_5 toxins in general.

X-ray crystallography shows that the holotoxin of CT or LT contains a symmetrical pentamer of five B subunits surrounding a central pore (Fig. 7.4). The A subunit is composed of two parts: a wedge-shaped A1 domain and an elongated A2 domain consisting of an α-helix plus a tail that extends through the pore formed by the B pentamer. The interactions between the A and B subunits are almost entirely

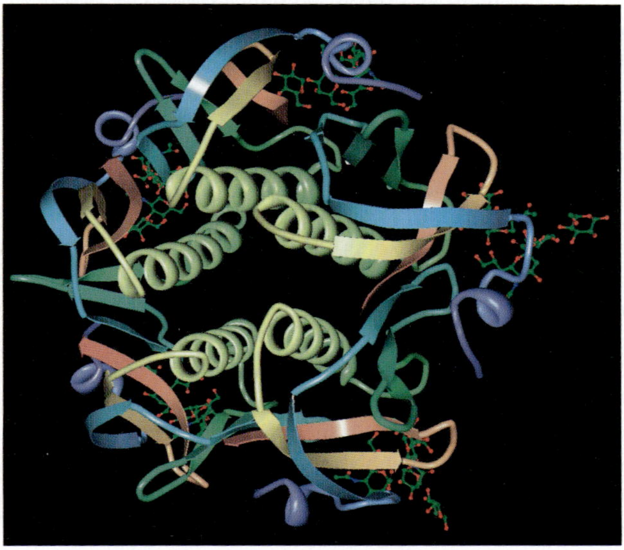

Fig. 7.4 The structure of cholera toxin B pentamer complexed with GM1 pentasaccharides, as derived by X-ray crystallography. (Reproduced from the Protein Data Bank (PDB), with PDB ID 2CHB submitted by Merritt et al. 1997; picture obtained through the kind courtesy of W.G. Hol)

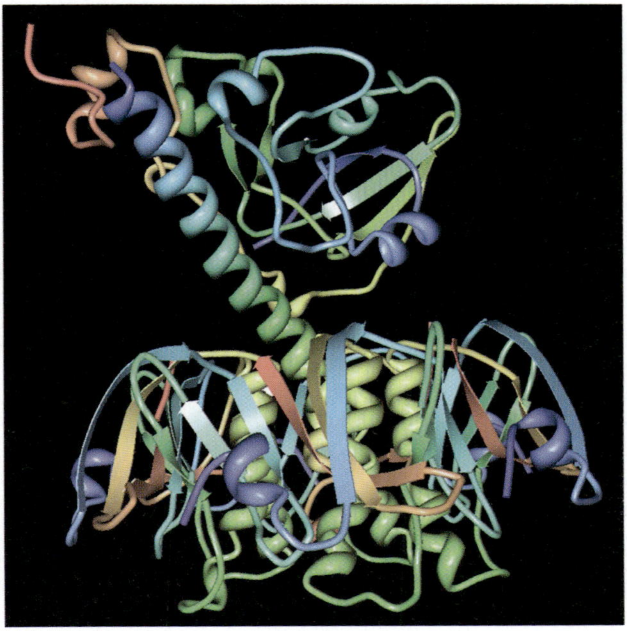

Fig. 7.5 The 3D structure (ribbon diagram) of the cholera toxin molecule, as derived by X-ray crystallography. (Protein Data Bank (PDB) ref. PDB ID 1XTC by Zhang et al. 1995a; picture obtained through the kind courtesy of W.G. Hol)

mediated by the A2 domain and the B-pentamer pore. The two domains (A1 and A2) of the A subunit are linked by an exposed loop containing a site for proteolytic cleavage (after arginine 192) and a disulfide bond that bridges the cleavage site. The pentameric B subunit (55 kDa) specifically binds five GM1 molecules with high affinity. Proteolytic cleavage within the exposed loop of the A subunit and reduction generates the enzymatically active A1 peptide (22 kDa), which enters the cytosol in the cell and activates adenylate cyclase by catalyzing the ADP-ribosylation of the subunit of the heterotrimeric GTPase, Gs. The A2 peptide (5 kDa) forms the scaffolding that tethers the A and B subunits together and contains a COOH-terminal K(R)DEL motif. The KDEL motif protrudes from the pentameric B subunit on the side that binds GM1 at the cell surface.

In the holotoxin, the wedge-shaped A subunit is loosely held high above the plane of the pentameric B subunits by the tethering A2 chain. However, the most striking difference between the two toxins (CT and LT) occurs at the carboxyl terminus of the A2 chain. Whereas the last 14 residues of the A2 chain of LT threading through the central pore of the B5 assembly form an extended chain with a terminal loop, the A2 chain of CT remains a continuous α helix throughout its length (Zhang et al. 1995a). The four carboxyl terminal residues of the A2 chain (KDEL sequence), disordered in the crystal structure of LT, are clearly visible in the electron density map of CT. The A subunit has higher temperature factors and less well defined secondary structure than the B subunits. It interacts with the B pentamer mainly via the C terminal A_2 fragment, which runs through the highly charged central pore of the B subunits. The pore contains at least 66 water molecules, which fill the space left by the A_2 fragment. A detailed analysis of the contacts between A and B subunits showed that most specific contacts occur at the entrance to the central pore of the B pentamer, while the contacts within the pore are mainly hydrophobic and water-mediated, with the exception of two salt bridges. Only a few contacts exist between the A1 fragment and the B pentamer, showing that the A_2 fragment functions as a "linker" for the A and B parts of the protein. Interactions of the B subunits with the A subunit do not cause large deviations from a common B subunit structure, and the fivefold symmetry is well maintained. The three-dimensional structure of the cholera toxin molecule (CT), as derived from X-ray crystallographic data, is shown in Fig. 7.6.

7.4.1 B Monomer and Pentamer

The B monomer is a small, compact, highly structured subunit with dimensions of 38 Å along the fivefold axis and 28 Å × 42 Å in directions perpendicular to the fivefold axis. The sequence and secondary structure of the cholera toxin B subunit presented in Fig. 7.2 show that the 103 amino acid residues of each B subunit are distributed among two α-helices and ten β-strands. Each subunit consists of a short helix (α1) at the N-terminus, a long helix (α2), two three-stranded antiparallel β-sheets, and another two β-strands. The overall fold consists of six antiparallel

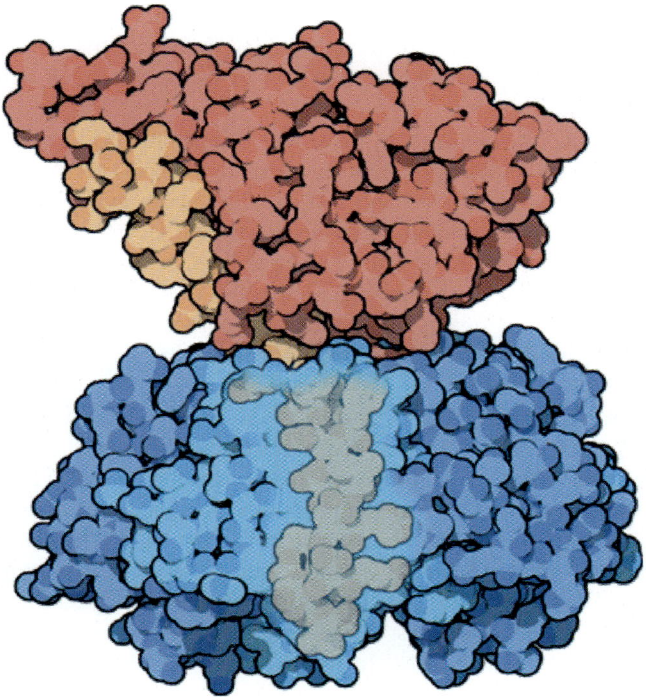

Fig. 7.6 The 3D structure of the cholera toxin molecule, where the A1 fragment is shown in *red*, A2 in *orange*, and the B subunits in *blue*. In this figure, the B subunit complex has been rendered partially transparent to show the tail of the A2 chain passing through the axial pore of the B pentamer. (Image reproduced from the Protein Data Bank (PDB); the image was derived from PDB ID 1XTC and presented as the September 2005 molecule of the month by D.S. Goodsell; it is reproduced here with his kind permission)

β-strands forming a closed β-barrel, capped by an α-helix. The long helical cap (residues 58–79) is gently curved, with its hydrophobic face contributing to the boundary wall of the central pore of the B_5 pentamer. The B subunit's sole disulfide bridge (Cys9–Cys 86) anchors a short, solvent-exposed amino terminal helix (residues 4–12) to an interior β-strand. The structure of the CTB (choleragenoid) is very similar to that of LTB, with a root mean square deviation of 0.57 Å/atom after superposition of the respective Cα backbones (Zhang et al. 1995b). This compares favorably with deviations calculated for the superposition of adjacent B subunits within the pentamer (0.42–0.47 Å/atom) and for the superposition of the B subunits of choleragenoid and choleragen (0.59 Å/atom). When viewed looking down the fivefold axis into the central pore (Fig. 7.4), the monomeric backbones are tightly packed to form a pentamer of interlocking subunits. The amount of surface area buried during oligomer formation is approximately the same for CT and LT and equal to 2700 A^2. When the pentamer is formed, the surface area of the monomer accounts for 39% of the total accessible surface. These structural features may account for the very high stability of the pentameric form. The residue B:Ala64 is in contact with the completely buried one, B:Met31, of the adjacent subunit, which explains why the mutation of B:Ala64 to valine interferes with pentamer formation.

The B pentamer is stabilized by a variety of interactions, among which the hydrogen bonds between the B subunits are important. The total number of hydrogen bonds between a single B subunit and its neighboring molecules is 30 for CT or at least 26 for LT. Each subunit forms seven salt bridges in CT or at least four in LT with each neighbor, giving a total of more than 20 inter-subunit salt bridges in the pentamer, with ten of these localized in the central pore. Further, there is extensive packing of subunits against each other. Choleragenoid or CTB pentamer is further stabilized by the tight interdigitation of hydrophobic groups at the subunit interface.

The inner surface of the ring of B subunits is hydrophobic, with a total of not fewer than 25 positive and 15 negative charges lining the central wall. The monomers that comprise the pentamer form a six-stranded antiparallel β-sheet with a sheet from the next subunit, giving the ring the appearance of a smooth outer surface, while long α-helices form a helical barrel in the center. The pentamer has an overall diameter of approximately 64 Å and a height of 40 Å. For all the toxins of the AB_5 group there is a central pore or channel that runs along the fivefold axis of the B_5 assembly. The boundaries of this conical pore are established by long, closely packed, parallel α-helices. In the case of CT, these helices gently bow inward during their course, reducing the effective diameter of the pore from 16 Å (amino end) to 11 Å (carboxyl end).

7.4.2 A1 Fragment of the A Subunit

The enzymatic A1 fragment consists of a single domain with a wedge shape, formed by numerous but not very regular secondary structure elements (Fig. 7.1), and consists of 192 amino acids. It is organized into three distinct structures. The first 132 amino acids form a compact globular unit comprising a mixture of α-helices and β-strands $(A1)_1$. The $(A1)_2$ structure (residues 133–161) forms an extended bridge between the compact $(A1)_1$ and $(A1)_3$ domains. The structure of $(A1)_2$ suggests that it acts as a molecular tether, like A2. The $(A1)_2$ linker extends 23 Å from the distal-free face of $(A1)_1$ near the catalytic site to the $(A1)_1$/A2 interface. The distal $(A1)_2$ chain is quite flexible and becomes increasingly disordered as it approaches the nick site located along its remote free edge. A third globular structure $(A1)_3$ is formed from the carboxy terminal 31 residues that surround the disulfide bridge linking the A1 and A2 fragments. The evidence that is available indicates that catalysis occurs in the well-defined cleft on the free surface of $(A1)_1$. This cleft is situated away from the A1/A2 and A/B interfaces and is probably the binding site for both NAD and substrate. This enables the A1 chain to act as both an

ADP-ribosyltransferase and a NAD-glycohydrolase (Gill and King 1975, Lai et al. 1983, Gill and Coburn 1987).

7.4.3 A2 Fragment of the A Subunit

The cholera toxin's A2 chain, which consists of a near-continuous α-helix broken only by a central 52° kink, anchors the enzymatic A1 chain to the B pentamer. In contrast, the A2 chain of LT is divided into three discrete segments, a long amino-terminal helix (residues 197–224), a length of extended chain that winds through the pore of the B pentamer (residues 225–231), and a small carboxyl terminal helix (residues 232–236). The sequences of the last four residues of the A2 chain of both CT (KDEL) and LT (RDEL) mimic that of an endoplasmic retention signal. The KDEL residues lie outside the ventral opening of the central pore with little or no stabilization by the B subunits. Although the tetrapeptide is clearly visible in the electron density map of CT, the corresponding residues are disordered in the crystal structure of LT. Deletion of these terminal four residues has little effect on B-subunit oligomerization but significantly reduces the stability of the holotoxin.

The A2 chain, however, shares an extensive interface with the A1 chain and the B pentamer. Also, the A2 chain intimately interacts with all five B subunits. The A2 subunit passes through the central pore of the B pentamer either as a continuous helix (CT) or as an extended chain anchored at its carboxy terminus by a short turn of the helix (LT). The pore diameter is just wide enough to accommodate the A2 chain as a helix. Stabilizing contacts within the pore between the A2 chain and the B subunits are largely hydrophobic, with very few specific hydrogen bonds.

7.5 Receptor Binding

In nature, the toxin CT binds by its B subunits to ganglioside GM1 in the apical membrane of polarized intestinal epithelial cell. The toxin receptor was first identified by King and Van Heyningen (1973), who observed that ganglioside GM1 prevented CT from increasing the capillary permeability of rabbit skin (skin blueing test), prevented CT from inducing the accumulation of fluid in ligated ileal loops of rabbit intestine, and inhibited the action of CT on the adenylate cyclase system in the small intestine of the guinea pig. It was subsequently shown by others that GM1, but not other gangliosides, had this effect on CT (Cuatrecasas 1973a, b, Holmgren et al. 1973). Both CT and LT bind ganglioside GM1. CT does not bind any other related ganglioside, such as GM2, GM3, etc., while LT can interact with a second class of receptors not recognized by CT. For CT, the recognition event is specific to GM1; cells not exhibiting the GM1 saccharide are not bound by the toxin, and conversely the endogenous addition of GM1 allows previously immune cells to be attacked and intoxicated (Eidels et al. 1983). Five molecules of GM1 on the membrane surface are bound by five identical binding sites on the toxin B pentamer (Fishman et al. 1978).

The saccharide moiety of GM1 is bound by the complete AB_5 hexamer and also by the B pentamer but not by the monomeric B subunits (de Wolf et al. 1981a). There are five binding sites on the toxin, and the binding of GM1 to the five sites is known to be cooperative (Schon and Freire 1989). The neuraminidase produced by *V. cholerae* can increase the number of receptors by acting upon higher-order gangliosides to convert them to GM1 (Galen et al. 1992).

Several key residues in the B-subunit have been found to be necessary for receptor binding. The single tryptophan residue (Trp-88) in each subunit is essential for binding, as shown by chemical modification and site-directed mutagenesis studies (de Wolf et al. 1981a, b, Ludwig et al. 1985, Jobling and Holmes 1991). Substitution of Gly-33 by negatively charged or large hydrophobic residues also abolishes binding and toxicity (Jobling and Holmes 1991). The two cysteine residues at positions 9 and 86 are also essential for the proper functioning of the B subunit.

The interaction of the B pentamer of LT or CT with GM1 caused a blue shift of about 12 nm in the fluorescence emission maximum of tryptophan (Moss and Vaughan 1977). Since there is only one tryptophan residue (Trp-88) in each B monomer, its involvement in this receptor binding was clearly indicated. This was further substantiated by a ^{13}C nuclear magnetic resonance study (Sillerud et al. 1981), where a resonance shift was observed and attributed to the ε_2 carbon of Trp-88 upon the binding of GM1 by the B pentamer subunit. It was confirmed that complex formation between CT and GM1 results in energy transfer between the indole moiety of the tryptophan residue and a dansyl derivative of GM1, indicating that the residue is in or near the receptor-binding site.

CT and LT bind specifically to the oligosaccharide portion of ganglioside GM1, a normal component of the outer leaflet of many cell membranes. Other AB_5 toxins recognize other gangliosides. Ceramide is linked to the GM1 sugar moiety, which is a branched pentasaccharide.

Although CT recognizes the terminal sequence Gal-Nac-Gal, the minimum requirement for LT binding is the terminal Gal alone (Fukuta et al. 1988, Smith 1984). The toxin B pentamers, however, all bind specifically to saccharides, either to the oligosaccharide moieties of gangliosides in the cell membrane or to the glycosylated proteins at the cell surface (Merritt and Hol 1995). Neither GM1 nor CT can pass through intercellular tight junctions. The B-pentamer of these AB_5 toxins is capable of target cell recognition and binding, even in the absence of the A-subunit, but actual intoxication follows only from exposure to the AB_5 holotoxin.

The B pentamer contains five equivalent GM1 binding sites with dissociation constants ranging from 10^{-9} to 10^{-10} M. Small-angle X-ray diffraction studies suggest that, in the absence of ligand, the negatively charged pentasaccharide of GM1 (Gal-β-1-3-GalNAc-β1-(NeuAc-α2-3)-4Gal-β-1-4-Glc-β1-ceramide) lies nearly 21° above the lipid/water interface. Co-crystals of LT with lactose (Gal-β1-4-Glc), and of CTB pentamer with galactose or with oligosaccharide have shown that the distal saccharides of GM1 bind to a small cleft adjacent to B:Trp 88 on the ventral flange of the B-pentamer. It is likely that binding to GM1 takes place with the B pentamer facing toward the membrane surface.

7.6 Structure at the Binding Site

The details of the structure of the cholera toxin B pentamer bound to the GM1 pentasaccharide were determined by X-ray crystallography (Fig. 7.4) (Merritt et al. 1994a, b). It involved the receptor-binding domain of the toxin and the specificity-determining portion of the corresponding cell surface receptor. The cholera toxin B pentamer (CTB) cocrystallizes readily with the GM1 pentasaccharide. Structure determination showed that the receptor essentially only interacts with toxin side chains belonging to a single monomer of the toxin at each binding site. Each receptor-binding site on the toxin is found to lie primarily within a single B subunit, while there is a single solvent-mediated hydrogen bond from residue Gly33 of an adjacent subunit. The large majority of interactions between the receptor and the toxin involve the two terminal sugars of GM1, galactose and sialic acid, with a smaller contribution coming from the *N*-acetyl galactosamine residue. The receptor–toxin binding interaction may be described as a "two-fingered grip" in which the Gal-GalNac forefinger of the longer branch of the pentasaccharide is fairly deeply buried in the toxin, and the sialic acid "thumb," which constitutes the shorter branch, lies along the toxin surface (Fig. 7.7). The terminal galactose residue of the GM1 is completely buried in the toxin–pentasaccharide complex; the binding site for this terminal is notable for comprising a complex net of hydrogen-binding interactions tying all of the galactose hydroxyl oxygens to the protein, either directly or via tightly associated water molecules. Each of the five identical receptor-binding sites on the toxin pentamer lies almost entirely within a single subunit, although the neighboring subunit contributes to the hydrogen-bonding network via a single solvent-mediated interaction. All of the binding sites are located at the base of the AB_5 toxin assembly, opposite rather than adjacent to the catalytic domain A1. This configuration makes it virtually certain that the orientation of the toxin upon binding to the cell surface is that shown in Fig. 7.7 (Merritt and Hol, 1995). This has the effect of bringing the C terminal residues of the A subunit, which extends all the way through the central pore of the CT or LT holotoxin structure, into the proximity of the cell membrane.

Cholera toxin molecules—both the B subunit and the complete molecule, randomly bound to the gangliosides—were imaged by atomic force microscopy with a resolution of better than 2 nm (Yang et al. 1993). A two-dimensional projected structure of cholera toxin B subunit–GM1 complex was also determined by electron crystallography (Mosser et al. 1992). Further, the orientation of cholera toxin bound to its cell surface receptor, ganglioside GM1, in a supporting lipid membrane was determined by electron microscopy of negatively stained toxin-lipid samples

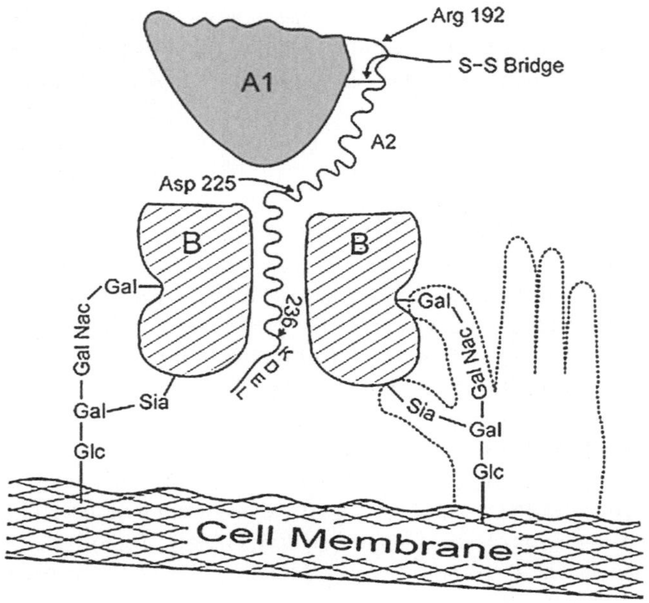

Fig. 7.7 Schematic diagram showing the binding of the cholera toxin molecule through its B subunits to the GM1 pentasaccharides of the epithelial cell membrane, drawn on the basis of the X-ray crystallographic data on receptor–toxin interactions. The binding between one B subunit and the GM1 pentasaccharide is illustrated here as a "two-fingered grip." (Adapted from Fig. 3 of Merrit and Hall 1995)

(Cabral-Lilly et al. 1994). These and other related studies (Ludwig et al. 1986, Reed et al. 1987, Ribi et al. 1988) more or less supported the picture derived by X-ray crystallography in respect to the binding of CT or its B subunit with the ganglioside GM1. Images of the pentamers showed fivefold symmetry and a central depression or pore; images of the holotoxin showed that it extends higher above the membrane surface than the B pentamer alone and is capped by a rounded protrusion with no visible symmetry.

7.7 Structural Basis of Toxicity

When the actions of CT and LT in polarized human (colonic) epithelial cells (T84) were monitored by measuring toxin-induced Cl^- ion secretion, CT was found to be the more potent of the two toxins. The structural basis for this difference in toxicity was examined by engineering a set of mutant and hybrid toxins (e.g., CTA^{RDEL} CTB, $CTA^{(1-225)}$ $LT^{(226-240)}$ LTB, CTA^{RDEL} LTB, $LTA^{(1-224)}$ $CTA^{(225-240)}$ CTB, as well as the wild-types CTAB and LTAB) and testing their activities in human colonic adenocarcinoma cell line T84 (Rodighiero et al. 1999). It was found that the differential toxicity of CT and LT was: (i) not due to differences in the A subunit's C terminal KDEL targeting motif (which is RDEL in LT), as a KDEL to RDEL substitution had no effect on cholera toxin activity; (ii) not attributable to the enzymatically active A1 fragment, as hybrid toxins in which the A1 fragment in CT was substituted for that of LT (and vice versa) did not alter the relative toxicity; and (iii) not due to the B subunit, as the replacement of the B subunit in CT with that of LT caused no alteration in toxicity, thus excluding the possibility that the broader receptor specificity of LTB is responsible for the reduced activity. However, the difference in toxicity could be mapped to a ten amino acid segment of the A2 fragment that penetrates the central pore of the B subunit pentamer. A comparison of the in vitro stability of two hybrid toxins, differing only in this ten amino acid segment, revealed that the CT A2 segment conferred a greater stability on the interaction between the A and B subunits than the corresponding segment from LTA2. This suggests that the reason for the relative potency of the CT compared with LT is the increased ability of the A2 fragment of CT to maintain holotoxin stability during uptake and transport into intestinal epithelia. However, the structural basis for the activation of CT was still not understood properly.

O'Neal et al. (2004), on the other hand, argued that the crystal structures of CT and LT reported to date represented only inactive, pro-enzyme toxin forms (Zhang et al. 1995a, Sixma et al. 1993), and that the attempts to obtain the crystal structure of activated CT or LT had been unsuccessful, as the fully activated toxins or free A1 subunits aggregated at low protein concentrations and had resisted crystallization. Also, the structure of a proteolytically nicked but unreduced LT showed no major conformational difference from the wild type, suggesting that conformational change must occur only upon full enzyme activation. Van den Akker et al. (1995) suggested that the A1 loop regions (25–36 residues of the activation loop and the 47–56 residues of the active loop) served as potentiators of activation-related conformational change. Following this, systematic mutagenesis of residues in the CT 25–36 loop led to the discovery of an A1:Y30S mutant that, while unnicked and unreduced, exhibited a similar level of activity to that of the wild-type, activated enzyme. Crystal structures of the CT:Y30S showed differences from the wild-type CT in the A subunit loop regions that had been previously implicated in activation. The 25–36 activation loop was disordered in CT:Y30S, while the 47–56 active site loop displayed varying degrees of order in the CT:Y30S structure, suggesting that disorder in the activation loop predisposes the active site loop to a greater degree of flexibility than that found in unactivated wild-type CT. O'Neal et al. (2004) thus proposed a new model to describe how the activational modifications experienced by wild-type CT are communicated to the active site. O'Neal et al. (2004) further determined the structure of the wild-type CT at higher resolution and accuracy than were previously available to get a better idea of the structure–function relation. Their study showed an elongated conformation of A2 (Fig. 7.8) through residues 226–236, in contrast to the helical-type tail proposed by the originally published structure of the wild-type CT by (Zhang et al. 1995a). Although this new CT tail model was elongated like that of LT, the conformations of the CT and LT tails are still clearly different. The authors argued that because the

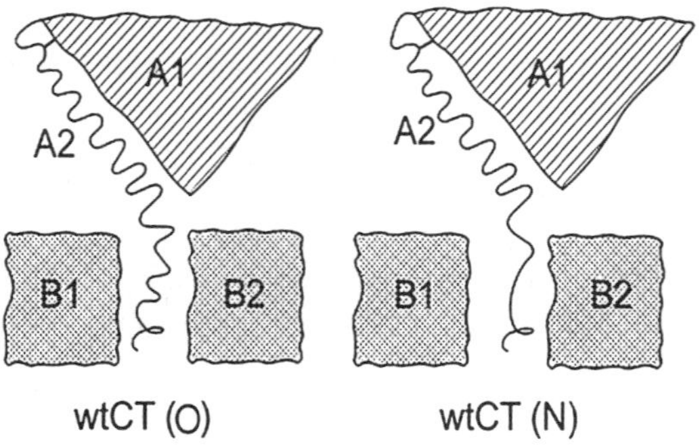

Fig. 7.8 Schematic diagram (not to scale) illustrating the secondary structure of the A2 fragment as derived from the X-ray study of Zhang et al. (1995a) [wtCT(O)] compared to that reported by O'Neal et al. (2004) [wtCT(N)]. See text for further discussion on these structures

A2 helix of CT extends two residues past that of LT, and because the CT tail exhibits gentle curvature around Asp229, the new CT also makes more interactions with B5 than the LT tail does; their new model was still compatible with conclusions (as stated previously) made by Rodighiero et al. (1999) that the greater potency of CT was due to the greater stability offered to the holotoxin by its A2 tail.

The structural basis for the activation of CT was further explored by O'Neal et al. (2005), who determined the 1.8 Å crystal structure of a CTA1 variant in complex with ARF6-GTP. The key event leading to this experiment was that the cholera toxin (CT) catalyzed the ADP-ribosylation of the human signaling protein, Gsα, and that this reaction is allosterically activated by the human ADP-ribosylation factor (ARF6). ARFs are guanine nucleotide-binding proteins, but only ARF-GTP, not ARF-GDP, can bind human effector proteins and, likewise, only ARF-GTP activates the CT. Again, the A1 fragment of CT and not the holotoxin can interact with ARF-GTP, and hence this fragment, CTA1, must separate from the holotoxin beforehand to enable its interaction with ARF-GTP. However, the manner in which ARF-GTP activates A1 had remained unknown. The crystal structure of CTA1:ARF6-GTP, as determined by O'Neal et al. (2005) (Fig. 7.9), has resolved many issues. (i) The crystal structure of the CTA1:ARF6-GTP complex revealed that the binding of the human activator elicits dramatic changes in CTA1 loop regions that allow NAD^+ to bind to the active site. (ii) The CTA1:ARF6-GTP heterodimer buries an extensive solvent-accessible surface of ~1900 $Å^2$ between the two proteins. The interface is predominantly hydrophobic, explaining the low solubility of CTA1 in the absence of a binding partner. (iii) CTA1 binds ARF6-GTP about 15 Å away from the floor of the active site where many different structural elements meet. (iv) About 70% of the ARF6-GTP binding surface of CTA1 also binds CTA2, or about 50% of the CTA2 binding surface interacts with ARF6-GTP. This explains why CTA1 must separate from the holotoxin in order to interact with ARF6-GTP. The CTA1 activation loop (residues 25–40) interacts with both partners but in different ways. The loop forms an ordered coil when bound by CTA2 in wild-type holo-CT; the active site loop (residues 47–60) occludes the active site and must

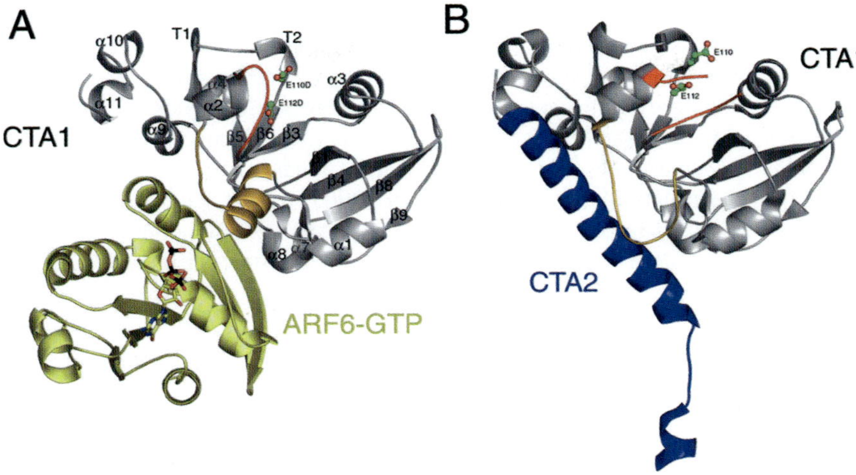

Fig. 7.9 **A** The structure of the CTA1 fragment complexed with human protein ARF6-GTP, as derived by the X-ray crystallographic study of O'Neal et al. (2005). The secondary structural elements of CTA1 are labeled according to the work of Sixma et al. (1991). In this figure, the CTA1 is colored *gray*, with important loop regions delineated in *gold* (activation loop) and *red* (active site loop); the ARF6 is *yellow* and the bound GTP molecule is shown as stacks. **B** CTA1 in complex with the CTA2 (*blue*), as in the structure of the holotoxin, CT. (Figure 1 of O'Neal et al. 2005, reproduced with the kind permission of the Editor of *Science* and W.G. Hol)

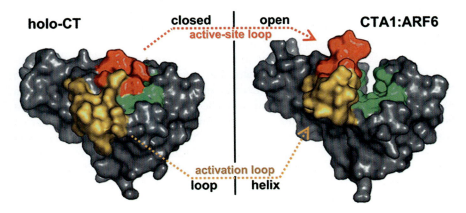

Fig. 7.10 Figure illustrating the changes in the loop regions of CTA1 between holotoxin, CT (*left*), and CTA1:ARF6:GTP (*right*), which led to the opening of the active site (*green*). The positions of the active site and the activation loop before and after binding with the ARF6:GTP are also shown here. (Figure 3 of O'Neal et al. 2005, reproduced with the kind permission of the Editor of *Science* and W.G. Hol)

move to allow substrate binding. When CTA1 is bound to ARF6-GTP, this loop swings out of the active site with a maximal displacement of about 11 Å. The open conformation of the active site loop exposes the active site residues Arg7, Ser61, etc., all of which are implicated in substrate binding and catalysis (O'Neal et al. 2005). The 2.0 Å crystal structure of the quaternary complex, CTA1:ARF6-GTP:NAD$^+$, revealed how this substrate, NAD$^+$, binds to the CTA1 active site (Fig. 7.10). The structural data of O'Neal et al. (2005) further suggested how the G protein, ARF, can alter the CTA1 structure to allow the toxin to adopt an enhanced conformation in order to attack its human G protein substrate, Gsα.

7.8 Structure-Based Inhibitor (Drug) Design: Possible Approaches

The design of drugs to inhibit the action of cholera toxin (CT) within the human system has been a challenging task. Elucidation of the crystal structure of CT has opened up new avenues in inhibitor design (Fan et al. 2004). The crystal structure of CT has revealed several sites where the inhibitor, if appropriately designed, may act. These include: (i) blocking of the active site of A1 in such a manner that it cannot catalyze the ADP-ribosylation reaction; (ii) blocking the A2–B5 interaction and thus the assembly of the holotoxin, and; (iii) preventing the binding of the B pentamer to the cell surface receptor GM1.

The blocking of the enzyme action of CTA1 demands a thorough knowledge of the structural basis of the CTA1 activation. A series of conformational changes that would lead to CTA1 activation has been proposed on the basis of biochemical and structural information about the enzyme. Competitive blocking of the active site of CTA1 by some natural or suitably designed molecules may inhibit the ADP-ribosylation activity of the CTA1 fragment of the holotoxin. If one can design a small molecule that can enter the bacterial periplasm and bind to the pore region of the B pentamer, it would compete with the A subunit for this site and inhibit holotoxin formation. One of the potential inhibitor compounds obtained after an extensive search of the databases was MDT (3-methylthio-1,4-diphenyl-1H-1,3,4-triazolium). Structural analysis indicated that not less than three MDTs bind in the pore region of toxin B pentamer, as targeted (as cited in Fan et al. 2004). Further research into these "pore-plug" inhibitors is being actively pursued. In respect to inhibiting the B pentamer–GM1 binding, significant attention has been directed to the structure-based design of a large number of receptor-binding antagonists. Investigations revealed that galactose, the terminal sugar of GM1 oligosaccharide, binds very specifically at a buried pocket of the receptor-binding site. Thus, the use of galactose as an "anchor" for inhibitor design was extensively explored, and a new molecule, MNPG (*m*-nitrophenyl-α-D-galactopyranoside) was found to be a much better inhibitor. A large number of potential inhibitors were found to be cocrystallized with toxin B pentamers.

The structural studies have thus provided insight into the mechanism of how (i) the CTA1 component of the cholera toxin, CT, gets activated upon complexing with ARF6-GTP, and (ii) the different ways of designing drugs or inhibitors to prepare or develop antagonists of the modes of action of the toxin molecule. However, there are many other aspects of the entire life cycle of CT that have largely remained unexplored so far. These other aspects, which include the synthesis of the toxin or its component parts in the bacterial cytoplasm, their assembly into the 3D structure of the holotoxin in the periplasmic space, its secretion into the extracellular medium by the complex secretory mechanism evolved by the bacteria, its entry and retrograde voyage into the epithelial cell of the host, etc., provide many opportunities to interfere with or intervene in the action of the toxin molecule. All of these possibilities must be investigated thoroughly by structural and other methods of molecular biology in order to enable us to design an effective method for inhibiting the physiological action of the toxin and thus preventing the onset of the disease cholera.

Chapter 8
Cholera Toxin (CT): Organization and Function of the Relevant Genetic Elements

Abstract The expression of virulence of the organism *Vibrio cholerae* greatly depends on a number of horizontally acquired genetic elements. Of these, the two virulence factors—cholera toxin (CT) and toxin-coregulated pilus (TCP)—are largely interrelated, and the expression of these two factors is primarily responsible for the virulence of the organism and disease onset. The two horizontally acquired genetic elements are the CTX genetic element containing the *ctxAB* gene and its associated 5′ sequences and the vibrio pathogenicity island (VPI) VPI-1, which encodes the TCP and other proteins. The CTX genetic element corresponds to the genome of an integrated filamentous bacteriophage, CTXφ, which uses the TCP as its functional receptor. Although the major structural subunit of TCP is encoded by the *tcpA* gene, the formation and the function of the pilus assembly require the products of a number of other genes located in the VPI-1. The pathogenicity island VPI-2 and the genomic islands VSP-I and VSP-II (vibrio seventh pandemic island I and II) are the other horizontally acquired gene clusters which have significant roles in the expression of virulence of the organism *V. cholerae*. The organization and function of these horizontally acquired gene clusters relevant to the expression of virulence of the organism *V. cholerae* are presented in this chapter. This chapter further presents a brief account of the biology of the filamentous bacteriophage, CTXφ, particularly its mode of infection and integration into the genome of *V. cholerae*, its mode of replication and its life cycle within the host, as a relevant topic of interest.

8.1 Introduction

Although cholera toxin (CT), the protein secreted by *V. cholerae* in the lower intestine of the infected host, is mainly responsible for causing the disease, its biosynthesis or expression is dependent on the expression and function of several other virulence genes of the organism. Primarily, the expression of the two virulence factors, cholera toxin (CT) and toxin-coregulated pilus (TCP), are largely interrelated and govern the onset and virulence of the disease. Both of these virulence factors are encoded by genes that are parts of larger genetic elements. The *ctxAB* genes encoding the holotoxin CT (comprising the subunits A and B) are part of a cluster of genes normally referred to as the CTX genetic element. This was subsequently shown to be the genome of the single-stranded DNA phage CTXφ. The genes encoding TCP are also part of the TCP-ACF genetic element, which is also referred to as the vibrio pathogenicity island (VPI-1). VPI-1 encodes the genes required for the expression of toxin-coregulated pilus (TCP), accessory colonization factors (ACF), and several other proteins with regulatory functions.

Both of these gene clusters are known to be acquired by *V. cholerae* by the horizontal gene transfer mechanism. This gene transfer mechanism allows the bacteria to instantly obtain a range of genetic traits that may increase virulence and fitness under different environmental conditions. These horizontally acquired gene clusters generally display several properties that mark them as being atypical compared to the overall genome of the organism in which they are found. These features include: (i) a large chromosomal region present in a subset of isolates of a species and absent from closely related isolates; (ii) the presence of mobility genes such as integrases and transposases; (iii) association with a tRNA gene; (iv) flanking direct repeat sites, which mark the sequence where the incoming DNA recombined with the genome; (v) a G + C content which differs significantly from that of the overall G + C content of the host genome, and; (vi) instability in their chromosomal insertion sites. Although the VPI-1 and the CTXφ are the principal gene clusters that are essential for the expression of the pathogenicity of the organism, *V. cholerae* strains are known to harbor several other gene clusters or pathogenicity islands (PAIs) that may play some roles in the expression of pathogenicity or increased fitness under adverse environmental conditions. These other PAIs include (i) VPI-2, (ii) VSP-I, (iii) VSP-II, (iv) integron islands, etc. While the organizations and functions of the two essential gene clusters, CTXφ and VPI-1, will be presented in detail, a brief account of the other horizontally acquired gene clusters will also be presented in this chapter in order to indicate their relative roles in the expression of virulence of the organism *V. cholerae*.

8.2 The CTX Genetic Element

The genetic region consisting of *ctxAB* and its associated 5′ sequences was termed the CTX genetic element (Pearson et al. 1993), as initial studies revealed that this region was present in toxinogenic *V. cholerae* but was absent in nontoxinogenic strains (Mekalanos 1983, Miller and Mekalanos 1984). The CTX genetic element comprises a 4.5 kb core region flanked by copies of directly repeated sequences that were initially named RS1 (Mekalanos 1983). Due to divergence between near-identical repeated sequences, the 2.4 kb sequence proximal to the core region is termed RS2 and the 2.7 kb sequence flanking RS2 was referred to as RS1; these RS1 and RS2 together are generically called "RS" sequences (Pearson et al. 1993). The variation in the size of the CTX genetic element in different strains is due to a difference in the copy number of RS. The core region carries the *ctxAB* operon (Mekalanos 1985), the *zot* gene coding for zonula occludens toxin (Zot) (Fasano

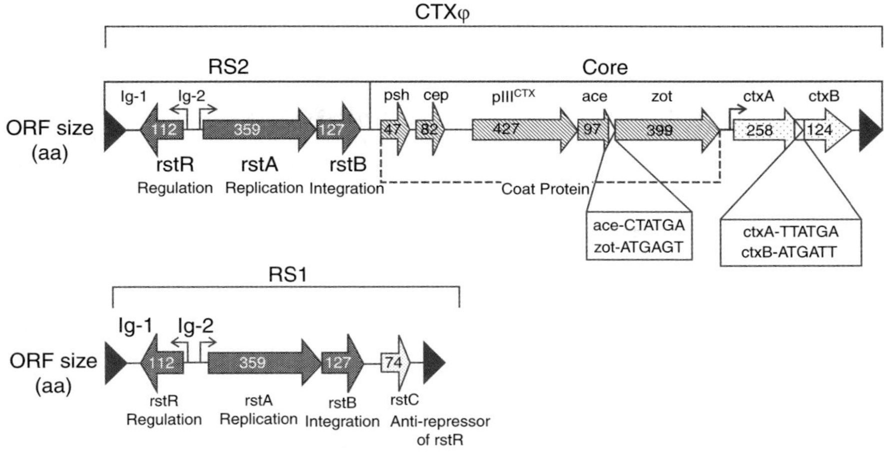

Fig. 8.1 Genetic organizations of CTXφ (comprising the core and the RS2 regions) and RS1 (shown separately below). The open reading frames are represented by *block arrows* showing the directions of transcription. The *numbers inside the block arrows* represent the numbers of amino acids in the respective ORFs. The *black triangles* represent the repeated sequences flanking the prophages. *Bent arrows* indicate the promoters for *rstR, rstA*, and *ctxAB*. The gene names and functions of the ORFs are indicated either below or above the block arrows. Genes with similar functions are shaded alike. The overlapping regions between *zot* and *ace* and also *ctxA* and *ctxB* are expanded below

et al. 1991), the *ace* gene encoding accessory cholera toxin (Ace) (Trucksis et al. 1993), a core-encoded pilin (*cep*) responsible for enhancing colonization (Pearson et al. 1993), and an open reading frame (ORF) of unknown function (*orfU*) (Waldor and Mekalanos 1996). This CTX genetic element was found to be duplicated in tandem in some strains of the El Tor biotype of *V. cholerae* and to undergo amplification upon growth in the intestine (Mekalanos 1983). The RS1 sequence encodes a site-specific recombination system, which can insert into a specific 18 bp sequence called attRS1 (Pearson et al. 1993). The organization of the CTX genetic element is presented in Fig. 8.1.

Naturally occurring CT-negative (CT$^-$) strains lack sequences homologous to the entire core region (Kaper and Levine 1981), but they do contain an 18 bp sequence called attRS1 (Pearson et al. 1993). In toxinogenic strains, RS sequences associated with the CTX genetic element apparently mediate *recA*-independent site-specific recombination between the element and attRS1.

8.3 Cloning and Sequencing of the *ctxAB* Operon

The CT structural genes *ctxAB* were initially cloned from different strains of *V. cholerae* by hybridization with heat-labile toxin LTI probes from enterotoxinogenic *Escherichia coli* (ETEC) (Pearson and Mekalanos 1982, Kaper and Levine

1981). LT and CT cross-react antigenically and have similar structural and biochemical properties (Dallas and Falkow 1979). DNA probes derived from cloned LT genes hybridized to toxinogenic but not to nontoxinogenic *V. cholerae* genomic DNA (Moseley et al. 1982). Complete nucleotide sequencing of *ctxAB* (Mekalanos et al. 1983) revealed 777 and 375 bp coding sequences for *ctxA* and *ctxB*, respectively; the primary translation products were 258 (18 + 240) amino acid and 124 (21 + 103) amino acid polypeptides, respectively. Like LT, both A and B subunits of CT have 18 and 21 amino-acid-long hydrophobic amino-terminal signal sequences. The A2 coding sequence lies at the carboxy terminal end of the A subunit. Sequence analysis revealed that *ctxA* and *ctxB* are in an operon, *ctxA* being upstream of *ctxB*, and both genes are transcribed as a single mRNA molecule. Some amino acid sequence variation was observed in *ctxB* genes from different strains (Lockman and Kaper 1983, Mekalanos et al. 1983). The open reading frames of *ctxA* and *ctxB* overlap; the first two bases of the *ctxA* translation termination signal (TGA) are the last two of the *ctxB* translation initiation codon (ATG), suggesting a translational coupling between these two genes (Fig. 8.1). Both *ctxA* and *ctxB* possess ribosome binding sites (RBSs) located immediately upstream of their start codons; the site for *ctxB* is located at the 3' end of the *ctxA* sequence. The higher expression of B subunits for every A subunit (5:1) is regulated at the translational level. Deletion and the dependence of the expression of in-frame fusion *ctxB* on *ctxA* initiation sequences showed that the RBS of *ctxB* is much (about ninefold) more efficient than the *ctxA* site, thus promoting efficient translation of the B subunit. The promoter region of *ctxAB* contains a number of tandem repeats (3–8, depending on the strain) of the sequence TTTTGAT located about 77 bp upstream of the start *ctxA*. Comparison of the nucleotide sequence of *ctx* and *elt* showed that the A and B cistrons of the two different toxin operons are 75% and 77% homologous at the nucleotide level. In contrast to the structural genes of CT and LT, the two toxin promoters do not show significant homology.

8.4 CTX Genetic Element Belongs to a Filamentous Bacteriophage

In 1996, Waldor and Mekalanos discovered that the genes encoding CT (the *ctxAB* operon) are not integral components of the *V. cholerae* genome, but instead correspond to the genome of an integrated filamentous bacteriophage designated CTXφ (Waldor and Mekalanos 1996). CTXφ is unusual among filamentous bacteriophages as it can either replicate as a plasmid or integrate into the *V. cholerae* chromosome. CTXφ integrates into the *V. cholerae* chromosome via a specific attachment site, attRS, forming a stable lysogen. CTXφ can replicate extrachromosomally as a plasmid in bacterial strains lacking appropriate integration sites (Waldor and Mekalanos 1996). The CTXφ prophage was first found to yield virions within an El Tor strain of *V. cholerae*. Since CTXφ is not a plaque-forming phage, virion formation was detected by the ability of supernatants from a strain containing a kanamycin resistance-marked prophage to transduce a recipient strain to Km^R. It has been demonstrated that, under appropriate conditions, toxinogenic *V. cholerae* strains can be induced to produce extracellular CTXφ particles. Cultures of *V. cholerae* harboring the RF of CTXφ produce high titers of the phage in their supernatants.

8.5 The CTXφ Genome

The CTXφ genome has two regions, the "core" and RS2. The core region contains the genes encoding CT and functions related to phage morphogenesis, located upstream of *ctxAB*. The genes in the RS2 region code for products that control phage replication and site-specific integration. RS1 elements flank the CTXφ genome (Fig. 8.1).

8.5.1 The Core Region

The gene products of the core region include the putative major coat protein (Cep), three putative minor coat proteins (Psh, OrfU or $pIII^{CTX}$, and Ace), and Zot (Fig. 8.1). The *zot* gene product was found to be homologous to the gene I product of M13 (Koonin 1992). When a plasmid containing a mutation in *zot* was introduced into *V. cholerae* strain Bah-2 (*recA*+, lacking a CTX prophage and a phage attachment site), it failed to produce detectable Km^R transducing particle, in contrast to pCTX-Km, which produced ~2.5×10^7 transducing particles/ml (Waldor and Mekalanos 1996), suggesting the involvement of *zot* in phage morphogenesis. Incidentally, Zot has been associated with toxic activity (discussed in detail in Chap. 13), and has been designated "zonula occludens toxin" (Fasano et al. 1991). The *cep* gene is predicted to encode a 47 amino acid polypeptide after cleavage of the N-terminal signal sequence. Cep closely matched the gene VIII product of M13, a major virion capsid protein, in the sizes and distributions of charged and nonpolar amino acids. Thus it was proposed that the *cep* gene product corresponds to the virion capsid protein of CTXφ. The *cep* gene product was earlier designated "core-encoded pilin" (Pearson et al. 1993) because of its homology to the mini-pilin subunit of the flexible pilus *fxp* of *Aeromonas hydrophila* (Ho et al. 1990).

The open reading frame *orfU*, initially described as being an ORF of unknown function, has been hypothesized to function like pIII of coliphage fd based on its size and position (Waldor and Mekalanos 1996). In the frame deletion of *orfU* from pCTX-Km^R (the plasmid form of a kanamycin-marked CTXφ) when electrophoresed into Bah-3 (*recA*$^-$, lacking a CTX prophage and a phage attachment site), no infectious particles could be detected in the supernatant compared to Bah-3(pCTX-Km^R). This defect was complemented by the expression of *orfU* in trans, thereby confirming a role for this gene in CTXφ assembly and/or release (Heilpern and Waldor 2003). The open reading frame *ace* is comparable in size to gene VI of M13, and shared ~61% similarity with the gene VI homolog of *Pseudomonas* filamentous phage Pf1. Thus, *ace* has been proposed to play a role in the assembly into the virion particle (Waldor and Mekalanos 1996).

8.5.2 RS2 Region

The RS2 region is located just upstream of the core CTXφ and contains three open reading frames, designated *rstR, rstA*, and *rstB*, and two apparently untranslated regions called intergenic regions Ig-1 and Ig-2 (Fig. 8.1). The ORFs *rstA* and *rstB* are transcribed in the same direction as other CTXφ genes, while *rstR* is divergently transcribed. The ORF *rstR* is predicted to encode a polypeptide of 112 amino acids, and exhibits significant similarity to repressor proteins, many of which are phage-encoded and play critical roles in maintaining lysogeny (Waldor et al. 1997). Characterization of function showed that *rstR* encodes a transcriptional repressor and represses *rstA*. The ORF *rstA* is predicted to encode a polypeptide of 359 amino acids, and its functional characterization suggests that RstA is required for the replication of CTXφ and perhaps for its site-specific integration. The *rstB* overlaps *rstA* by 15 bp and is predicted to encode a polypeptide of 126 amino acids. The *rstB* gene function has been shown to be required for the site-specific integration of CTXφ into the *V. cholerae* chromosome, but not for the replication of the RF form of CTXφ (Waldor et al. 1997). Ig-2 appears to encompass the *rstA* promoter and the *rstR* operator; this region contains two short dyad repeats (5 bp and 6 bp), which might be within operator sites for RstR binding. No role has yet been established for Ig-1, although it has been suggested to form stem–loop structures that might play role in establishing the phage packaging signal and phage DNA replication (Waldor et al. 1997).

8.6. RS1 Element: The Flanking Region

The CTXφ genome is flanked (5' and/or 3') by RS1 elements (2.7 kb) in toxinogenic *V. cholerae* El Tor or O139 strains, but is absent in the genomes of classical biotype (Waldor et al. 1997, Davis et al. 2000a). The RS1 element encodes four polypeptides and a similar intergenic region to that of RS2. Like RS2, it contains the genes for replication (*rstA*), integration (*rstB*) and transcription regulation (*rstR*). In addition, RS1 also includes *rstC*, a gene not found in CTXφ, which encodes a 74 amino acid polypeptide, an anti-repressor that counteracts the activity of the phage repressor protein RstR and facilitates CTXφ gene expression (Davis et al. 2002).

RS1 has been shown to be another filamentous phage, although not an autonomously transmissible one, as it lacks the genes for phage coat proteins, assembly, and CT. Instead, RS1 utilizes the coat and assembly proteins of other *V. cholerae*-specific filamentous phages such as CTXφ, VGJφ, and KSF-1φ for packaging and transmission of its genome (Campos et al. 2003a, b, Faruque et al. 2002, 2005b, Davis and Waldor 2003).

8.7 TLC: Another Upstream Element

A low copy number plasmid, pTLC (4.7 kb), has been found in all *ctxAB*-positive strains studied by Rubin et al. (1998). Hybridization analysis demonstrated that sequences highly related or identical to this plasmid were present in the chromosomes of all toxinogenic strains of *V. cholerae* but were notably absent in all non-toxinogenic environmental isolates lacking CT and TCP. This element was designated the TLC (toxin-linked cryptic) element because it is adjacent to the CTX prophage on the *V. cholerae* chromosome. The TLC element can replicate autonomously, and integrate as well. The largest ORF of pTLC is predicted to encode a protein similar to the replication initiation protein (pII) of *E. coli* F-specific filamentous phages. The chromosomal location, the existence of the filamentous phage replication gene and the strain distribution suggest that the TLC element might play some role in the biology of CTXφ.

8.8 Variation of CTXφ Among Classical, El Tor, O139, and Non-O1 Non-O139 Strains

The CTX prophage-RS1 arrays differ widely among various pathogenic isolates of *V. cholerae* belonging to O1, O139, and non-O1 non-O139 serogroups. These differences include: (i) variation in copy number and location within chromosome; (ii) variation in sequence of *rstR* and Ig-2; (iii) variation in the integration site; (iv) variation in the flanking region, and; (v) production of infectious particles.

8.8.1 Variation in Copy Number and Location Within Chromosome

In O1 El Tor genomes, CTXφ prophage may be found as either a single copy or multiple copies arranged in tandem (Mekalanos 1983, Bhadra et al. 1995, Nandi et al. 2003). In sharp contrast, the classical vibrios contain CTXφ prophage in two copies, which are widely separated on the chromosome (Mekalanos 1983). *V. cholerae* was found to harbor two unique chromosomes, one large and one small (Trucksis et al. 1998). Genetic mapping revealed that the two copies of the CTXφ prophage in the classical biotype strain O395 are present as one copy on each chromosome (Trucksis et al. 1998). However, the whole-genome sequencing of an El Tor strain N16961 revealed the integration of only one copy of the CTX prophage in the large chromosome (Heidelberg et al. 2000). Like O1 El Tor, O139 strains have just one chromosomal locus within which the phage genome is integrated (Nandi et al. 2003). A few of these strains belonging to the serogroups of *V. cholerae* other than O1 and O139, collectively termed non-O1 non-O139, also harbor CT. The organization of CTX prophage in these strains was different from those of O1 or O139 (Bhattacharya et al. 2006, Maiti et al. 2006). The copy number and chromosomal locations of CTXφ in O1 El Tor, classical, O139, and non-O1 non-O139 are summarized in Table 8.1.

8.8.2 Variation in Sequence of *rstR* and Ig-2

Distinct variations of CTXφ have been distinguished on the basis of sequence variation of *rstR* genes and also of Ig-2, the *rstR* promoter (Kimsey and Waldor 1998) (Fig. 8.1). The existence of at least four different *rstR* genes carried by different CTX phages has been recognized. These are termed $CTX^{ET}φ$, mostly produced by *V. cholerae* El Tor; $CTX^{Class}φ$, derived from classical biotypes; $CTX^{Calc}φ$, found in some O139 strains isolated at Calcutta, India (Davis et al. 1999); and $CTX^{Env}φ$, found in non-O1 non-O139 (Mukhopadhyay et al. 2001). Another variant, $CTX^{Var}φ$, has also been distinguished in pre-O139 El Tor isolates (Nandi et al. 2003). Genetic hybrids of classical and El Tor biotypes that cause cholera have been shown to exist (Nair et al. 2002) on the basis of *rstR* genes and other phenotypic traits.

8.8.3 Variation in the Integration Site

The integration of CTXφ genome is site specific, but the integration sites differ between the two biotypes of *V. cholerae*. Within El Tor biotype strains, CTXφ prophages are found at a chromosomal site known as attRS (Pearson et al. 1993). Integration of CTXφ DNA into attRS occurs via recombination between an 18 bp sequence in the phage genome and a near-identical sequence in attRS. Classical strains have two integration sites; one site is identical to the attRS integration site found in El Tor strains. The second site has not been well characterized but is localized to a different chromosome.

8.8.4 Variation in the Flanking Region

RS1 generally flanks 5' and or 3' of CTXφ prophage genome in toxinogenic El Tor vibrios, but is absent from the classical strains (Davis et al. 2000a).

Table 8.1 No. of copies of CTXφ prophage in different serogroups and biotypes of *V. cholerae*

V. cholerae strain	Serogroup (biotype)	Year of isolation/country	Origin	Chromosomal location of CTXφ prophage	References
569B	O1 (Classical) Inaba	1948[a]/India	Clinical	Two copies, large (1), small (1)	Sharma et al. (1997)
O395	O1 (Classical) Ogawa	1964/India	Clinical	Large (1), small (1)	Trucksis et al. (1998)
N16961	O1 (El Tor) Inaba	1975/Bangladesh	Clinical	Large (1)	Heidelberg et al. (2000)
P27459	O1 (El Tor)	1976/Bangladesh	Clinical	One copy	Mekalanos (1983)
E7946	O1 (El Tor) Ogawa	1978/Bahrain	Clinical	Large (2)	Mekalanos (1983)
VCE232	O4	1984–1985/Calcutta, India	Environmental	Large (2)[b], small (2)	Maiti et al. (2006)
VCE233	O27	1984–1985/Calcutta, India	Environmental	Small	Bhattacharya et al. (2006)
VCE228	O27	1984–1985/Calcutta, India	Environmental	Small	Bhattacharya et al. (2006)
C6709	O1 (El Tor) Inaba	1991/Peru	Clinical	Large (1)	Waldor and Mekalanos (1994)
VC44[c]	O1 (El Tor)	1992/India	Clinical	Small (2)	Nandi et al. (2003)
VC20[c]	O1 (El Tor)	1992/India	Clinical	Small	Nandi et al. (2003)
VC106[c]	O1 (El Tor)	1992/India	Clinical	Small	Das et al. (2007)
SG24	O139	1992/India		Large (2)	Bhadra et al. (1995)
CO471	O1 (El Tor)	1994/India	Clinical	Large (1)	Nandi et al. (2003)
CO473	O1 (El Tor)	1994/India	Clinical	Large (1)	Nandi et al. (2003)
CO457	O1 (El Tor)	1994/India	Clinical	Large (1)	Nandi et al. (2003)
AS197	O139	1996/India	Clinical	Large (3)	Khetawat et al. (1999)
AS207	O139	1996/India	Clinical	Large (3)	Khetawat et al. (1999)
B33	O1 (El Tor)	2004/Mozambique	Clinical	Small	Das et al. (2007)

[a]Not known with certainty, but believed to be as listed
[b]CTXφ prophages are devoid of *ctxAB* genes
[c]Isolated in India just prior to the O139 outbreak; called pre-O139 El Tor isolate

8.8.5 Production of Infectious Particles

CTX prophage in El Tor and O139 strains generally give rise to infectious phage particles at a low but measurable rate, even in the absence of induced stimuli such as mitomycin C. These virions are secreted, rather than released through bacterial lysis, and virion production is not known to limit growth of the host bacterium (Davis and Waldor 2000). CTX prophage within classical strains does not give rise to infectious phage particles. CTXφ derived from El Tor strains does not integrate following CTXφ infection of classical strains. The phage DNA replicates as a plasmid in classical strains rather than recombining into either of the two integration sites.

8.9 Overview of CTXφ Biology

This section briefly presents recent advances in our knowledge of CTXφ biology pertaining to its infection, integration, replication, phage gene expression, assembly, and secretion. Unlike the well-studied F pilus-specific filamentous coliphages, CTXφ integrates site specifically into its host chromosome and forms stable lysogens. CTXφ has a remarkable dependence on host-encoded factors, and at the same time plays a critical role not only in pathogenesis but also in the evolution of the pathogenicity of *V. cholerae*.

8.9.1 Infection of CTXφ

Like several other filamentous bacteriophages which use pili as receptors for infection of host bacterial cells, CTXφ uses toxin-coregulated pili (TCP) as its functional receptor (Waldor and Mekalanos 1996). This is based on the following evidence: (a) CTXφ adsorbs efficiently to TCP-piliated cells of *V. cholerae* but not to nonpiliated cells; (b) antisera directed against TCP blocks phage infection, and; (c) TCP defective mutants were resistant to transduction by CTX-Kmφ particles. However, biochemical or electron microscopic evidence of CTXφ binding to TCP has not been demonstrated.

The infection of CTXφ has been studied by Heilpern and Waldor (2000, 2003), and the following model has been proposed. At first, the CTXφ virions bind to TCP, a receptor that protrudes some distance out from the outer membrane. The virions are then drawn to the bacterial outer membrane, possibly by pilus retraction, as found in the case of infection of Ff phages. However, the retraction of TCP has not been demonstrated experimentally. From the outer membrane, the virions are translocated to the periplasmic space with the help of TolRQA proteins. The coat protein $pIII^{CTX}$ binds to the C-terminus of TolA and mediates the uncoating of the phage

DNA. Finally, the phage DNA (ss) enters the cytoplasm; once inside the cell, the CTX DNA can integrate into the chromosome by some mechanism, as discussed below.

8.9.2 Integration of CTXφ

A proper understanding of the mode of integration of phage DNA into the host chromosome has been a problem, and two different views have been proposed. One view (McLeod and Waldor 2004) states that after the phage DNA (ss) enters the host cytoplasm, its complimentary strand is synthesized to generate a dsDNA replicative form, and that this form acts as the substrate for integration. A model explaining the details of the integration steps has been put forward. The host-encoded tyrosine recombinases, XerC and XerD, were shown to take part in the recombination process leading to integration of phage DNA into host chromosome. However, this model raised several questions (Blakely 2004) that needed to be resolved. In an attempt to resolve the issue, a second view of the integration mechanism was proposed by Val et al. (2005). These authors showed that integration of the CTXφ genome into the host chromosome does not require its prior conversion to a double-stranded replicative form. Rather, CTXφ integration directly uses the (+) ssDNA genome of the phage as a substrate. Their model shows that the two dif-like sequences present on the (+) ssDNA genome and the 90 bp sequence separating them fold into a double-forked hairpin, which unmasks a new Xer recombination site. Only one strand exchange is performed, which is mediated by XerC catalysis. These authors claimed that their integration strategy explains the complex structure of the integration region found on the replicative form of the phage. It also explains why the prophage cannot be excised from its host genome. For details, the readers are referred to the original publications cited above.

8.9.3 Replication of the CTXφ Genome

Unlike the known filamentous phages, the chromosomally integrated CTXφ prophage acts as a template for synthesis of viral DNA. The replication process requires the presence of tandem elements CTXφ-CTXφ or CTXφ-RS1 within the chromosome and the function of the phage-encoded protein RstA (Moyer et al. 2001). The predicted amino acid sequence of RstA contains motifs similar to those found in RC replication initiation proteins (Waldor et al. 1997). RstA is believed to make a site-specific cleavage (nick) in the CTXφ prophage at the phage origin of replication located at Ig-1, yielding a 3'-end that can serve as a substrate for new DNA synthesis by host-encoded DNA polymerase. DNA synthesis displaces one strand of the phage genome as a new strand of phage DNA is synthesized. Replication has been studied extensively by Moyer et al. (2001). The apparent advantage of this replicative process is that phage DNA (and subsequently virions) can be produced without loss of the CTXφ prophage from the chromosome. Thus it allows for both horizontal transmission of CTXφ DNA via phage particles and vertical transmission of the CTXφ prophage DNA to daughter cells.

Classical strains of *V. cholerae*, however, lack RS1, and CTXφ is inserted into their chromosomes as either a solitary prophage or an array composed of two truncated prophages (Davis and Waldor 2000). Because the CTXφ prophage array constitutes the template for phage DNA production, the arrangement of CTX elements in classical strains would not be expected to produce extrachromosomal CTXφ phage DNA, a critical step in the production of virions (Davis and Waldor 2000).

8.9.4 CTXφ Gene Expression

The phage CTXφ enters into a lysogenic state following its integration into the host chromosome. In this state, the expression of phage genes controlling the replication and virion production are repressed. CTXφ lysogeny depends on the phage-encoded transcriptional repressor RstR, which binds to the promoter of *rstA*, thereby repressing the expression of *rstA, rstB*, and other morphogenesis genes. The gene *rstR* encodes a 13 kDa protein with an N-terminal helix-turn-helix (HTH) DNA-binding element similar to the HTH present in the Cro/cI superfamily of repressors (Waldor et al. 1997), whereas the short C-terminal region is unrelated to the oligomerization domain of cI repressor. Like lambda lysogens, treatment of CTXφ lysogens with either mitomycin C (MMC) or UV light, DNA-damaging agents known to provoke the SOS response in *E. coli*, results in an increased transcription from *rstA* promoter and increased production of CTXφ virions in a *recA*-dependent fashion (Waldor and Mekalanos 1996, Quinones et al. 2005). In addition, SOS induction of CTXφ production requires RecA-stimulated autocleavage of LexA, the global SOS regulator in *E. coli* (Quinones et al. 2005). Upon activation of RecA and DNA damage, LexA undergoes autoproteolysis, which relieves the repression of numerous genes involved in DNA repair (Little 1984). Quinones et al. (2005) has shown that LexA autoproteolysis is also required to induce the expression of the RstA promoter. For more detail on this topic, readers may refer to Kimsey and Waldor (2004) and Quinones et al. (2005).

8.9.5 Assembly and Secretion of CTXφ

CTXφ, like other filamentous phages, is secreted from *V. cholerae* without cell lysis. In Ff phages, the virion assembly and secretion occur simultaneously (Russel 1995). The CTXφ-encoded protein Zot is homologous to the Ff protein pI (Koonin 1992, Waldor and Mekalanos 1996) and, like pI, is thought to be an inner membrane protein that acts in both phage assembly and secretion. As Zot contains a putative ATPase domain, it may provide energy needed for these processes. Zot may also interact with phage coat proteins, which are also predicted to reside in the inner membrane, to facilitate the assembly of phage particles.

Ff phages are known to encode a secretin, an outer membrane pore that serves as a channel for secretion of virions out of the cell. CTXφ, however, is secreted through a chromosome-encoded outer membrane channel, EpsD (Davis et al. 2000b). EpsD also serves as the secretion component of the *V. cholerae* Eps type II secretion system (T2SS). This apparatus consists of approximately 15 proteins that mediate secretion of CT, hemagglutinin-protease, chitinase, and other proteins (discussed in Chap. 10). CTXφ appears to require only EpsD among T2SS components for its secretion. Although it is possible that other factors compete with CTXφ for access to EpsD, *V. cholerae* produces such low titers of CTXφ that it seems unlikely that phage production significantly impairs protein release.

8.10 The Vibrio Pathogenicity Island 1 (VPI-1)

Although the major structural subunit of TCP is encoded by the *tcpA* gene, the formation and function of the pilus assembly require the products of a number of other genes located in a larger genetic region adjacent to the *tcpA* gene referred to as the TCP gene cluster, and which was shown to be a part of a large pathogenicity island termed the TCP-ACF element, or the vibrio pathogenicity island (VPI) (Brown and Taylor 1995, Karaolis et al. 1998, Kovach et al. 1996), and was later designated VPI-1 (Jermyn and Boyd 2005). VPI-1 is a 41 kb region that encodes the toxin-coregulated pilus (TCP), an essential colonization factor, the accessory colonization factor (ACF), and the virulence regulators ToxT and TcpH. In addition, the VPI-1 contains several open reading frames with unknown function.

The VPI-1 has many features that are typical of pathogenicity islands (Blum et al. 1994, Hacker et al. 1997, Hacker and Kaper 1999). It has a low percent G+C content (35%) compared to the rest of the genome (48%) (Heidelberg et al. 2000), has phage-like attachment (att) sites at its termini, is inserted site specifically into the chromosome of epidemic *V. cholerae* strains downstream of a tRNA-like locus (*srrA*), and has its left- and right-end genes with potential roles in DNA mobility, including a transposase-like gene (*vpiT*) and a phage-like integrase gene (*int*), which belongs to the family of site-specific recombinases (Karaolis et al. 1998, Kovach et al. 1996). It was shown that VPI-1 can be transferred via generalized transduction between *V. cholerae* serogroup O1 strains (O'Shea and Boyd 2002). VPI-1 was further shown to have the ability to excise from its chromosomal insertion site and circularize to form a circular intermediate; however, the cognate integrase was not essential for excision (Rajanna et al. 2003).

Transposon mutagenesis identified another set of genes important for host colonization (Peterson and Mekalanos 1988). These accessory colonization factor (*acf*) genes, *acfA, acfB, acfC*, and *acfD* are also located within the VPI-1. The *acf* genes

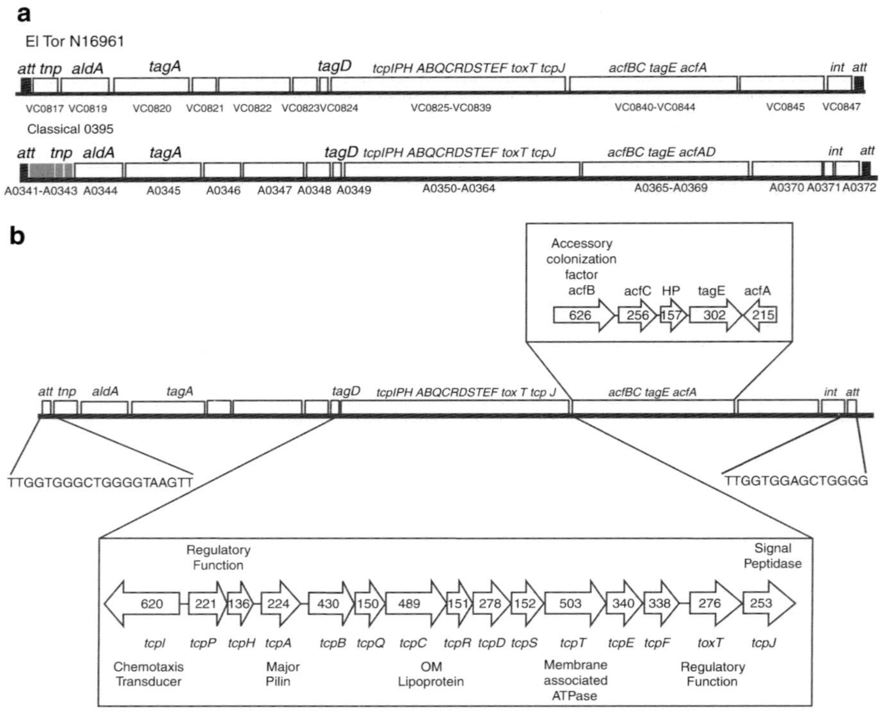

Fig. 8.2 **(a)** Genetic organizations of the TCP pathogenicity islands (VPI-1) of *V. cholerae* El Tor N16961 and classical O395. The genes or gene clusters are represented by *open boxes* and their names are indicated above the boxes. The *att* sites at each end are represented by *black boxes*. The putative integrase gene (*int*) is near the right end of the island. The transposase sequence (*tnp*) is near the left end of the island. **(b)** Expanded view of the TCP and ACF gene cluster present in VPI-1 of El Tor N16961, showing the positions of the individual genes and directions of their transcription. The functions or the proteins translated are also shown. The nucleotide sequences at the two *att* sites are also shown by enlarging the regions. *HP*, hypothetical protein

are required for enhanced intestinal colonization, but the functions of the encoded proteins are yet to be fully elucidated. AcfB shares homology with methyl-accepting chemotaxis receptors (MCP) (Everiss et al. 1994).

The organization of genes in VPI-1 from the complete genome sequence of *V. cholerae* El Tor N16961 and classical O395 is represented in Fig. 8.2a. VPI-1 is flanked on both sides by a putative phage-like att-site (elaborated in Fig. 8.2b for N16961). The left and right ends of VPI-1 contain genes with potential roles in DNA mobility, which include a transposase-like gene (*tnp*) and a phage-like integrase gene (*int*) belonging to the family of site-specific recombinases (Fig. 8.2a). These genes could be involved in the transfer and integration of the VPI-1 region (Karaolis et al. 1998). VPI-1 encodes proteins with essential roles in virulence, such as (i) the *tcp* gene cluster encoding toxin-coregulated pili (TCP), an essential colonization factor, and proteins that regulate virulence such as ToxT, TcpP, and

TcpH, and (ii) the *acf* gene cluster, which is assumed to encode an accessory colonization factor; the exact nature of this colonization factor is unclear (Karaolis et al. 1998, Kovach et al. 1996) (Fig. 8.2b). Like other prokaryotic pathogenicity islands, VPI-1 is associated with pathogenic strains of *V. cholerae* (Karaolis et al. 1998, 2001). The structure, GC content and codon usage within the VPI suggests that it was recently acquired by *V. cholerae*. As the CTXφ encoding CT uses TCP as its receptor for infecting new strains (Waldor and Mekalanos 1996), these two horizontally moving elements are proposed to be linked evolutionarily (Faruque and Mekalanos 2003).

It has been proposed by Karaolis et al. (1999) that the VPI-1 constitutes the genome of a novel filamentous bacteriophage designated VPIφ and is responsible for inter-strain gene transfer, although the transfer of VPIφ between the serogroups O1 and O139 has not been shown. Later studies could not detect VPIφ particles in TCP-positive strains by repeated assays (Faruque et al. 2003). O'Shea and Boyd (2002) demonstrated that VPI-1 can be transferred via generalized transduction between *V. cholerae* serogroup O1 strains. It was proposed that the mechanism is simply generalized transduction of a chromosomal marker rather than a specific, VPI-mediated process, and VPI-1 was hypothesized to correspond to a satellite element (Faruque et al. 2003).

8.11 Toxin-Coregulated Pili (TCP)

In order to cause disease, *V. cholerae* must attach to the intestinal epithelium, where it colonizes and secretes cholera toxin, which is ultimately responsible for the massive fluid efflux characteristics of cholera. The colonization process in bacteria is mediated by surface appendages like fimbriae or pili. The toxin-coregulated pili (TCP) in *V. cholerae* were first described by Taylor et al. (1987) and were shown to play a critical role in colonization in the infant mouse cholera model and subsequently in human volunteers (Herrington et al. 1988). Antibodies to TCP are sufficient to confer protection in the infant mouse cholera model (Sun et al. 1990). These initial studies were performed with strains of the classical biotype. Interestingly, the expression of TCP parallels the synthesis of cholera toxin (Taylor et al. 1987).

TCP-deficient mutants are unable to colonize when tested in suckling mouse small bowel (Taylor et al. 1987) and in gut mucosa of human (Herrington et al. 1988). The elaboration of TCP confers several properties upon *V. cholerae* that may be relevant to its role in colonization. When grown in vitro, production of TCP causes the bacteria to autoagglutinate in liquid culture, manifesting as granular clumps of cells that fall out of suspension after overnight growth (Taylor et al. 1987). This phenomenon provides an easily discernible phenotype and suggests a possible in vivo role for autoagglutination in intestinal microcolony formation. In addition, TCP mutants show increased serum sensitivity, indicating a potential role for pili to mediate resistance to a complement-like activity possibly associated with the intestinal environment (Chiang et al. 1995). Although no specific enterocyte receptor has yet been identified for TCP, it has been suggested that TCP mutants adhere less well than wild-type strains to intestinal mucosa.

TCP has been shown to be important for colonization by *V. cholerae* O1 of the classical biotype (Taylor et al. 1987, Herrington et al. 1988, Sharma et al. 1989, Sun et al. 1990). The expression of TCP on the bacterial surface, as well as the importance of this pilus as a colonization factor of vibrios of the El Tor biotype, on the other hand, has been questioned by several investigators (Attridge et al. 1993, Osek et al. 1994). Antibodies raised against classical biotype TCP, or TcpA, exhibit strong protection against challenge by strains of the same biotype, but provide weaker protection, or none at all, against El Tor strains (Sharma et al. 1989, Sun et al. 1990, Osek et al. 1994). Jonson et al. (1991) have demonstrated that most or all strains of *V. cholerae* of the El Tor biotype instead express another pilus structure called the mannose-sensitive hemagglutinin (MSHA) pilus. Specific antibodies against MSHA pili were able to protect against experimental cholera caused by El Tor vibrios (Osek et al. 1992), suggesting an important role for MSHA pili in the pathogenesis of cholera caused by the El Tor biotype of *V. cholerae* O1. The recent isolation and characterization of the *tcpA* gene from El Tor and O139 strains has helped to clarify some of these issues. The *tcpA* sequence from El Tor strain N16961 is identical to that of O139 strain MO3 (Rhine and Taylor 1994), but shows significant deviation from the classical biotype gene, especially in the portion encoding the C-terminal region of the pilin, where epitopes recognized by protective monoclonal antibodies map (Iredell and Manning 1994, Sun et al. 1991, Rhine and Taylor 1994). Knowledge of the sequence has also facilitated the construction of El Tor *tcpA* insertion mutants. As in the case of classical biotype strains, such mutants exhibit a marked decrease in their ability to colonize (Attridge et al. 1993, Rhine and Taylor 1994).

TCP is a member of the type IV pili, which are 1000–4000 nm long, 60–90 Å diameter fibers formed by the ordered association of thousands of identical pilin subunits plus a few copies of pilus-associated proteins. TcpA, a 20 kDa protein, is the major component of TCP (Strom and Lory 1993), and serves as the receptor for CTXφ (Waldor and Mekalanos 1996). The amino acid sequences of the major pilin protein TcpA are known to differ considerably between the El Tor and classical biotypes (75% similarity at the nucleotide level) (Iredell and Manning 1994, Rhine and Taylor 1994). Several variant TcpA sequences have been observed among non-O1 non-O139 serogroup isolates (Ghosh et al. 1997, Nandi et al. 2000, Mukhopadhyay et al. 2001). The genetic variability of *tcpA* is illustrated in Fig. 8.3a and b. The alignments of the deduced amino acid sequences of TcpA from different strains of *V. cholerae* obtained from NCBI database are presented in Fig. 8.3a, which shows the distribution of conserved and polymorphic amino acid sites in TcpA among classical, El Tor, O139, and non-O1-non-O139 strains. These sequence comparisons indicate that the genetic variability in the *tcpA* locus occurs mostly in the carboxy-terminal region of the protein. A dendogram of the *tcpA* locus based on pairwise comparisons of the deduced amino acid sequences is presented in Fig. 8.3b and reveals the presence of three distinct clusters. *V. cholerae*

a

```
VC_O395        MQLLKQLFKKKFVKEEHDKKTGQEGMTLLEVIIVLGIMGVVSAGVVTLAQRAIDSQNMTK
V52            ------------------------------------------------------------
MAK757         ------------------------------------------------------------
MO10           ------------------------------------------------------------
2740-80        ------------------------------------------------------------
C6706_ElTor    ------------------------------------------------------------
N16961         ------------------------------------------------------------
B33            ------------------------------------------------------------
NCTC8457       ------------L-----------------------------------------------
V51            ------------------------------------------------------------
VC_O27         ----------------R-------------------------------------------
VCE232         ------------------------------------------------------------
               ************:***:*******************************************

VC_O395        AAQSLNSIQVALTQTYRGLGNYPATADATAASKLTSGLVSLGKISSDEAKNPFIGTNMNI
V52            ----------------------------------T-------------------------
MAK757         ---N---V-IAM-----S--------N-N--TQ-AN-------V-A-------T--A-G-
MO10           ---N---V-IAM-----S--------N-N--TQ-AN-------V-A-------T--A-G-
2740-80        ---N---V-IAM-----S--------N-N--TQ-AN-------V-A-------T--A-G-
C6706_ElTor    ---N---V-IAM-----S--------N-N--TQ-AN-------V-A-------T--A-G-
N16961         ---N---V-IAM-----S--------N-N--TQ-AN-------V-A-------T--A-G-
B33            ---N---V-IAM-----S--------N-N--TQ-AN-------V-A-------T--A-G-
NCTC8457       ---N--T----------S-----V----A--AA-T----------A-------T---L--
V51            ---N--T----------S-----T----A--AA-T----------A-------T---L--
VC_O27         ---N--T----------S-----T----N--AA-TA-----------------T-S-L--
VCE232         ---N--T----------S-----T----N--AA-TA-----------------T-S-L--
               ***.**::*:::*****.*****.**:* **: *: *******:*:******* *: :.*

VC_O395        FSFPRNAAANKAFAISVDGLTQAQCKTLITSVGDMFPYIAIKAGGAVALADLGDFENSAA
V52            F-----------------------------------------------------------
MAK757         F-----S--------T-G----------V--------F-NV-E-AFA-V-------T-V-
MO10           F-----S--------T-G----------V--------F-NV-E-AFA-V-------T-V-
2740-80        F-----S--------T-G----------V--------F-NV-E-AFA-V-------T-V-
C6706_ElTor    F-----S--------T-G----------V--------F-NV-E-AFA-V-------T-V-
N16961         F-----S--------T-G----------V--------F-NV-E-AFA-V-------T-V-
B33            F-----S--------T-G----------V--------F-NV-E-AFA-V-------T-V-
NCTC8457       W-------G-------------------V----------NVQQKTSI-L---N---TN--
V51            W-------G-------------------V----------NVQQKS-I-L---N---TN--
VC_O27         W-----G-G-------------------V----------NVQQKASMPL-------TGV-
VCE232         W-----G-G-------------------V----------NVQQKASMPL-------TGV-
               :*****.*.******:*.**********:********:* ::     .:***.***...*

VC_O395        AAETGV-VIKSIAPASKNLDLTNITHVEKLCKGTAPFGVAFGNS
V52            ----------------------------E---------------
MAK757         D-A--A--------G-A--N-----------T-----T------
MO10           D-A--A--------G-A--N-----------T-----T------
2740-80        D-A--A--------G-A--N-----------T-----T------
C6706_ElTor    D-A--A--------G-A--N-----------T-----T------
N16961         D-A--A--------G-A--N-----------T-----T------
B33            D-A--A--------G-A--N-----------T-----T------
NCTC8457       N-AA-T-I-----TG-V--N--E----QN--A--GT-S------
V51            N-AA-T-I-----TG-V--N--E----QN--A--GT-S------
VC_O27         --G--T-I------T-V--N--E----QN--A--GT-S------
VCE232         --G--T-I------T-V--N--E----QN--A--GT-S------
                * :*.*:*****. * **:**:****::** **..* ******
```

Fig. 8.3 (a) Multiple alignment of predicted amino acid sequences of TcpA from different *V. cholerae* strains, as performed with ClustalW. Strain designations are on the *left hand side* of the alignment. *Dashes* represent amino acids identical with respect to TcpA of the *V. cholerae* classical O395 strain. *indicates positions where different strains have identical amino acids; **:** indicates positions where not all of them have same amino acid; **.** indicates where unrelated amino acids exist

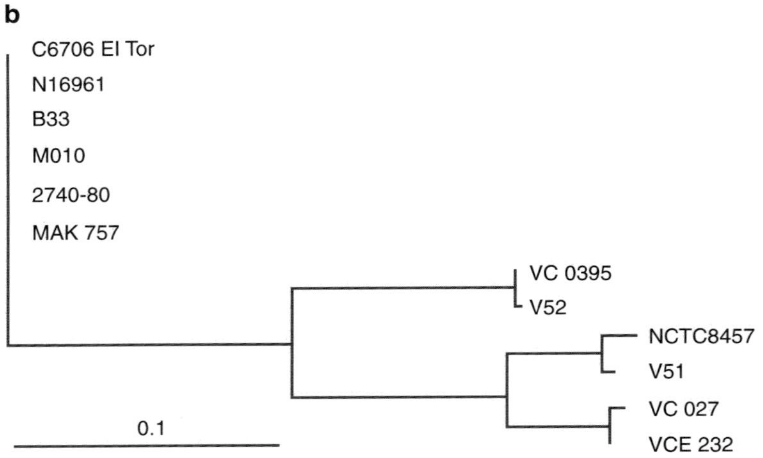

Fig. 8.3 (b) Phylogenetic tree showing the relatedness of predicted TcpA sequences from 12 *V. cholerae* strains. The unrooted tree was constructed based on the multiple alignment of TcpA sequences presented in Fig. 8.3 a using a neighbor-joining algorithm. *Labels* indicate strain designations. See text for further details

classical biotype strain O395 is grouped with V52, a O37 serogroup clinical isolate responsible for the epidemic in Sudan; the El Tor biotype strains N16961, C6706, 2740-80, Mak757, and B33 (a hybrid between the classical and El Tor biotypes, isolated in Mozambic 2004) converge with O139 strain MO10. On the other hand, an early El Tor NCTC8457 isolated in Saudi Arabia in 1910 grouped with V51, a serogroup O141 clinical isolate from the United States, VCO27 (an El Tor progenitor) (Li et al. 2002) and VCE232, a non-O1-non-O139 environmental toxinogenic strain.

8.12 Other Horizontally Acquired Gene Clusters

8.12.1 VPI-2

The pathogenicity island VPI-2 is a 57.3 kb island confined predominantly to toxinogenic *V. cholerae* O1 and O139 serogroup isolates. It encompasses 52 ORFs spanning VC1758 to VC1809 in El Tor N16961 (Jermyn and Boyd 2002) (Fig. 8.4). VPI-2 has several characteristic features of a pathogenicity island, including low G+C content (42%), presence of a bacteriophage-like integrase (P4-like) (*int*)

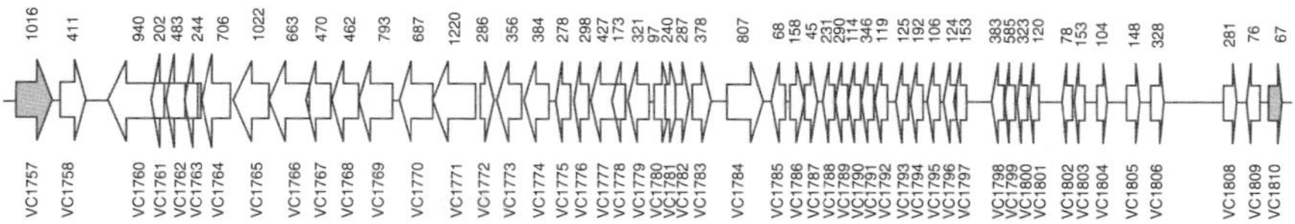

Fig. 8.4 Genetic organization of the pathogenicity island VPI-2 in *V. cholerae* El Tor N16961 based on the complete genome sequence (Heidelberg et al. 2000). ORFs are represented by *open arrows* showing the directions of transcription. The locus numbers of and the number of amino acids in each ORF are presented *below* and *above*, respectively

(VC1758), insertion in a tRNA gene (serine) (VC1757.1), and the presence of direct repeats at the chromosomal integration sites (Jermyn and Boyd 2002). In addition, VPI-2 encodes a type 1 restriction-modification (RM) system, a region that shows homology to Mu phage and genes required for the utilization of amino sugars as well as neuraminidase. The enzyme neuraminidase is known to convert higher-order gangliosides to GM1-gangliosides, releasing sialic acid, and is therefore proposed to facilitate or increase the binding of cholera toxin to its GM1 receptor (Galen et al. 1992). *V. cholerae* neuraminidase may also form part of the mucinase complex that hydrolyzes intestinal mucus, enabling the bacterium to move more freely to the epithelium.

8.12.2 VSP-I and VSP-II Islands

The genomic islands VSP-I and VSP-II were identified by microarray analysis by Dziejman et al. (2002). A *V. cholerae* genomic microarray containing spots corresponding to over 93% of the predicted genes of strain N16961 was constructed. These spots were hybridized to labeled genomic DNA from classical, prepandemic El Tor, pandemic El Tor, and two nontoxinogenic strains, in order to compare the gene content of N16961 to those of other *V. cholerae* isolates. Although a high degree of conservation was observed among the strains tested, two unique regions, termed VSP-I and VSP-II (vibrio seventh pandemic island-I and II), were found to be specifically present in seventh pandemic-specific El Tor and related O139 serogroup strains.

VSP-I is a 16 kb region spanning a block of 11 genes (VC0175-VC1085) (Fig. 8.5), and it has an atypical G+C content of 40% (compared to 47% for the entire genome), suggesting that it could be acquired by horizontal transfer. Of these 11 genes, seven encode hypothetical proteins and VC0185, which defines the 3′ end of the region, is predicted to encode a putative transposase; VC0175 shows similarity in its C-terminal domain to deoxycytidylate deaminase-related proteins; VC0176

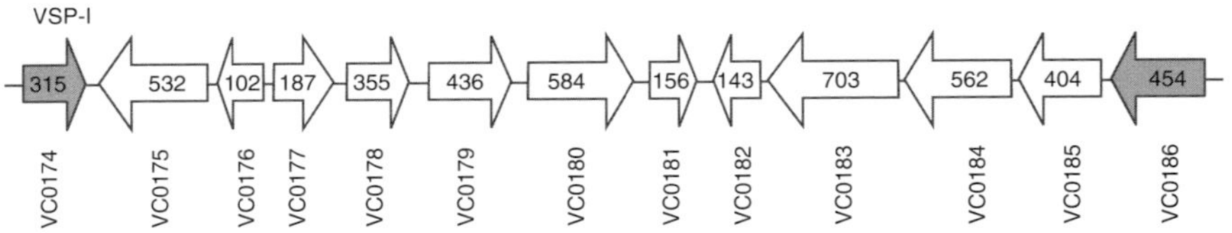

Fig. 8.5 Schematic representation of the vibrio seventh pandemic island VSP-I from *V. cholerae* El Tor N16961, based on the complete genome sequence (Heidelberg et al. 2000). ORFs are represented by *arrows* showing the directions of transcription of each individual gene. The chromosomal genes flanking the island are indicated by *shaded arrows* and the island region genes are indicated by *open arrows*. The locus number for each ORF is presented below. The *numbers inside the arrows* represent the number of amino acids of the respective ORFs

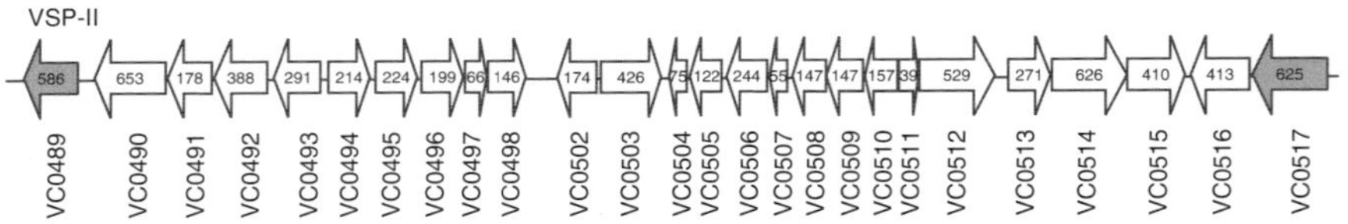

Fig. 8.6 Genetic organization of the vibrio seventh pandemic island VSP-II in *V. cholerae* El Tor N16961, based on the complete genome sequence (Heidelberg et al. 2000). ORFs are represented by *arrows* showing the directions of transcription of individual genes. The chromosomal genes flanking the island are indicated by *shaded arrows*, and the island region genes are indicated by *open arrows*. The locus number for each ORF is presented below. The *numbers inside the arrows* represent the number of amino acids in the respective ORFs

encodes a product belonging to a paralogous family that includes the lysogeny repressor protein for CTXφ and RstR; VC0178 product shows significant homology to a variety of phospholipases.

VSP-II was initially identified as a 7.5 kb region encompassing eight genes spanning VC0490-VC0497 and an average G + C content of 41%, although the boundaries of VSP-II were not defined (Dziejman et al. 2002). Subsequent studies revealed that VSP-II is a much larger region (27 kb) spanning ORFs VC0490 to VC0516 (Fig. 8.6) that integrates at a tRNA-methionine locus, is flanked by direct repeats, and encodes a P4-like integrase (O'Shea et al. 2004). A recent study by Murphy and Boyd (2008) has demonstrated that the PAIs, VPI-2, VSP-I and VSP-II, can excise from the tRNA attachment site and form extrachromosomal circular intermediate molecules, which is possibly a first step in the horizontal transfer of these PAIs. The characteristic features of many of the PAIs acquired by *V. cholerae* under different environmental conditions are summarized in Table 8.2.

All of the pathogenicity islands detected so far in *V. cholerae* show uniquely different G+C contents compared to the rest of the genome, suggesting that these

Table 8.2 Horizontally acquired gene clusters in *V. cholerae*

Designation	Location in El Tor N16961	G+C content (%)	Characteristics
CTXφ prophage	VC1456–VC1464	44	Chromosomally integrated phage genome, carries the cholera toxin genes (*ctxAB*), other genes include *zot, ace*, and *cep*
TCP island or VPI-1	VC0817–VC0847	35	A 41 kb chromosomal region integrated at the tmRNA site (*ssrA*) comprising TCP and ACF gene clusters, a putative integrase, and a transposase
VPI-2	VC1758–VC1809	42	A 57.3 kb gene cluster encoding genes for neuraminidase (*nanH*) and amino sugar metabolism, inserted at a tRNA-serine locus flanked by direct repeats. Encodes a type 1 restriction modification system, region homologous to Mu phage, silialic acid transport, and catabolism genes
VSP-I	VC0175–VC0185	40	A 16 kb region present only in O1 El Tor and O139 strains. The genes encode hypothetical proteins, putative XerCD recombinase
VSP-II	VC0490–VC0516	40	A 27 kb region integrated at a tRNA-methionine site, and which encodes homologs of RNaseH1, a type IV pilus, a DNA repair protein, transcriptional regulators, methyl-accepting chemotaxis protein, and P4-like integrase
Integron islands	VCA0291–VCA0506	42	A 125.3 kb gene cluster that constitutes an efficient gene capture system

clusters were acquired from exogenous sources. The four PAIs, VPI-1, VPI-2, VSP-I and VSP-II, have been found associated with *V. cholerae* epidemic and pandemic isolates. Studies of these gene clusters are expected to throw light on their significant role in the emergence and reemergence of pathogenic bacteria. However, significant genetic heterogeneity has been found within the VPI-2, VSP-I and VSP-II islands, the proper significance of which remains to be elucidated.

Chapter 9
Cholera Toxin (CT): Regulation of the Relevant Virulence Genes

Abstract The regulatory cascade responsible for the induction of cholera toxin is typically referred to as the ToxR regulon. The expression of genes in the ToxR regulon is controlled by a cascade of regulatory proteins, ToxR, ToxT, and TcpP. In general, the ToxR regulon genes fall into four classes: cholera toxin (CT) genes, toxin-coregulated pilus (TCP) biosynthesis genes, accessory colonization factor (ACF) genes, and the ToxR-activated genes of unknown function. The primary and direct transcriptional activator of the *Vibrio cholerae* virulence genes is the ToxT protein, and the ToxR regulon consists of the complex pathway regulating ToxT expression and, consequently, downstream virulence genes including the *ctxAB* and *tcp* operons. In general, the products that function to promote intestinal colonization, toxin production and survival within host, and which are coordinately expressed, are dependent on the ToxR regulon. The story of this coordinated regulation of the virulence genes of *V. cholerae* in response to diverse environmental and other signals is presented here.

9.1 Introduction

The organization and function of the cholera toxin genes (*ctxAB*) belonging to the CTXφ genetic element and the genes encoding the toxin-coregulated pilus (TCP) and the accessory colonization factors (ACF) belonging to the large genetic island VPI-1 have already been described in the previous chapter (Chap. 8) of this book. The expression of these genes or the expression of virulence of the organism under different environmental conditions is dependent on the interplay of a complex regulatory network. *V. cholerae* enters its host orally, survives passage through the gastric acid barrier of the stomach, uses motility and chemotaxis functions to reach and penetrate the mucus coat of the upper intestinal epithelium, colonizes the epithelium, survives in the intestinal environment, and produces the potent enterotoxin CT (Kaper et al. 1995). To do this, the vibrios must coordinately express two critical sets of genes, the *tcp* and *ctx* operons, which encode the toxin-coregulated pilus and cholera enterotoxin, respectively. While CT is responsible for the severe

K. Chaudhuri and S.N. Chatterjee, *Cholera Toxins*.

diarrhea that is the hallmark of the disease, TCP pili are essential for colonization in mice and human (Taylor et al. 1987).

CT production in *V. cholerae* is a part of a complex regulatory network in which several other virulence factors in addition to CT are expressed and coordinately regulated in response to specific environmental signals. The molecular mechanism of action of this regulatory network is presented in this chapter.

9.2 Modulation of Cholera Toxin Expression by Environmental Factors

Early studies showed that in vitro CT expression is influenced by growth medium, temperature, and other environmental parameters (Evans and Richardson 1968, Richardson 1969, Callahan and Richardson 1973). Several investigators (Craig 1966, Finkelstein 1970, Finkelstein and LoSpalluto 1970) have attempted to enhance in vitro cholera toxin production by modifying the culture medium and conditions and have defined the environmental parameters that control CT expression in vitro. The early in vitro studies uncovered several variables like temperature, pH, osmolarity, presence of amino acids, aeration, bile, lincomycin, etc. that regulate CT synthesis.

It was shown that the CT production is favored in cultures grown at 30 °C, pH 6.6, 66 mM NaCl compared to those grown in 37 °C, pH 8.5, 300 mM NaCl—conditions prevailing in the human intestine. However, the true environment of the human intestine cannot be reproduced in vitro, as the in vivo conditions are not precisely known. The type and level of nutrients, the human intestinal tissue secretory products, the presence of bile, and the presence of commensal microorganisms add to the complexity of the in vivo environment, which cannot be mimicked in vitro. Further characterization showed that amino acids also promote CT production. Toxin production occurs in minimal medium only when it is supplemented with amino acids like asparagines, arginine, glutamate, and serine (Richardson 1969, Callahan and Richardson 1973). Miller and Mekalanos (1988) proposed that this might have some physiological relevance; the action of *V. cholerae* proteases might release amino acids from mucus during colonization of the intestinal epithelium.

Other parameters that increase CT production include the presence of 5% CO_2 in cultures grown at 37 °C, despite slight suppression of cell growth (Shimamura et al. 1985). Bile salts (0.1% sodium deoxycholate) increase the CT production in cultures grown aerobically at 37 °C (Fernandes and Smith 1977). However, CT expression was repressed in the presence of crude bile (ox bile) in the culture medium (Gupta and Chowdhury 1997). Unlike crude bile, pure bile acids—unconjugated as well as glyco- and trans-conjugated—do not repress but rather stimulate CT production, while the unsaturated fatty acids (arachidonic, linolic, and oleic acids) drastically reduced the expression of *ctxAB* genes. Lincomycin, an inhibitor of protein synthesis, increases the rate of synthesis of the toxin, increasing the extracellular and intracellular CT levels significantly (Levner et al. 1980).

Interestingly, the laboratory conditions for the expression of CT differ between the two major disease-causing biotypes of *V. cholerae*: O1, classical and El Tor (Iwanaga et al. 1986). There are distinct differences in the compositions of the media and the culture conditions required by these two biotypes; the media and culture conditions that permit high-level CT expression in the classical biotype do not favor its expression in strains of the El Tor biotype, which are responsible for the current world pandemic. The classical strains are relatively permissive; they produce CT under a variety of laboratory media and conditions, but the induction of CT is more demanding and requires defined conditions (AKI conditions) for the El Tor biotype. AKI is composed of bacto-peptone (1.5%), yeast extract (0.4%) and NaCl (0.5%), and cultures are grown in tubes under static conditions for 4h and then transferred to flasks with approximately tenfold higher capacities where they are grown with shaking up to 16h (Iwanaga et al. 1986). Although a variety of media and culture conditions have been developed, it is difficult to define a perfect combination of medium and culture conditions that would yield optimal cholera toxin production by these two biotypes of *V. cholerae*. The reasons for the differential regulation of CT in these two biotypes are not fully understood; there could be metabolic differences in the genetic regulation of CT in these two biotypes.

9.3 The Regulation of Virulence: An Overview

Regulation of the virulence factors of *V. cholerae* has been an area of great concern recently. The three proteins ToxR, TcpP and ToxT coordinately regulate transcription of the structural genes for CT, TCP, and other virulence factors (Faruque et al. 1998, Lee et al. 1999, 2001). These proteins together form the *V. cholerae* virulence gene regulatory cascade (DiRita et al. 1991, Carroll et al. 1997, Hase and Mekalanos 1998). ToxR is a transcription factor located in the inner membrane that directly regulates the expression of CT, ToxT, and the porins OmpU and OmpT (Miller and Mekalanos 1984, Champion et al. 1997, Crawford et al. 1998). The function of ToxR is greatly enhanced by a second inner membrane protein, ToxS, which associates with ToxR (Miller et al. 1989). ToxT is a soluble transcriptional activator that amplifies its own expression and directly regulates the expression of CT, TCP, and other virulence factors (DiRita et al. 1991). Recently, TcpP and TcpH, which exhibit homology to ToxR and ToxS, respectively, have been shown to regulate CT and TCP expression, presumably through ToxT (DiRita 1992, Hase and Mekalanos 1998). *V. cholerae* has further developed a mechanism of sampling and responding to the environment in order to monitor temperature variations (Parsot and Mekalanos 1990), depletions in iron concentration (Sigel and Payne 1982), presence of bile salt (Gupta and Chowdhury 1997), etc. Also, the expression of CT, TCP, and other virulence factors has been shown to differ between the classical and El Tor biotypes of *V. cholerae* (Jonson et al. 1990, DiRita et al. 1996). Up to the late

1990s, investigations of these regulatory networks of *V. cholerae* were mostly limited to in vitro culture conditions (Lee et al. 2001). Although in vitro assays contribute greatly to our understanding of bacterial pathogenesis, they frequently cannot reproduce the complex environment encountered by pathogens during infection (Chiang et al. 1999). Recently, several methods have been developed that greatly simplify the in vivo analysis of large number of strains (Chiang et al. 1999). Of these, in vivo expression technology (IVET) and signature-tagged mutagenesis (STM) (Chiang and Mekalanos 1998) have been used to identify possible regulators of virulence genes during infection. The requirements of the virulence regulators, ToxR, TcpP, and ToxT, for the expression of *tcpA* and *ctxA* were found to differ significantly during infection versus growth in vitro (Lee et al. 1999).

Several genes that are differentially expressed following infection have also been identified using two other methods, global transcription response and RNA arbitrarily primed (RAP)-PCR fingerprinting techniques (Chakrabortty et al. 2000, Das et al. 2000). Further, a role for quorum sensing in the regulation of virulence gene expression has recently been established (Zhu et al. 2002). Readers are referred to some of the recently published reviews and papers on this topic for a detailed account of the in vivo genetic analysis of virulence (Lee et al. 1999, 2001, Chiang and Mekalanos 1999, Banerjee et al. 2002, Das et al. 2002, Krukonis and DiRita 2003a, b, Nag et al. 2005).

9.4 The ToxR Regulon

Much attention has been focused on the dissection of the regulatory cascade responsible for the induction of cholera toxin, and this cascade is typically referred to as the ToxR regulon. The reason for this nomenclature is purely historical; ToxR was the first regulatory protein identified in this cascade that was required for CT expression in *E. coli* and whose deletion in *V. cholerae* results in the inability to produce CT (Miller and Mekalanos 1984). However, direct ToxR activation of the *ctxAB* promoter in *V. cholerae* has not been demonstrated. Instead, a regulatory cascade, the ToxR regulon, has been defined, and multiple positive and negative transcriptional regulators have been fed into this cascade to ensure virulence factor regulation in response to environmental cues (DiRita 1992, Kaper et al. 1995, Skorupski and Taylor 1997, Faruque et al. 1998, Klose 2001, Peterson 2002, Childers and Klose 2007).

The expression of a large number of *V. cholerae* genes in response to specific environmental conditions within the human small intestine is required for successful colonization and expression of the cholera toxin, CT. The ToxR regulon contains these genes as members, and their expression is controlled by a cascade of regulatory proteins (ToxR, TcpP, and ToxT) (Skorupski and Taylor 1997). In general, the ToxR regulon genes fall into four classes: cholera toxin (CTX) genes; toxin-coregulated pilus (TCP) biosynthesis genes; accessory colonization factor (ACF) genes; and ToxR-activated genes of unknown function (Peterson and

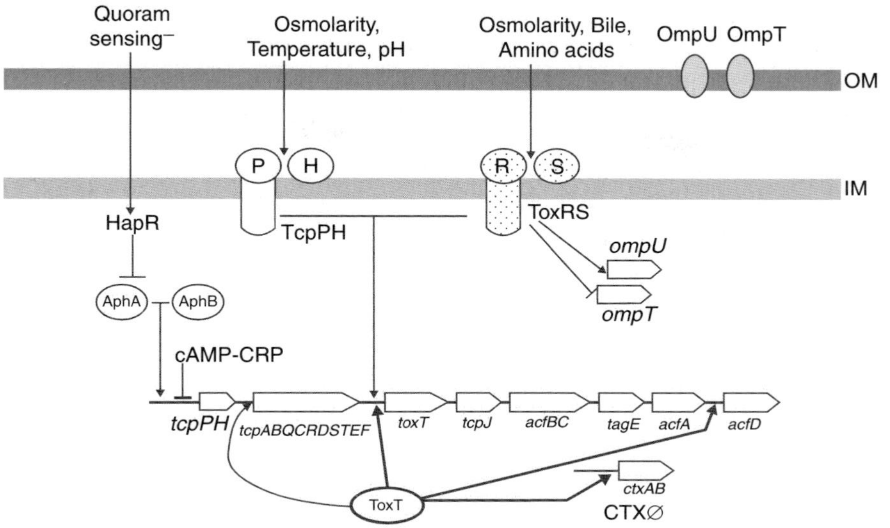

Fig. 9.1 Schematic diagram illustrating how the various environmental factors influence the regulatory cascade (ToxR regulon) that controls the *ctx* and *tcp* transcription in *V. cholerae*. See text for details of the regulatory processes described in this scheme

Mekalanos 1988). In fact, the primary and direct transcriptional activator of the *V. cholerae* virulence genes is the ToxT protein, and the ToxR regulon consists of the complex pathway regulating ToxT expression and, consequently, downstream virulence genes including *ctxAB* and *tcp* operon. In general, the products that function to promote intestinal colonization, toxin production and survival within host, and which are coordinately expressed, are dependent on the ToxR regulon (Fig. 9.1). The coordinated regulation of these virulence genes by the ToxR regulon in response to diverse environmental and other signals will be described in the following.

9.5 ToxT-Dependent Transcription

ToxT is a soluble transcription factor that directly activates transcription of the *ctx, tcp*, and *acf* genes, as well as additional genes within the VPI-1. The *toxT* gene is also located within the VPI-1, downstream of the *tcp* operon, so ToxT can possibly influence its own expression as well. *V. cholerae* strains lacking *toxT* do not express CT or TCP and are highly attenuated for virulence.

ToxT, a 32 kDa protein 276 amino acids in length, is composed of two domains: an N-terminal domain (NTD) of approximately 176 amino acids involved in dimerization (presumably forming a secondary domain within ToxT), and the 100 amino acid AraC/XylS family domain at its C-terminal, necessary for DNA binding and transcriptional activation (Prouty et al. 2005). ToxT directly activates the transcription of the vast majority of *V. cholerae* virulence genes, including the genes encoding cholera toxin and TCP. ToxT binds to the upstream of the *ctxAB* genes and the upstream of *tcpA*, the first gene in a large operon encoding the components of TCP (DiRita et al. 1991, Higgins et al. 1992, Bina et al. 2003). *V. cholerae* lacking *toxT* express neither CT nor TCP and are highly attenuated for virulence (Champion et al. 1997).

The function of the secondary domain within ToxT is not clear. The ToxT N-terminal domain (NTD) sequence did not show any homology to any other protein, and significant variation is found within the NTD in environmental *V. cholerae* strains (Mukhopadhyay et al. 2001). On the other hand, the AraC family domain is nearly invariant among all known ToxT sequences. Usually these secondary domains are involved in effector binding and/or multimerization. Bile happens to be a natural effector of ToxT and causes a decrease in the expression of CT and TCP (Gupta and Chowdhury 1997, Schuhmacher and Klose 1999). That bile interacts with the ToxT NTD is indicated by mutational analysis (Prouty et al. 2005). Similarly, virstatin, a synthetic compound, acted as an inhibitor of *ctx* gene expression and also inhibited ToxT activity by interacting with the ToxT NTD, suggesting that there is an effector binding patch in this portion of the NTD (Shakhnovich et al. 2007).

The presence of a predicted coiled-coil domain characteristic of dimerization in the N-terminus of ToxT suggests the possibility of dimerization in this protein. Prouty et al. (2005), utilizing both a LexA-based reporter system and dominant/negative studies with isolated ToxT domains, showed that ToxT is able to dimerize. The ToxT N-terminus fused to the DNA binding domain of LexA is able to repress *sulA* transcription, which is consistent with dimerization determinants being located within the N-terminal domain. The ToxT N-terminus shares little sequence homology with the AraC N-terminus, but it is predicted to share secondary and tertiary structural similarities, as determined by threading programs. Moreover, the ToxT C-terminus region cannot bind *tcpA* promoter, as determined in an electrophoretic mobility shift assay using Cy5-labeled *tcpA* promoter, unless it is dimerized by a heterologous dimerization domain from C/EPB. However, the dimerized C-terminus is unable to activate transcription, suggesting that—unlike AraC and several other family members—there may be additional determinants in the ToxT N terminus required for transcriptional activation. DNAse I footprinting analysis of ToxT bound to the *tcpA* promoter suggested that ToxT protects two direct repeats, raising the possibility that at least two monomers of ToxT bind this promoter, and that these two monomers might in turn dimerize and generate stable binding to this promoter.

Evidence for dimerization of ToxT prior to transcriptional activation was further presented by Shakhnovich et al. (2007), who used a bacterial two-hybrid system to study ToxT and its truncation mutants. Here ToxT, its truncated form or a mutant was used drive a protein–protein interaction in a bacterial two-hybrid system in *E. coli*, which was induced with isopropyl-β-D-thio-galactoside (IPTG) and the activity was measured by β-galactosidase assay. The N-terminal domain (amino acids 1–167) of ToxT dimerized in this system, in contrast to full-length ToxT (amino acids 1–267), while the full-length ToxT mutant (L113P) was able to dimerize effectively. A series of N-terminal truncation mutants of ToxT were constructed by Shakhnovich et al. (2007), and they measured the ability of these mutants to activate *ctx* transcription by CT ELISA. The deletion of the first five residues was tolerated, but the deletion of the first nine amino acids abrogates activity. This finding is reminiscent of AraC, in which the N-terminus is required to bind arabinose to induce a conformational change resulting in transcriptional activation, and suggests that the N-terminus of ToxT may similarly bind a small molecule effector. It was found that the same amino acids (6–9) that are critical for activity had a similar effect on dimerization, drawing a strong correlation between the ability to dimerize and the activity of ToxT.

A comprehensive scanning alanine mutagenesis study was used to identify amino acids critical for ToxT function. Specific amino acids were identified as being involved in dimerization, DNA binding, and environmental modulation of ToxT (Childers et al. 2007). The Ala substitution at F151, located in a region of predicted alpha-helical structure of ToxT, most adversely affected the dimerization in the LexA-based assay system. Two other mutants, F152A and L107A, also appeared to contribute to dimerization to a lesser extent, suggesting the involvement of this region of ToxT in dimerization. Ala substitutions in two putative helix-turn-helix (HTH) recognition helices that caused differential promoter activation (K203A and S249A) did not appear to alter specific DNA binding, suggesting that these residues contribute to other aspects of transcriptional activation. A number of Ala substitutions (18 residues) were also found that result in a higher level of ToxT transcriptional activity compared to that produced by the native protein, and these mutations were almost exclusively found within the N-terminus (17 residues), consistent with this domain being involved in the modulation of ToxT activity.

ToxT directly activates the transcription of the ToxT-dependent genes by binding to their promoter regions (Table 9.1). The ToxT binding sites upstream of the promoters of ToxT-activating genes have been determined using a combination of nested *lacZ* fusions, directed mutagenesis, and copper-phenanthroline footprinting experiments. ToxT binds to specific sites within the ToxT-activated promoters, characterized as "toxbox" sequences (yrTTTTwwTwAww), a fairly degenerate 13 base pair AT-rich motif. ToxT binds to two toxbox sequences in a direct repeat configuration to activate transcription of *ctxAB, tcpA, tcpI-2*, and binds to two sites in inverted repeat orientation for activation of *acfA, acfD, tagA, tcpI-1*, while only a single ToxT binding site is present in *aldA* promoter (Withey and DiRita 2006). The fact that ToxT uses a variety of binding site configurations to activate the transcription of different genes raises the question of whether ToxT binds independently to its two sites. EMSA coupled with the copper-phenanthroline footing technique showed that occupancy of both sites is required for ToxT-directed transcription of *acfA, acfD*, and *tcpA*. Mutation of one of the two binding sites does not significantly affect occupancy by ToxT of the unaltered binding site. Moreover, as ToxT is indeed able to bind to pairs of DNA sites with both direct and inverted orientations, the protein

Table 9.1 ToxT binding site sequences and their arrangements in ToxT-regulated gene promoters

Gene[a]	Toxbox sequence[b]	ToxT binding site				References
		Location[c]	No. of sites	Spacing	Orientation[d]	
ctxAB	GATTTTTGATTTT GATTTTTGATTTT	–80 to –115	2	8 bp	Direct	Yu and DiRita (2002)
tcpA	CATTTTTTGCTGT TATTTTTTTAATA	-44 to –76	2	7 bp	Direct	Yu and DiRita (2002)
tagA	AATTTTAAGTTAA TGTTTTTTTAATG	–45 to –79	2	9 bp	Inverted	Withey and DiRita (2005a)
aldA	TGTTTTTTTAAAT	–46	1			Withey and DiRita (2005a)
acfA/acfD	AATTTTTAAAAAT CATTTTGTTAAAT	–51 to –79/–46 to –74	2	2 bp	Inverted	Withey and DiRita (2005b)
tcpI-1	TATTTTCCTAAAG CGTTTTAAATAGT	–62 to –96	2	8 bp	Inverted	Withey and DiRita (2006)
tcpI-2	TGTTTTTTTTAAT TATTTTTTTTAAC	–44 to –76	2	7 bp	Direct	Withey and DiRita (2006)
Consensus[e]	yrTTTTwwTwAww					

[a]*Gene* indicates the name of the gene upstream of which the ToxT binding sites are identified
[b]*Toxbox sequence* signifies the ToxT binding site
[c]The numbers represent the start/end of toxbox1 and toxbox2. The values refer to the distances upstream of the transcription start site
[d]Relative orientation of the two toxbox sequences
[e]The consensus toxbox sequence; *r* represents the nucleotide A or G; *w* is A or T, and *y* stands for C or T

must be remarkably flexible to form or maintain protein–protein interactions between ToxT monomers in these different orientations. This question was addressed by Withey and DiRita (2006) by altering the spacing between the two binding sites by +5 bp or +10 bp and examining whether ToxT would footprint the altered DNA. If ToxT requires interactions between the monomers bound at either site, or binds as a dimer, rotating the binding sites one half-turn of the helix relative to each other should disrupt any interactions between ToxT molecules, and the footprinting of one or both ToxT sites should be affected. However, their results strongly suggested that ToxT was able to footprint both binding sites regardless of their positions relative to each other when tested for *acfA, acfD*, and *tcpA* transcription. The authors therefore proposed that ToxT binds as independent monomers to the two toxbox sites in the promoter. However, DNA insertions between these binding sites did abrogate activation of transcription by ToxT, suggesting that the spacing of the binding sites relative to the promoter is an important factor in activation. Both ToxT monomers probably contact RNA polymerase, presumably via α-CTDs bound to the promoter.

The *ctx* transcription regulation by ToxT is more complex than *tcpA* (Yu and DiRita 2002). ToxT was found to have dual roles in activating the expression of *ctx*. First, ToxT acts as an anti-repressor of *ctx*. It does so by competing for binding at DNA sites upstream of *ctx* with heat-stable nucleoid-structural (H-NS) protein, which has multiple binding sites located in this region of *ctx* promoter and acts to reduce the expression of *ctx*. Second, ToxT acts as a direct activator of *ctx* transcription, presumably by interacting with RNA polymerase. In contrast to its negative effect upon *ctx* expression, H-NS has minimal effect upon *tcpA* expression. The reason for H-NS having a stronger effect on *ctx* than *tcpA* is probably that *ctx* has more potential binding sites for H-NS. The general consensus binding site for H-NS is TNTNAN, in which N is any nucleotide. It has also been shown by Nye et al. (2000) that H-NS exerts a stronger negative effect on the *ctx* than *tcpA* promoter in *V. cholerae*. Therefore, each direct repeat TTTTGAT includes a binding site for H-NS (TTTGAT). At *tcpA*, ToxT acts not as an anti-repressor, but only as a direct activator of transcription. This is most probably through interaction between ToxT and the C-terminal domain(s) of the alpha-subunits of RNA polymerase (α-CTD).

In the case of the *ctx* promoter, binding of H-NS may alter DNA conformation to an unfavorable topology for the formation of active transcription complexes, as in the case of H-NS binding to the *E. coli rrnB* p1 promoter. It was predicted by Yu and DiRita (2002) that ToxT displaces H-NS from the promoter and binds first to the high affinity sites (–111 to –41), which may initiate an oligomerization process, and the low-affinity sites that overlap the –35 hexamer are subsequently occupied. ToxT may then enhance the binding of RNAP to the promoter and/or help the isomerization of the RNAP promoter closed to open complexes.

The transcriptional activity of ToxT is found to be repressed by specific environmental signals, including temperature and bile, suggesting that ToxT might have environmentally responsive elements (Schuhmacher and Klose 1999). A natural variant of ToxT with only 60% identity in the N-terminus, as well as a mutant form of ToxT with an altered amino acid in the N-terminus (L107F), exhibited altered transcriptional responses to bile, suggesting that the N-terminus is involved in environmental sensing (Childers et al. 2007). Additional evidence obtained in vitro demonstrated different spatio-temporal patterns of *ctx* and *tcp* transcription, despite both of these promoters being activated by ToxT (Lee et al. 1999), suggesting that *toxT* can preferentially activate specific promoters. In addition, cyclic diguanylate levels affect ToxT-dependent transcription of *ctx*, but not *tcpA* (Tischler and Camilli 2004, 2005). These data hint at an ability of ToxT to preferentially activate certain promoters, possibly in response to environmental cues encountered within the intestine, facilitating finer regulation of spatiotemporal expression of virulence factors.

A recent study examining ToxT-dependent virulence factor expression revealed that several ToxR regulon genes are modulated by bile. ToxT-dependent expression of *ctxA* and *tcpA* was significantly decreased by the addition of 0.4% bile to the medium (Schuhmacher and Klose 1999, Provenzano and Klose 2000, Provenzano et al. 2000). It was suggested that the presence of bile within the intestinal lumen, where concentrations may be as high as 2% of individual bile salts, would prevent ToxT-dependent transcriptional activation of virulence genes. The authors further proposed that since the concentration of bile presumably decreases at the surface of the epithelial cells lining the intestinal tract, the repressive activity of bile would dissipate, thereby allowing for ToxT transcription of virulence genes.

9.6 Regulation of ToxT Transcription

Induction of *toxT* transcription occurs in the intestine, and this is an essential prerequisite for virulence factor expression. The initial induction of ToxT transcription is under the control of two integral membrane regulatory proteins, ToxR and TcpP (Krukonis et al. 2000). Both proteins have cytoplasmic domains that share homology with the OmpR family of winged membrane regions and periplasmic domains with little homology to other proteins (Miller and Mekalanos 1985, Miller et al. 1987, Hase and Mekalanos 1998). Unlike many other OmpR family activators, both ToxR and TcpP contain the DNA-binding motif in the N-terminal domain rather than in the C-terminal domain, and both are predicted to reside in the inner membrane of the bacterium with substantial periplasmic domains: 96 amino acids for ToxR and 52–60 amino acids for TcpP (depending on the location of putative transmembrane domain of TcpP). ToxR is necessary but not sufficient for *toxT* activation (Higgins and DiRita 1994). Mutant Δ*toxR* failed to activate *toxT*, showing that ToxR is required for *toxT* expression. However, overexpression of TcpP from a plasmid could restore an intermediate level of *toxT* activation to the delta *toxR* strain, suggesting that ToxR is required for *toxT* activation at wild-type TcpP expression levels.

Mutations in the *tcpPH* locus lying upstream of the TCP biosynthetic operon led to a loss of *toxT* transcription. Additionally, overexpression of the TcpP protein activates *toxT* transcription, as measured by both expression of a *toxT-lacZ* gene fusion and primer extension analysis in *V. cholerae*. The expression of TcpPH is sufficient to activate a *toxT-lacZ* fusion construct, and co-expression with ToxRS enhances this activation in *E. coli* (Murley et al. 1999). Enhancement of TcpP-mediated activation of *toxT* by ToxR has also been shown in *V. cholerae* (Hase and Mekalanos 1998).

The co-dependence of maximal *toxT* expression on TcpP and ToxR may indicate a direct interaction between these two proteins, although attempts to demonstrate such an interaction have failed. The *toxT* promoter sequences required for TcpP or ToxR interaction was determined by two different strategies: (i) a series of *toxT-lacZ* promoter deletion constructs was analyzed in the *V. cholerae* Δ*tcpP*Δ*toxR* double mutant for the ability of either TcpP or ToxR expressed from a plasmid to activate transcription; (ii) *V. cholerae* membrane extract derived from strains expressing ToxR or TcpP or both or neither protein were mixed with *toxT* promoter (−172 to +45) construct in which the top strand or bottom strand had been labeled, and this was analyzed by DNAse I footprinting (DNAse protection assay). Both a promoter activation experiment and electrophoretic mobility shift analysis (EMSA) using *V. cholerae* membranes demonstrated that TcpP interacts with a region of the *toxT* promoter in close proximity to the predicted RNAP consensus binding site. TcpP binds the *toxT* promoter in close proximity to the RNAP consensus binding site encompassing nucleotides −51 to −32. ToxR occupies a more distal site from −100 to −69, possibly involving more than one ToxR molecule (DiRita and Mekalanos 1991, Harlocker et al. 1995).

ToxR binds the *toxT* promoter at −100 to −69 with respect to the transcriptional start site, while TcpP binds the −51 to −32 region (Krukonis et al. 2000). ToxR alone is unable to activate the *toxT* promoter, while TcpP must be overexpressed to activate *toxT* in the absence of ToxR (Hase and Mekalanos 1998, Murley et al. 1999, Krukonis et al. 2000). The evidence suggests that ToxR serves at this promoter as an enhancer for TcpP binding, and that TcpP alone makes contact with RNA polymerase but requires interaction with ToxR in order to activate transcription (Krukonis and DiRita 2003 a, b). Membrane localization of ToxR is required for its ability to stimulate TcpP-dependent *toxT* transcription, presumably because this facilitates interaction with membrane-bound TcpP.

ToxR interacts with another co-transcribed membrane protein, ToxS (DiRita and Mekalanos 1991). ToxS does not appear to be required for stable DNA-binding activity of ToxR, but ToxR requires ToxS for maximal transcriptional activation (Pfau and Taylor 1998, Beck et al. 2004). TcpP interacts with another co-transcribed membrane protein, TcpH (Beck et al. 2004). In the absence of TcpH, TcpP is targeted and degraded by proteases, including the membrane-localized YaeL (Matson and DiRita 2005), and targeting is dependent on the periplasmic domain of TcpP (Beck et al. 2004). Furthermore, shifting *V. cholerae* from permissive to nonpermissive conditions for virulence factor expression in vitro leads to the degradation of TcpP even in the presence of TcpH (Matson and DiRita 2005), and prolonged growth of *V. cholerae* under conditions permissive for virulence factor expression leads to the accumulation of spontaneous inactivating mutations in the *tcpPH* gene (Carroll et al. 1997). These results suggest that degradation of TcpP is a mechanism for turning off virulence factor expression after the system has been induced.

The *tcpPH* genes are encoded within VPI-1, which is only found in pathogenic strains, whereas the *toxRS* genes are found in all *Vibrio* species and are therefore part of the ancestral vibrio genome. The *tcpPH* genes are only transcribed under permissive conditions for virulence factor expression, while *toxRS* appears to be constitutively expressed (Murley et al. 1999). ToxR, in the absence of TcpP, regulates the expression of numerous genes (Bina et al. 2003). Thus, the horizontally acquired VPI-1 encodes a regulatory factor TcpP, which, under virulence-inducing conditions, is expressed and redirects the host regulator ToxR to facilitate the transcription of another regulatory factor within the VPI-1, ToxT, which in turn activates additional VPI-1 genes as well as the phage-encoded CT genes. The incorporation of the ancestral protein ToxR into this otherwise horizontally acquired regulatory system suggests that ToxR may respond to environmental conditions that could provide additional control over *toxT* transcription (DiRita 1992, Lin et al. 1993, Reich and Schoolnik 1994).

9.7 Transcriptional Regulation of *tcpPH*

Recent studies have made it clear that the induction of ToxR regulon under permissive conditions is dependent on the induction of *tcpPH* transcription (Kovacikova and Skorupski 1999, Skorupski and Taylor 1999). The transcription of *tcpPH* is controlled synergistically by two proteins, AphA and AphB, which are located on the large chromosome and are not encoded on the VPI-1 or CTX elements. *V. cholerae* strains deficient in either *aphA* or *aphB* show reduced expression of the *tcpPH* operon and as a result do not produce virulence factors such as CT or TCP. Double mutants of *aphA* and *aphB* shows lower expression of *tcpPH* compared to either of the single mutants, suggesting that AphA and AphB do not act sequentially but instead activate *tcpPH* transcription synergistically.

AphA is a member of a novel and largely uncharacterized transcriptional regulator family comprising at least 30 proteins with mostly unknown functions that show homology to PadR, a repressor that controls the expression of genes involved in the detoxification of phenolic acids. The N-terminal end of AphA has a conserved domain architecture (CDART) predicted by BLAST to resemble strongly the helix-turn-helix (HTH) domain of MarR, a repressor that controls the expression of a variety of genes involved in multiple antibiotic resistance. In addition, a number of MarR homologs (SlyA in *Salmonella*, RovA in *Yersinia*) play roles in pathogenesis.

The crystal structure of AphA has been determined. AphA was found to be a dimer; each AphA subunit consists of an N-terminal DNA binding domain that adopts a winged helix fold architecture similar to that of the MarR family of transcriptional regulators. Unlike this family, however, AphA has a unique C-terminal antiparallel coiled coil domain that serves as its primary dimerization interface. AphA monomers are highly unstable by themselves and form a linked topology, requiring the protein to partially unfold to form the dimer. AphB is a 33 kDa protein which exhibits significant homology to the LysR family of transcriptional regulator. AphB is unique among the LysR family members in its requirement for a second protein to activate transcription.

Both AphA and AphB are believed to interact directly with the *tcpPH* promoter. AphA binds to the *tcpPH* promoter from positions –101 to –71 from the start of transcription, while AphB binds from positions –69 to –53, a region of interrupted dyad symmetry (5′-TGCAA-N7-TTGCA), with partial overlap between the two binding sites. A systematic mutational analysis of the AphA binding site from positions –75 to –98 has identified nine base pairs that are absolutely critical if AphA is to bind to the *tcpPH* promoter and activate transcription in the presence of AphB. These nine base pairs (located at –95 to –90, –83, –81, –79, –78) lie within a region of partial dyad symmetry that has the sequence TATGCA-N6-TNCNNA.

A variety of evidence suggests that AphB is the primary activator while AphA plays a more indirect role in *tcpPH* transcription; it enhances the ability of AphB to activate transcription. For example, (a) overexpression of *aphB* fully complements an *aphA* null mutant for *tcpPH* expression, whereas AphA only partially complements an *aphB* null mutant; (b) gel shift and DNAse I footprinting indicate that the presence of AphA enhances the binding of AphB to its recognition site; (c) insertion of half a helical turn between the AphA and AphB binding sites, which blocks this interaction by shifting the proteins to opposite faces of the DNA, prevents transcriptional activation of *tcpPH*. How this occurs is not yet clear. As there are only 8 bp between the AphA and AphB binding sites and their DNAse I footprints partially overlap, the two proteins may interact directly. In this model, AphB is believed to contact RNA polymerase, and AphA may increase the activator binding site occupancy of AphB through protein–protein interactions. The other hypothesis could be that AphA might alter the conformation of DNA to facilitate the contact between AphB and RNA polymerase.

The differential expression of virulence genes between the two disease-causing biotypes, classical and El Tor, has been shown to be the result of a single base-pair difference in the classical and El Tor *tcpPH* promoters, which dramatically influences the ability of AphB to activate transcription. The A at the –65 position on the *tcpPH* promoter that is critical for transcriptional activation by AphB lies within the dyad, and the presence of G in this position in the El Tor promoter disrupts the symmetry of this site. The reduced expression observed with the El Tor *tcpPH* promoter may be the result of a decreased affinity of AphB for this site, or it may influence the conformation or orientation of the protein on the DNA.

9.8 Direct Transcription of *ompU* and *ompT* by ToxR

ToxR regulates the production of two outer membrane proteins, OmpU and OmpT, independently of TcpP and ToxT (Miller and Mekalanos 1988, Champion et al. 1997, Childers and Klose 2007). ToxR positively activates transcription of OmpU (Crawford et al. 1998) but represses transcription of OmpT (Li et al. 2002a). OmpU is a porin (Chakrabarti et al. 1996), it confers resistance to some antimicrobial peptides, it may also function as an adhesin (Sperandio et al. 1995), and it protects the cell against organic acids (Mathur and Waldor 2004). OmpT is maximally expressed in cells lacking ToxR, and its role in virulence is not clear. Expression of these membrane proteins by ToxR is important for conferring bile resistance to *V. cholerae* cells, for intestinal colonization in mice, and for resistance to organic acids (Provenzano and Klose 2000). The bile salt (deoxycholic acid) blocks OmpT porin activity but not OmpU (Duret and Delcour 2006). Also, bile induces ToxR-dependent transcription of OmpU (Duret and Delcour 2006), suggesting the interplay of a regulatory cycle. *V. cholerae* cells expressing OmpT enter the intestine and encounter bile, which crosses the outer membrane through OmpT and induces ToxR in the inner membrane to express OmpU, which then makes the cells resistant to bile. *V. cholerae* cells expressing only OmpT express less CT and TCP and colonize the intestine poorly compared to cells expressing only OmpU (Provenzano and Klose 2000).

ToxR binds to three regions (–238 to –139, –116 to –58 and –53 to –24) from the transcription start site of *ompU*, and activates transcription (Crawford et al. 1998). At the *ompT* promoter, ToxR binds to a region located at –95 to –30 from the transcription start site. This *ompT* promoter is also activated by the cyclic AMP receptor protein (CRP), which requires a region centered at -310, and ToxR binding may disrupt contact between RNA polymerase and CRP, thus repressing *ompT* expression (Li et al. 2000). CRP is a global regulator that activates *ompT* expression in carbon and energy source limiting conditions. Moreover, ToxR-dependent *ompU* transcription occurs at high levels under laboratory conditions, but can be increased by the presence of bile (Provenzano and Klose 2000) and can apparently also be modulated by osmotic conditions (Miller and Mekalanos 1988), indicating that ToxR activity is regulated by environmental signals. Alteration of ToxR-dependent modulation of OmpU and OmpT reduces the ability of *V. cholerae* to colonize the intestine, probably by reducing the transcription of *toxT*, (Provenzano and Klose 2000), indicating that the porins in the outer membrane may facilitate the recognition of inducing environmental signals that stimulate the ToxR/TcpP/ToxT cascade.

9.9 Environmental Regulation of *tcpPH* Transcription

The ToxR virulence regulon, which controls CT and other virulence gene expression, is strongly influenced by environmental conditions such as temperature, presence of bile, glucose availability, and cell density or quorum sensing. The mechanism for this regulation is not completely understood, especially in relation to how the environmental stimuli are sensed. The environmental signals are known to exert their effects at different levels of the regulatory cascade. The role of bile has been discussed in previous sections, and only a brief outline of the modes of regulation of virulence genes by osmolarity, temperature, cell population density, and glucose availability is presented below.

9.9.1 Osmolarity

Osmolarity is sensed by ToxR periplasmic domain, which then adapts to a conformation that promotes transcriptional activation of ToxR. A role for osmolarity in this process comes from a study in which ToxR periplasmic domain was replaced with alkaline phosphatase. The ToxR-alkaline phosphatase fusion protein was able to activate *ctxAB* expression but was insensitive to high osmolarity, an in vitro growth condition that normally represses *ctxAB* expression. This observation corroborates with the fact that ToxR is an ancestral gene that is present in other vibrio species and is involved in the modulation of OmpU and OmpT outer membrane protein levels in response to different salinity levels found in the environment. Thus, ToxR/ToxS act as direct mediators of signal transduction via their ability to recognize environmental signals with their periplasmic domains and subsequently control transcription of ToxR regulon genes with the ToxR cytoplasmic domain. The exact nature of the osmolarity signal recognized by this system during intraintestinal infection remains to be determined.

9.9.2 Temperature

Transcription of *toxR* is negatively regulated by *htpG*, a gene that encodes a member of the HSP90 family of heat shock proteins. The genes *toxR* and *htpG* are divergently transcribed from overlapping promoters. The *htpG* transcription negatively impacts *toxR* expression. These data suggest that conditions of stress, such as those encountered by *V. cholerae* within the stomach and small intestine (low pH, anoxia, bile salts) induce the expression of *htpG*, which in turn inhibits the transcription of *toxR* and subsequent ToxR-dependent virulence gene activation. Negative regulation of virulence gene expression during a maximum heat shock response may be biologically relevant, since virulence gene activation during the experimental infection of mice does not occur until 4–10 h postinfection.

9.9.3 Quorum Sensing

Quorum sensing (QS) refers to a cell-to-cell signaling mechanism through which bacterial cells respond to chemical molecules, called autoinducers (AIs), in their environment in a cell density-dependent manner. The AIs can be produced by bacteria of the same species or by bacteria belonging to different genera. Depending on the concentration of AIs, the bacteria detect and respond to the signal by altering their gene expression. Through this mechanism the bacteria exhibit a collective living process—as if they are in a society.

In the *V. cholerae* system, QS represses the expression of virulence genes, *ctx* and *tcp*, at high cell density, but allows their expression at low cell density using three parallel signaling systems. These three systems apparently function in parallel to control the expression of the virulence cascade through the activity of the response regulator, LuxO. Signaling system 1 includes CqsA, a putative synthase for the CAI-1 signal molecule that has remained unidentified, and CqsS, a hybrid sensor/kinase that responds to the CAI-1 signal. System 2 in *V. cholerae* is the LuxS/AI-2 system, which uses LuxO and LuxP as sensors of AI-2. Genetic and other types of evidence suggest that there is also a third system, system 3, whose components have remained largely unidentified. However, all three systems converge at the response regulator LuxO. Mutation of LuxO in *V. cholerae* results in severe intestinal colonization defects (Zhu et al. 2002). All three systems involve a LuxR homolog called HapR (Jobling and Holmes 1997) that serves as a repressor of virulence genes and biofilm formation and as an activator of the Hap protease. The LuxO regulation is activated by phosphorylation and in turn activates transcription of small RNAs, which, together with Hfq, mediate destabilization of the *hapR* mRNA, thereby repressing expression of *hapR* posttranscriptionally (Lenz et al. 2004). The functioning of the QS in the *V. cholerae* system is described briefly below and presented schematically in Fig. 9.2.

At low cell densities, when the concentration of autoinducers (AIs) is low, LuxO is phosphorylated by a relay from the sensor proteins. Information from system 1 and system 2 flows in the form of phosphate to the histidine phosphotransfer protein LuxU and finally to the response regulator LuxO. The activated form of LuxO interacts with sigma-54, activating the expression of four genes encoding small regulatory RNAs (sRNAs). These sRNAs (called Qrr sRNAs for quorum regulatory RNA), in conjunction with the sRNA chaperone Hfq, bind to *hapR* mRNA transcript and destabilize it; thereby preventing HapR from binding to the *aphA* promoter, thus permitting a high level of AphA expression with consequent high-level expression of *tcpPH*, leading to high-level expression of CT.

At high cell densities, when the autoinducer concentrations reach their critical thresholds, they are bound by the cognate sensors. This event switches the sensors from kinases to phosphatases, leading to dephosphorylation and inactivation of LuxO. LuxO then fails to transcribe *qrr* genes, *hapR* mRNA is stabilized, and HapR is translated. This leads to diminished LuxO activity and consequently enhances *hapR* expression. HapR binds to a site –85 to –58 from the start of transcription in the promoter of *aphA* and represses its expression, thus preventing the activation of the *tcpPH* and the rest of the virulence cascade (Kovacikova and Skorupski 2002).

HapR binds to its own promoter and can repress its own transcription at high cell densities. The binding site for HapR is located downstream of its own promoter located between +8 and +36 from the start of transcription, as determined by gel mobility shift assay and DNAse I footprinting. A single A to G mutation at +18 prevents HapR from binding to its site in vitro in gel mobility shift assays and eliminated autorepression in vivo. The recognition site at the *hapR* promoter is weakly conserved with that at the *aphA* promoter; therefore, these promoters are temporarily regulated by HapR as its intracellular level increases.

HapR was initially characterized as an activator of hemagglutinin (HA) protease, a member of a family of zinc metalloproteases in *V. cholerae*, and has been

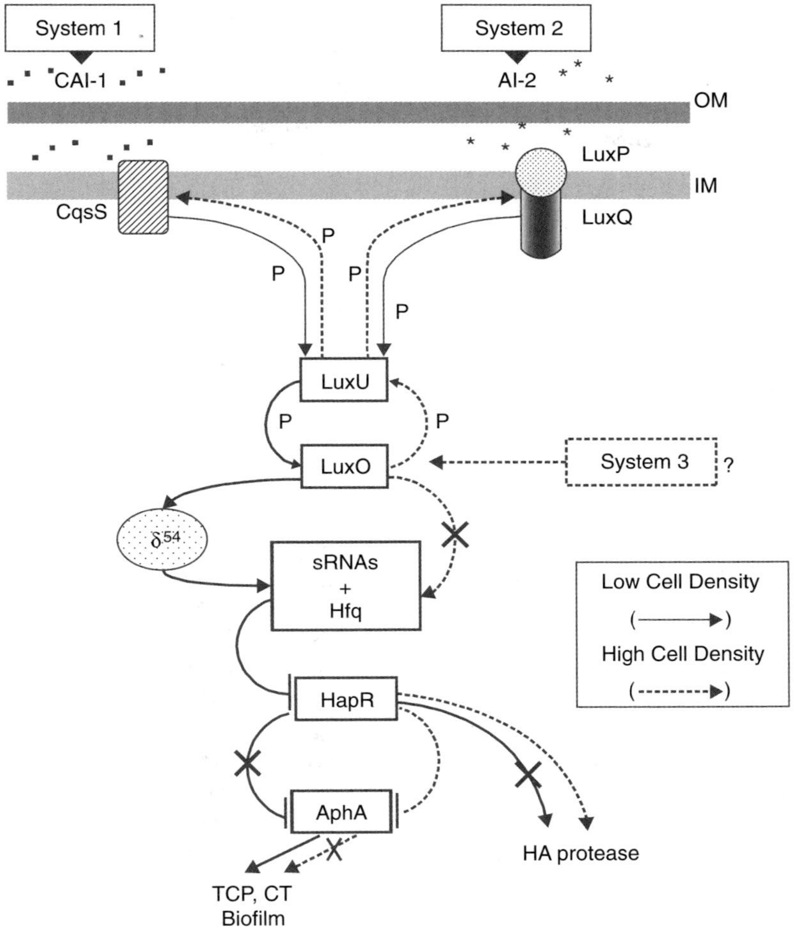

Fig. 9.2 Regulation of *V. cholerae* virulence by quorum sensing. For a detailed description of the model see text

shown to play a role in repressing biofilm formation. HapR can act as both an activator and a repressor. HapR directly upregulates the expression of *hapA*, which produces secreted hemagglutinin (HA) protease, responsible for detaching the vibrios from the intestinal epithelium. HapR is the *V. cholerae* homolog of LuxR and activates luciferase genes also. In addition to its function as a repressor for virulence factors, HapR represses *vps* (vibrio polysaccharide synthesis) operon, thus negatively regulating biofilm formation.

In some toxinogenic strains, like *V. cholerae* El Tor N16961 and classical O395, the cell density dependent virulence regulation does not operate because there is a naturally occurring frameshift mutation in the *hapR* gene (Heidelberg et al.

2000); other strains, like classical CA 401, have a naturally occurring point mutation in the *aphA* promoter preventing the binding of HapR (Kovacikova et al. 2004).

9.9.4 Glucose Availability

Glucose availability influences cAMP levels and the cAMP receptor protein (CRP). Inactivation of CRP in *V. cholerae* leads to increased CT and TCP expression. CRP represses *tcpPH* transcription by binding to a region of the *tcpPH* promoter that is located within the AphA–AphB binding region, thus explaining CRP's ability to negatively regulate CT and TCP expression.

Our knowledge of the complexity of regulation of the virulence genes by *V. cholerae* is by no means complete. This organism has the unique ability to devise newer methods or means for coping with the changing environment, which poses a great challenge to researchers.

Chapter 10
Cholera Toxin (CT): Secretion by the Vibrios

Abstract The secretion of cholera toxin (CT) by the organism *Vibrio cholerae* involves different sets of proteins spanning the inner and outer membrane of the bacterial cell. In the bacterial cytosol, the subunits A and B are synthesized as unfolded chains with N-terminal signal peptide. These unfolded chains are then translocated across the inner membrane via the Sec-dependent pathway to reach the periplasmic space. Here, the chains are freed of the signal peptides; they then fold to achieve their respective 3D structures and ultimately assemble into the AB_5 type of holotoxin structure. This fully formed CT holotoxin is then translocated across the outer membrane via another secretory pathway spanning the inner and outer membranes, which is known as the type II secretion system (T2SS) or the Eps (extra cellular protein secretion) system of *V. cholerae*. Both of these secretion channels consist of multiprotein complexes, whose organizations and functions govern the secretion of CT by *V. cholerae*.

10.1 Introduction

The preceding chapters have presented brief accounts of our current knowledge of (i) the structure of the cholera toxin (CT) molecule, (ii) the genetics of the biosynthesis of the CT molecule along with other virulence factors, and (iii) the regulation of these virulence factors by environmental cues. Early works clearly indicated that CT is an exotoxin that is liberated or secreted by the organism during its active growth phase. The secretion machinery of the organism *V. cholerae* and the mechanism of secretion of the CT molecule, as demonstrated by biochemical, genetic and structural studies, will be discussed in this chapter.

10.2 Early Electron-Microscopic Enquiries

Ever since the discovery in 1959 that cholera toxin is an exotoxin (De 1959, Dutta et al. 1959) with a size that was subsequently estimated to be about 88 kDa, it had been argued that a biomolecule of this size cannot simply leak through the membranes or envelopes of the *V. cholerae* cell, and that its secretion by the organism must involve an (unknown) active process. An attempt to understand the mechanism of toxin secretion by *V. cholerae* cells by electron microscopy was initiated almost immediately after De's discovery, and the microanatomy of a novel secretory activity of *V. cholerae* cells was discovered (see Chap. 4, Fig. 4.3) (Chatterjee and Das 1966, 1967, Chatterjee and Sur 1974, Chatterjee 1990). This secretion mechanism involved the formation and release of surface blebs of dimension 400–1100 Å by the actively growing cells. Some finer particles of size 40–100 Å were also found to be released in association with the blebs. The authors argued that bleb formation represented a novel mechanism by which the actively growing cells extracellularly secreted not only the endotoxins but also materials residing in the periplasmic space of the bacterial cell. It was further observed that the finer particles (40–100 Å) released in association with the blebs were dimensionally similar to the cholera toxin molecule. Subsequently, a correlated biochemical and electron microscopic study (Chatterjee et al. 1974) further revealed that the finer particles were also released independent of the blebs. Besides the lipopolysaccharides, several different proteins were detected in the filtrate (membrane filtration) of the log phase culture of *V. cholerae*. No further characterization of the finer particles, particularly their functional and chemical identities with respect to the cholera toxin molecule, was performed, and the mechanism of the secretion of the cholera toxin molecule remained unresolved.

However, the interesting points to be noted in this context are: (i) the surface bleb-associated secretion mechanism discovered in *V. cholerae* was subsequently found by different investigators to be operative in *V. cholerae* (Kondo et al. 1993a) and many Gram-negative bacteria in general (Li et al. 1998, Kuehn and Kesty 2005) and it has already been proposed that this mechanism represents a type VI secretion process for a bacterial cell (for a recent review, see Kuehn and Kesty 2005), and; (ii) in *Escherichia coli*, the heat-labile enterotoxin molecule (LT) was found to be secreted in association with the surface blebs, and it was proposed that vesicles should be considered a specific secretion mechanism for virulence factors (Horstman et al. 2004). A genetic study of several vesiculation mutants of *E. coli* revealed that vesiculation was not a consequence of bacterial lysis, could not be correlated with membrane instability, and may be considered a fundamental characteristic of Gram-negative bacterial growth (McBrown et al. 2006). A recent study further demonstrated that the vesiculation is a novel stress-response mechanism of Gram-negative bacteria and fulfils the proposed key requirements of a genuine prokaryotic secretion process (Economou et al. 2006, McBrown and Kuehn 2007). In the context of these observations, further investigations are definitely needed to decide whether any of the different toxin molecules secreted by *V. cholerae* cells under diverse environmental conditions are done so through the surface bleb formation mechanism, particularly since nature never operates without purpose.

10.3 The Secretion Mechanism: An Overview

The mechanism of secretion of cholera toxin (CT) by the *V. cholerae* organism is now largely understood as involving different sets of proteins spanning the inner and outer membranes of the bacterial cell. It is in fact a multistage process. In the first phase, the CT subunits, A and B, are synthesized in the bacterial cytoplasm as unfolded chains. These chains are produced with attached N-terminal signal peptides (Mekalanos et al. 1983), and as such are translocated across the cytoplasmic or inner membrane via the Sec-dependent pathway to reach the periplasmic space. Here, the chains are freed of the signal peptides and are folded to form the 3D structures of the CT subunits (Hirst et al. 1984a). Further, in the periplasmic space, the folded subunits of CT interact with each other and assemble into the complete AB_5 type of CT holotoxin (Hirst et al. 1984a, Hirst and Holmgren 1987a, b). The fully formed CT holotoxin is then translocated across another secretory pathway involving both the inner and the outer membranes, the type II secretion system (T2SS) (Sandkvist 2001a, b). This type II pathway is quite specific and can easily distinguish proteins to be secreted from the resident periplasmic proteins (Sandkvist 2001a, b). Both secretion channels—the Sec-channel and the type II secretion system (T2SS)—are composed of multiprotein complexes that are described in the following section. A schematic view of these essential steps in the secretion of cholera toxin, CT, is presented in Fig. 10.1. The details of the mechanism of secretion of CT are presented in the following.

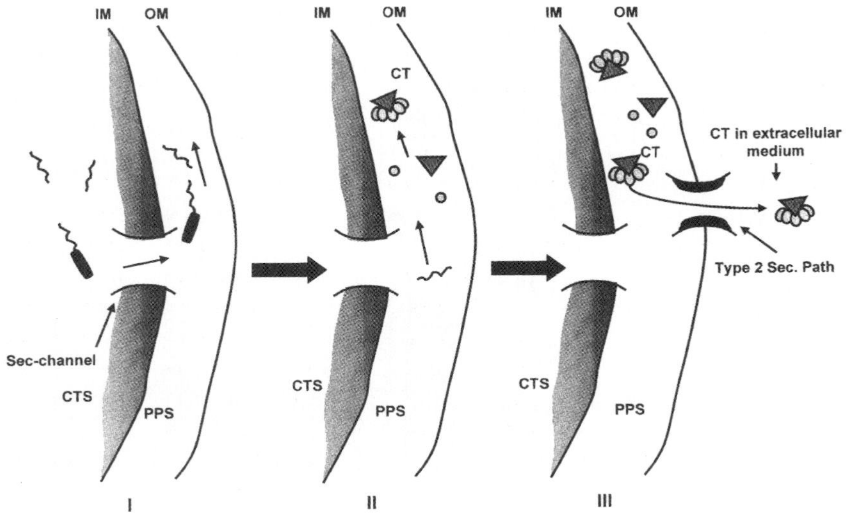

Fig. 10.1 Schematic diagram showing the different stages of cholera toxin secretion by the organism *V. cholerae*. Cholera toxin, synthesized in the bacterial cytosol in an unfolded state, is first translocated through the Sec channel of the inner membrane to the periplasmic space (I). Folding of cholera toxin subunits and their assembly into the holotoxin structure (AB_5) takes place at the periplasmic space (II). The fully assembled CT is finally secreted into the extracellular milieu by the type II secretion pathway (T2SS). Abbreviations are: *CTS*, cytosol; *PPS*, periplasmic space; *IM*, inner membrane; *OM*, outer membrane; *CT*, cholera toxin molecule

10.4 Translocation Across the Inner Membrane

In CT export, the crossing of the inner membrane is mediated by secretion machinery that constitutes a protein-conducting channel. Our knowledge of inner membrane translocation has developed due to studies on the translocation of LT in *E. coli* and the fact that the proteins constituting this translocation machinery are highly conserved across species of Gram-negative organisms.

Among the Gram-negative bacteria, the Sec translocase is the general translocase that transports newly synthesized proteins across the cytoplasmic membrane in an unfolded state, i.e., before they acquire their final structures (Dalbey and Chen 2004, Luirink et al. 2005). The essential *E. coli* translocase constituents include a heterotrimeric integral inner membrane protein complex (SecYEG), the cytoplasmic ATPase (SecA), and several integral membrane proteins: SecD, SecF, and YajC. The process of the translocation of secretory proteins across the inner membrane is illustrated schematically in Fig 10.2. The Sec machinery recognizes the preprotein with its signal sequence. The preprotein then utilizes SecB, a molecular chaperone, to target and stabilize the unfolded state prior to translocation. SecB actively targets the bound precursor to the translocase through its ability to bind SecA. The SecA ATPase interacts dynamically with the heterotrimeric complex,

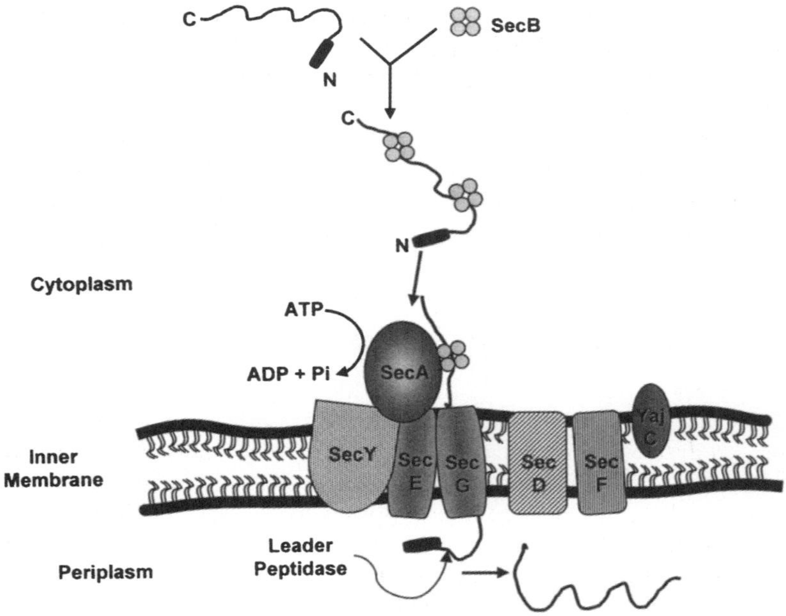

Fig. 10.2 Schematic diagram illustrating the translocation of the CtxA and CtxB components of CT, synthesized in the unfolded state in the cytosol, through the Sec translocase pathway into the periplasmic space of the *V. cholerae* organism. See text for details of the translocation process

SecYEG (formed from SecY, SecE and SecG, and constituting a channel), to drive the transmembrane movement of newly synthesized preproteins. The SecD and SecF subunits form a subcomplex, together with a small protein YajC. The SecDF–YajC complex is not essential for secretion, but it stimulates secretion up to tenfold, particularly at lower temperatures. An intact SecYEGDFYajC complex has been isolated under appropriate solubilization conditions (Duong et al. 1997). Translocation is thought to occur in a stepwise fashion; the initiation and continuation requires cycles of ATP hydrolysis and/or proton motive force across the membrane. The completion process is known to occur on the periplasmic side, leading to the cleavage of N-terminal signal peptide and the release of the substrate. The nascent inner membrane proteins, however, are targeted to the translocase by the signal recognition particle and its membrane receptor (Mori and Ito 2001).

Another translocase has recently been discovered that has been called the twin arginine translocation (Tat) machinery, and is responsible for translocating exported proteins that are folded before translocation and typically have bound cofactors (reviewed in Muller and Klosgen 2005, Lee et al. 2006). Proteins are targeted to the Tat pathway by N-terminal signal peptides harboring consecutive, essentially invariant, arginine residues within an S-R-R-X-F-L-K consensus motif, thus giving the nomenclature. Several genes have been shown to be components of the Tat export pathway in *E. coli*, and four are thought to be integral membrane proteins (Sargent et al. 2001). The functions of these four genes (*tatA, tatB, tatC*, and *tatE*) have not been fully elucidated yet; TatA is proposed to be the translocation channel, while TatC is the receptor to which preproteins bind. All four of these *tat* genes are present in the genome of *V. cholerae* (Heidelberg et al. 2000). Secretion of cholera toxin does not normally require the Tat pathway. It was shown that holotoxin-like chimeras in which the CTA1 domain was replaced by several other antigenic protein domains were transported to the periplasm of *E. coli* by the Tat system, although the corresponding B subunits were transported by the Sec system. The fluorescent fusion proteins, however, spontaneously assembled with the B subunits at the periplasmic space, similar to native CT (Tinker et al. 2005).

10.5 Protein Folding in the Periplasm

An understanding of the mechanism of assembly of holotoxin in the periplasm prior to secretion through the outer membrane emerged from studies on CT and cloned LT in *V. cholerae* CT$^-$ strain. LTs are similar to CT in immunological (Holmgren and Svennerholm 1979), structural (Dallas and Falkow 1980) and functional (Moss and Richardson 1978) properties. However, CT is secreted from *V. cholerae* (Finkelstein and LoSpalluto 1970), but LTs in *E. coli* are cell associated (Kunkel and Robertson 1979a, b) and are located in the periplasmic space of *E. coli* (Hirst et al. 1984a, b), despite the fact that many of the steps in the export of these toxins are similar, including synthesis of subunit precursors, translocation across the inner membrane, and maturation and subsequent assembly (Mekalanos et al. 1983, Hirst et al. 1983). Interestingly, the plasmid containing *E. coli* LT, when introduced into *V. cholerae*, resulted in the secretion of LT to the external milieu, whereas when plasmid containing cloned CT was introduced into *E. coli*, the accumulation of cell-associated CT resulted (Pearson and Mekalanos 1982). LT-producing *V. cholerae* strains (CT^-) were subsequently used to study the secretion mechanism of *V. cholerae*.

The kinetics of enterotoxin efflux from *V. cholerae* was determined by pulse-labeling the cells with [^{35}S]-methionine and chasing experiments. The number of labeled B subunits in the periplasmic fraction was then quantitated via the densitometric scanning of sodium dodecyl sulfate-polyacrylamide gel electrophoresis (SDS-PAGE). The results revealed that enterotoxin B subunits were exported in the periplasm, where they transiently reside before being secreted through the outer membrane (Hirst and Holmgren 1987a, b). The next obvious question was whether the B subunits passing through the periplasmic space remain monomeric or attain an oligomeric conformation before translocation. To examine this, Hirst and Holmgren (1987a, b) monitored the subunit flux through the periplasmic space by pulse-labeling cells with [^{35}S]-methionine and then isolating periplasmic fractions at different time intervals after the cessation of radiolabeled uptake. The B subunits were initially released into the periplasm as monomers; however, they assembled into oligomers with a half-time of ~1 min. The half-time for toxin efflux from periplasm was slow (half-time ~13 min) compared to the rate of oligomerization. It was concluded that all of the B subunits that enter the periplasm assemble prior to their secretion across the outer membrane. The assembled toxin was detected by sodium dodecyl sulfate-polyacrylamide gel electrophoresis (SDS-PAGE) as well as by GM1-enzyme-linked immunosorbent assay (GM1-ELISA) using a monoclonal antibody against the A1 subunit. No assembly intermediates (e.g., B-subunit dimers) were detected; the major toxin pools in the periplasm were A- and B-subunit monomers and fully assembled holotoxin.

The assembly of cholera toxin into AB_5 complexes in the periplasm was investigated both in vitro and in vivo (Hardy et al. 1988, Lesieur et al. 2002). The use of in vitro conditions to study the disassembly and reassembly of CT was first reported by Finkelstein (Finkelstein et al. 1974), who showed that purified CT could be denatured in acid urea and subsequently reassembled into active toxin upon neutralization. Investigations of the in vivo pathway proved difficult due to the complex periplasmic environment (Mekalanos 1985, Hardy et al. 1988). When purified B-subunit pentamers of CT or LT were denatured in acid and subsequently neutralized, the B monomers reassembled into stable pentameric complexes (Hardy et al. 1988, Ruddock et al. 1995, 1996). The reassembly of CTB pentamer formation was further probed by (i) GM1-ELISA, which captures assembled B subunits on GM1-immobilized microtiter plates; (ii) SDS-PAGE, which can detect monomers, pentamers, and or other intermediates, and; (iii) tryptophan fluorescence spectroscopy, with the single Trp residue (at position 88) in CTB being located at the subunit interface in native CTB_5 thus making it a useful probe for studying CTB assembly (Lesieur et al. 2002). The results suggested that a multimolecular event rather than intramolecular refolding as dimers, trimers, and tetramers could be detected in a reassembly mixture, which declined with time as stable pentamer formation occurred.

The synthesis of B subunits in bacterial strains that are unable to express A subunits results in the assembly of B subunits both in vitro and in vivo (Hirst et al. 1984a, b). The B pentamers are secreted to the extracellular environment as efficiently as assembled holotoxin (Hirst et al. 1984a, b, Hirst and Holmgren 1987a, b). This observation would suggest that the pathway of holotoxin assembly involved the formation of B pentamers followed by association with an A subunit to yield the native holotoxin complex. However, mixtures of purified CTB pentamer and CTA do not spontaneously associate in vitro to form an assembled holotoxin (AB_5). Instead, evidence of the direct association of A subunits with B subunits in intermediates has been presented. Furthermore, the A subunit was able to accelerate B subunit assembly in vivo (Hardy et al. 1988), which may involve conformational changes in the intermediate or the formation of salt bridges between A and B intermediates favoring subsequent interactions with B monomers.

It has been established that the proteins destined for the extracellular medium often contain disulfide bonds, which provide an additional level of stabilization for secreted proteins. Folding of cholera toxin subunits involves the formation of intrachain disulfide bonds in the A subunit (between Cys187 and Cys199) and each B subunit (between Cys9 and Cys86). Addition of dithiothreitol (DTT), a sulfhydryl reagent inhibiting disulfide bond formation in the B subunit, prevented the assembly of CTB or ETB monomers to pentamers (Hardy et al. 1988). Disulfides can form spontaneously in vitro in the presence of an oxidizing agent such as molecular oxygen or oxidized glutathione; however, this process is typically slow and inefficient. In vivo, disulfide bond formation is dependent on cellular enzymes, which catalyze the formation of new disulfides (oxidation) and the rearrangement of non-native disulfides (isomerization), both of which are necessary to form the full complement of native disulfide bonds. A battery of enzymes (called Dsb proteins) located in bacterial periplasm mediate disulfide bond formation. Dsbs are the prokaryotic homologs of the eukaryotic protein disulfide isomerase (PDI), and they can perform various thiol–disulfide exchanges, oxidation, reduction, or isomerization of disulfide bonds. Thus, these Dsb redox proteins help to accelerate the slow steps of folding.

In *V. cholerae*, a homolog of *E. coli* DsbA (having 40% identity at the amino acid level) has been identified, which was later shown to be identical to the TcpG protein present in the VPI-1 pathogenicity island (Peek and Taylor 1992, Yu et al. 1992, 1993). A mutation in the gene encoding DsbA severely affected the production of B-subunit pentamers in a *V. cholerae* strain engineered to express high levels of B subunit of *E. coli* LT. Like *E. coli* DsbA, *V. cholerae* DsbA contains the dithiol motif, -Cys-X-X-Cys-, which is typically found at the active site of dithiol/disulfide oxidoreductases such as thioredoxin and protein disulfide isomerase. The sequence homology between *V. cholerae* and *E. coli* DsbA rises to 90% in this putative active site region from Glu43 to Cys52 containing a Cys-Pro-His-Cys motif, similar to the active site of thioredoxin (Cys-Gly-Pro-Cys) and protein disulfide isomerase (Cys-Gly-His-Cys), which catalyze thiol–disulfide interchange reactions (Yu et al. 1992).

The functional equivalence of *V. cholerae* DsbA to *E. coli* DsbA was established through the complementation of the defect in EtxB biogenesis in *V. cholerae dsbA*::Tn*phoA* mutants upon the introduction of a plasmid encoding *E. coli* DsbA. Periplasmic fractions from the wild-type $DsbA^+$ strain exhibited high levels of PDI activity, whereas extracts taken from the mutant $DsbA^-$ strain failed to show PDI activity, and this could be restored and enhanced by approximately twofold by expressing the cloned *dsbA* gene in the mutant strain. Furthermore, oligonucleotide-directed mutagenesis of either of the two Cys residues to Ala in the putative active site of DsbA abolished both its isomerase activity and its capacity to promote enterotoxin biogenesis (Yu et al. 1993).

The crystal structure of the oxidized form of *V. cholerae* TcpG, the homolog of *E. coli* DsbA, has been refined at a resolution of 2.1 Å (Hu et al. 1997). TcpG has the same fold as DsbA and also shows surface features characteristic of DsbA, suggesting functional similarity between the two. However, unlike DsbA, TcpG expression is intimately related to *V. cholerae* colonization and toxin production, and this factor is coordinately regulated with the expression of CT and TCP (Peek and Taylor 1992, Peterson and Mekalanos 1988). While the overall architectures of TcpG and DsbA are similar, significant differences have been found at the kinked active site helix, the proposed peptide-binding groove of TcpG is substantially shortened, and the hydrophobic pocket of TcpG is shallower and the acidic patch is more extended. The divergent features may partially account for the additional pathogenic role of TcpG.

A model of the oxidation–reduction cycle of DsbA is presented schematically in Fig 10.3. DsbA catalyzes the formation of intrachain disulfide bonds, possibly through the formation of a mixed disulfide intermediate (Yu et al. 1992). The regeneration of oxidized DsbA from reduced DsbA probably requires the involvement of DsbB, an inner membrane protein (Kadokura et al. 2003).

10.6 Secretion Across the Outer Membrane

The different aspects of the secretion of the fully assembled AB_5 type of cholera toxin (CT) across the outer membranes of *V. cholerae* cells are presented below.

10.6.1 Genetics of the Extracellular Protein Secretion (Eps) Apparatus

In *V. cholerae*, the extracellular protein secretion process (T2SS) is generally termed the Eps system, and the corresponding genes are known as *eps* genes. Secretion of cholera toxin across the outer membrane requires a large set of accessory proteins, the components of T2SS, designated EpsA to EpsN and VcpD (PilD) (Sandkvist et al. 1997, Sandkvist 2001a, b, Russel 1998, Johnson et al. 2006).

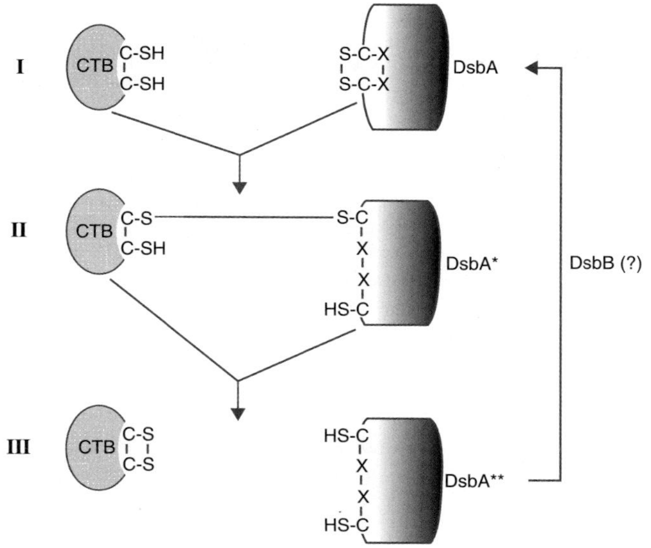

Fig. 10.3 Proposed model of intrachain disulfide formation in CTB by DsbA during its folding in the periplasmic space. It is proposed that the formation of intrachain disulfide bonds in CTB (stage III) is catalyzed by DsbA and occurs via the formation of a mixed disulfide intermediate (stage II). DsbA* and DsbA** are partially and fully reduced states of DsbA. The conversion of DsbA** to DsbA by interaction with another membrane protein DsbB and stage II (intermediate stage) of the reaction are both yet to be established

Mutations in the *eps* genes result in the accumulation of assembled cholera toxin in the periplasm, and also affect the secretion of hemagglutinin/protease (HA/protease) and a chitinase. In addition, the cell division and biogenesis of some outer membrane proteins appear to be altered. Gene loci similar to *eps* have also been described for protein secretion across the outer membranes of many other organisms, including *Klebsiella oxytoca, Pseudomonas aeruginosa, Aeromonas hydrophila, Erwinia carotovora, E. chrysanthemi, Yersinia enterocolitica, Burkholderia pseudomallei*, and *Xanthomonas campestris* (Filloux et al. 1990, Sandkvist 2001a, b), and are generally termed the Gsp (general secretion pathway) system.

Earlier studies were focused on determining the set of *eps* genes coding for functions involved in the extracellular secretion of CT. Two strategies were used to identify functions responsible for Eps (Sandkvist et al. 1997). First, a *V. cholerae* gene library was constructed and individual clones from this library were mobilized into a secretion-defective mutant (accumulating CT in periplasm), and the transconjugants were screened for protease secretion. Clones positive for protease secretion were then tested for their ability to secrete the toxin by GM1-ganglioside ELISA of culture supernatant and sonicated cells. The recombinant plasmids, which could rescue the secretion defect, were analyzed to identify the gene product mediating the secretion process. Transposon mutagenesis was then carried out and the

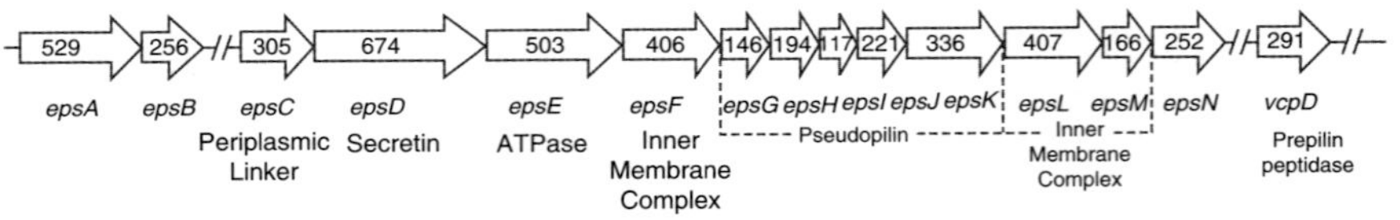

Fig. 10.4 Organization of the extracellular protein secretion gene cluster in *V. cholerae*. Genes are indicated as *unshaded boxes with arrowheads* indicating the direction of their transcription. The *number inside each box* represents the number of amino acids in the corresponding ORF. The cluster *epsAB* is located away from the cluster *epsCDEFGHIJKLMN*, as is the gene *vcpD*. The functions of the genes that have been established so far are also indicated

transposition mutants defective in protease secretion were tested for CT secretion as well as for periplasmic accumulation of CT. Mutants showing periplasmic accumulation of CT were complemented by the gene library as described above. Using these strategies, the investigators identified a set of genes termed as *eps* genes that were recognized to be the genetic determinant of the type II secretion system in *V. cholerae* (Sandkvist et al. 1997). These are mostly clustered and constitute the large operon *epsC-N*, the operon *epsAB*, and the *vcpD (pilD)* gene. The organization of the *eps* gene cluster in *V. cholerae* N16961 is shown in Fig. 10.4. DNA composition analysis of the secretion genes of *V. cholerae* suggests that the *eps* genes were present in the ancestral chromosome, while the *vcpD* gene has been acquired recently by a horizontal gene transfer (Heidelberg et al. 2000) mechanism.

10.6.2 Molecular Architecture of the Secretory Machinery

The assembled CT in the periplasm thus traverses the outer membrane in a second step that requires the *eps* gene products. Based on data from many different experimental approaches, including subcellular localization, genetic and biochemical techniques, X-ray crystallography, and protein–protein interactions between individual components of the secretion system, the Eps secretion apparatus (T2SS) has been proposed as being distinctly arranged into several subcomponents which function in a concerted manner.

The T2SS can be considered to consist of three subcomplexes (Fig. 10.5): (i) the "inner membrane platform" consisting of the "secretion ATPase" EpsE and the inner membrane proteins EpsC, EpsF, EpsL, and EpsM; (ii) the "pseudopilus" consisting of the major pseudopilin EpsG and the minor pseudopilins EpsH, EpsI, EpsJ, and EpsK, and; (iii) the "outer membrane complex" consisting of large "secretin" EpsD and, in many species, the "pilotin" EpsS. The prepilin peptidase, generally called EpsO, also belongs to T2SS. A schematic view of the architecture of the type II secretion system (T2SS) is presented at the center of Fig. 10.5. The crystal structures of the different proteins involved in this secretion mechanism as components of the T2SS are shown in the surrounding periphery.

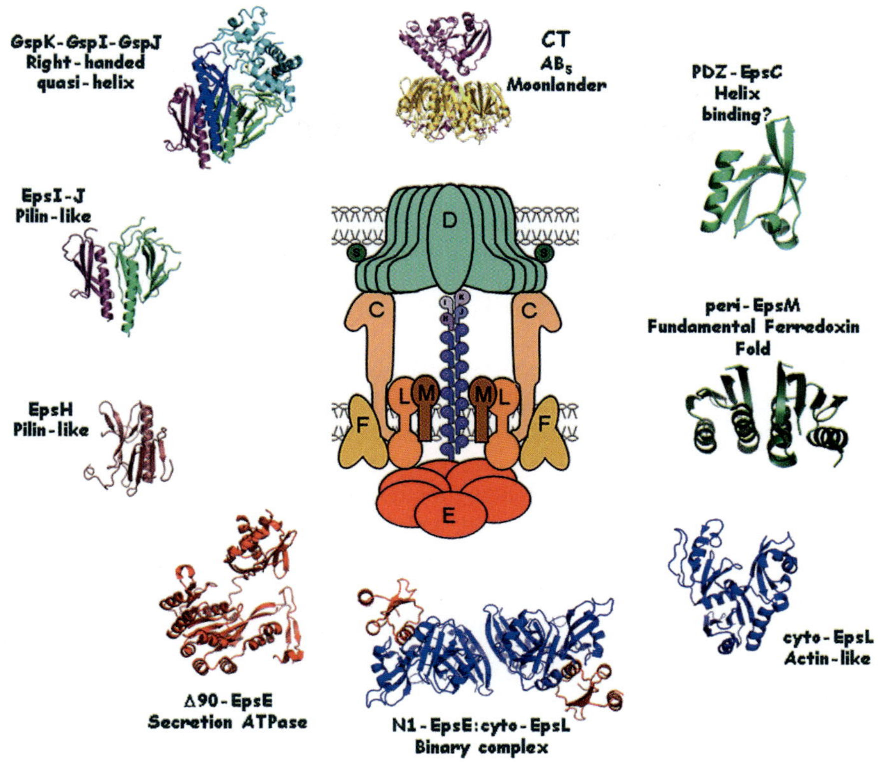

Fig. 10.5 Towards an understanding of the architecture of the type II secretion system (T2SS) in *V. cholerae* responsible for the transport of cholera toxin across the outer membrane to the extracellular environment. The derived structure of the T2SS system with all of the components assembled is shown at the *center* of the figure. The crystal structures of the individual components are shown in the *periphery* surrounding the central structure (clockwise from CT): (*i*) cholera toxin molecule (PDB: ILTT, ISSE; Sixma et al. 1991, O'Neal et al. 2004); (*ii*) PDZ-domain of GspC from *V. cholerae* or EpsC (PDB: 214S, 216V, Korotkov et al. 2006); (*iii*) the periplasmic domain of GspM from *V. cholerae* or EpsM (PDB: 1UV7, Abendroth et al. 2004b); (*iv*) the cytoplasmic domain of GspL from *V. cholerae* or EpsL (PDB: 1YF5, Abendroth et al. 2004a); (*v*) the cytoplasmic domain of GspL (*blue*) in complex with the N1 domain of the secretion ATPase GspE from *V. cholerae* or EpsE (*red*) (PDB: 2BH1, Abendroth et al. 2005); (*vi*) the secretion ATPase GspE from *V. cholerae* or EpsE, with the N1 domain removed (PDB: 1P9R, 1P9W, Robien et al. 2003); (*vii*) the globular domain of the pseudopilin GspH from *V. cholerae* or EpsH (PDB: 2QV8, Yanez et al. 2008a); (*viii*) the globular domains of pseudopilins GspI or EpsI (purple) from *V. cholerae* and GspJ or EpsJ (*green*) from *V. cholerae* (PDB: 2RET, Yanez et al. 2008b); (*ix*) the ternary complex of GspK (*light and dark blue*), GspI (*purple*), and GspJ (*green*), all from enterotoxigenic *E. coli* (ETEC) (PDB: 3CIO, Korotkov and Hol 2008). Details on the function of the T2SS system and its components are presented in the text. (Figure obtained through the kind courtesy of Konstantin Korotkov and W.G. Hol, University of Washington, Seattle, USA)

10.6.3 The Inner Membrane Complex

The inner membrane platform comprises of a subcomplex formed by EpsE, a cytosolic protein associated with the cytoplasmic face of the inner membrane, the cytoplasmic membrane proteins EpsL and EpsM, together with EpsF and the inner membrane–outer membrane linker EpsC. This subcomplex is assumed to either initiate the signal transduction to gate the pore or to provide the energy for active transport via conformational changes. X-ray crystallographic studies provide an atomic-level picture of the interactions between different components of the Eps system, and mutational studies have added to our understanding of this fascinating complex.

EpsE belongs to the subfamily of "type II secretion ATPases," which are thought to power bacterial protein secretion through nucleoside triphosphate (NTP) hydrolysis. EpsE contains Walker A and B boxes that likely participate in the binding and hydrolysis of ATP. A mutation (conserved lysine to alanine substitution, K270A) introduced in the Walker A ATP-binding motif of EpsE reduces the specific ATPase activity in vitro about two- to threefold, and is unable to support secretion in vivo (Sandkvist et al. 1995). Purified preparations of EpsE were predominantly monomeric but contained a small fraction of oligomers, specifically hexamers. The latter had increased specific activity compared to the monomers, suggesting that the functional form of EpsE proteins in vivo is most likely hexameric (Camberg and Sandkvist 2005). During subcellular fractionation of *V. cholerae* cells, a cytoplasmic soluble form representing the monomer and a cytoplasmic membrane-associated form of EpsE representing multimerization states of EpsE were detected. A hexameric ring model was proposed for EpsE based on X-ray crystallographic studies (Robien et al. 2003). The transitions from hexameric toroidal to helical assemblies and vice versa of ATPases like EpsE have been predicted to play a key role in maintaining the dynamic arrangements needed to push CT-like proteins through the outer membrane pores.

Biochemical and genetic studies revealed that recruitment of EpsE to the cytoplasmic membrane and to the rest of the Eps system is mediated by EpsL, a 403-residue bitopic inner membrane protein (Sandkvist et al. 1995). EpsL consists of a large cytoplasmic domain (Val1-Lys253), a single transmembrane helix (Val254-Phe271), and a smaller periplasmic domain (Gln272-Lys403). The cytoplasmic domain binds to and localizes EpsE to the cytoplasmic membrane, whereas either the transmembrane helix, the periplasmic domain or both interact with EpsM, another biotopic inner membrane protein (Sandkvist et al. 2000), resulting in a polar localization of EpsL (Scott et al. 2001).

The 28 kDa cyto-EpsL consists of three β-sheet-rich domains: domains I and III are similar to the RNaseH fold; domain II has the topology of an SHS2-fold module (Abendroth et al. 2004a). Subcellular fractionation and immunoblot analysis have shown that EpsL interacts with EpsE, and mutational studies have shown that the N-terminal 90 residues of EpsE form the interface between EpsE and EpsL. The EpsE binding domain of EpsL was mapped using a protein hybrid approach (Sandkvist et al. 2000), which revealed that an internal portion spanning residues 57 through 296 of EpsL determines species specificity, and may contain the region interacting with EpsE. The first structural view of these interacting partners comes from the structure of the N-terminal domain of EpsE (1-96) in complex with the cytoplasmic domain of EpsL (1-241), referred to as the N1E-cytoL complex (Abendroth et al. 2005). The structure of the complex reveals that the N1-EpsE and the cyto-EpsL form a heterotetramer in which EpsL is the central dimer and EpsE binds on the periphery. The structural configuration suggests that other residues of EpsE or another member of the T2SS machinery, such as EpsF, may also bind in this pronounced cleft (Abendroth et al. 2005).

Another T2SS component, EpsM, forms a stable complex with EpsL and EpsE at the cytoplasmic membrane. EpsM, a small, 165-residue, inner-membrane protein, is predicted to contain a short N-terminal cytosolic domain (residues 1–23), followed by the transmembrane domain (residues 24–43) and the C-terminal periplasmic domain (residues 44–165). EpsM inherently localizes to the cell pole, independent of other Eps proteins (Scott et al. 2001). A protein hybrid approach mapped the EpsM binding region in EpsL between residues 216 and 296, which includes the membrane-spanning region (254–271), suggesting that EpsL and EpsM interact with each other in the cytoplasmic membrane (Sandkvist et al. 2000).

Full-length EpsM and the soluble periplasmic domain are capable of forming dimers in solution, and two monomers of periplasmic EpsM are present in each asymmetric unit of the crystal (Abendroth et al. 2004b). A third cytoplasmic membrane protein, EpsF, has also been implicated in the cytoplasmic platform of the T2SS machinery in addition to EpsE, EpsL, and EpsM from co-immunoprecipitation studies from other Gram-negative organisms (Py et al. 2001). Another protein, EpsC, has been attributed the role of a regulator linking the two membranes, which will be discussed in a later section.

10.6.4 The Pseudopilus

The five T2SS proteins EpsG, EpsH, EpsI, EpsJ, and EpsK of *V. cholerae* are termed pseudopilins because of their homology with the proteins involved in type IV pilus biogenesis, and are proposed to form a pilus-like structure called pseudopilus. Type IV pilins are characterized by (Craig et al. 2004, Hansen and Forest 2006): (i) a prepilin leader sequence which is cleaved off by a prepilin peptidase, yielding a mature protein; (ii) an extended hydrophobic N-terminal α-helix, which is thought to anchor pilins to assembled pili and into a membrane when pili are not assembled; and (iii) a conserved glutamate at position +5 of the mature pilin sequence, which is thought to be involved in the assembly of pilins into the pilus. Type IV pilins can be subdivided into type IVa and type IVb pilins; the distinguishing criteria are summarized in Table 10.1. Hydrophobic sequences at the N-termini of EpsG, EpsH, EpsI, and EpsJ were similar to those recognized by the prepilin peptidase present in the type IV prepipilin subunits (Hansen and Forest 2006). These signal sequences consist of a short leader sequence, then a conserved region G↓FT(L/I)Q, followed by a hydrophobic region of 15–20 amino acid residues,

Table 10.1 Comparison of type IVa and type IVb pilins

Characteristic	Type IVa pilins	Type IVb pilins
Occurrence	Present in a variety of bacteria with a broad host range	Almost exclusively found among enteric pathogens
Length of the signal peptide	5–6 amino acids	15–30 amino acids
Average length of mature protein	150 amino acids	190 amino acids
N-terminal residue	N-methylated phenylalanine	N-methylated methionine, leucine or valine
Average length of D region	22 amino acids	55 amino acids
Topology of β-sheet	The β-sheet follows the pilin sequence with a N to N + 1 nearest-neighbor connectivity	The β-sheet connectivity is complex, with the most C-terminal segment forming the central strand of the β-sheet
N-terminal αβ loop	A sugar loop or minor β-strands	Well-defined α-helix
C-terminal disulfide bonded D regions	Two type I β-strands joined together, or one type I joined to one type II β-turn	Pair of α helices packed on top of a pair of antiparallel β-strands

which is conserved in all pseudopilins (Fig. 10.6). The cleavage between G↓F by the type IV prepilin leader peptidase followed by the methylation of the N-terminal amino acid of the processed pseudopilin is required for their normal functioning (Craig et al. 2004). The *V. cholerae* prepilin peptidase is encoded by the gene *pilD* (*vcpD*), located outside the cluster (Marsh and Taylor 1998). Genetics and biochemical evidence indicate that this peptidase is responsible for processing the Eps prepilin-like proteins required for cholera toxin secretion, and that it also processes the prepilins of mannose-sensitive hemagglutinin (MSHA) pilus biogenesis. On the other hand, the processing of the TcpA prepilin into its mature form is mediated by TcpJ, another type IV prepilin peptidase, present in the TCP gene cluster. TcpJ is distinctly different from PilD (25% sequence homology) and does not have an influence on toxin secretion.

Like T2SS of other organisms, many of the proteins of the *V. cholerae* type II secretion system show homology with many of the genes coding for TCP, MSHA biogenesis and assembly, as well as with other pili and fimbriae involved in twitching motility and which belong to type IV pilus biogenesis systems (Table 10.2). Due to similarities between the pseudopilins and type IV pilus biogenesis systems, assembly of the T2SS pseudopilus is proposed to occur in a manner analogous to the assembly of type IV pili, although the true pili extend outside the bacterial cell wall and pseudopili function in the periplasmic space, between the inner and outer membrane subcomplexes of the T2SS. A similarity in the organization has been

```
                                        -1 +1   +5
EpsG    ----------------MKKMRKQTG   FTLLEVMVVVVILGILASFVV
EpsH    -------------------MTATRG   FTLLEILLVLVLVSASAVAVI
EpsI    -------------------MKSKRG   FTLLEVLVALAIFATAAISVI
EpsJ    ----------MWRTNQVSSRQNMAG   FTLIEVLVAIAIFASLSVGAY
EpsK    ------------------MRAKQRG   VALIVILLLLAVMVSIAATMA
MshA    ----------------MVIMKRQGG   FTLIELVVVIVILGILAVTAA
TcpA    MQLLKQLFKKKFVKEEHDKKTGQEG   MTLLEVIIVLGIMGVVSAGV-
PilT    ----------------MELNQYLDG↑  MLTHKASDLYITVGAPILYRV
```

Fig. 10.6 Alignment of the N-terminal sequence of pseudopilins of the T2SS system and the proteins MshA, TcpA, and PilT of *V. cholerae*, demonstrating the high degree of homology and the important features of this region. The position of the leader sequence cleavage site is indicated with an *arrow*. Positions in the proteins relative to the prepilin peptidase cleavage site are shown *above the alignment*. The residue G at −1 is conserved; the conservation of F at 1 and E at +5 is also apparent. The leader peptide and hydrophobic regions are shown

assumed between T2SS pseudopilus and type IV pilus, with EpsG being recognized as the major pseudopilin, while EpsH, EpsI, EpsJ, and EpsK are called minor "pseudopilins" (Craig et al. 2004). This is supported by the fact that when the expression of proteins encoded by the pseudopilin genes of *V. cholerae* was monitored through a T7 promoter polymerase system, EpsG protein was synthesized in large amounts, whereas EpsH to J showed lower expression (Sandkvist et al. 1997).

Much of our understanding of pseudopilin function has been obtained from recent biochemical studies as well as from the crystal structures of these proteins, notably PulG (Kohler et al. 2004) (the EpsG homolog of *V. cholerae*, sharing 80% sequence similarity), EpsH of *V. cholerae* (Yanez et al. 2008a), and the EpsI:EpsJ heterodimer of *V. vulnificus* (Yanez et al. 2008b), a close relative of *V. cholerae*. The crystal structure of GspG (PulG) from *K. oxytosa* reveled that this major pseudopilin adopts the type IVa pilin fold defined by an N-terminal hydrophobic α-helix, a variable region and a "conserved β-sheet" consisting of four antiparallel β strands. However, PulG lacks a highly variable loop region containing a disulfide bond, as found in all type IV pilins (Kohler et al. 2004).

The crystal structure of EpsH of *V. cholerae* revealed a hydrophobic N-terminal α-helix (Asp30-Leu54) characteristic of all type IV pilins in each monomer, followed by nine β-strands forming two β-sheets (Yanez et al. 2008a). The pairwise protein sequence identity within the EpsH/GspH family varies between 19 and 50%, much less than that observed for EpsG/GspG homologs in the same species. Most of the conserved residues are hydrophobic and are clustered around a specific region of the EpsH structure, forming a solvent-exposed hydrophobic groove.

EpsI of *V. cholerae*, a 117 amino acid protein, includes a propeptide (1–6), a mature chain (7–117), and an *N*-methylphenylalanine at the residue 7 position. A recent study on the interaction between the *V. cholerae* minor pseudopilins EpsI and EpsJ suggests that EpsI and EpsJ are likely to form a heterodimer (Vignon et al. 2003).

Table 10.2 Homology between the type II secretion system and type IV pilus biogenesis proteins in *V. cholerae* El Tor N16961

T2SS protein[a]	Locus no.[b]	Size (amino acids)	Function	Type IV pilus biogenesis homologs[c]		
				TCP	MSHA	Others
EpsA	VC2444	529	ATPase	–	–	–
EpsB	VC2445	256	–	–	–	–
EpsC	VC2734	305	–	–	–	–
EpsD	VC2733	674	Secretin	TcpC (VC0831, 489)	MshD (VC0411, 203), MshL (VC0402, 559)	Fim (VC2630, 578)
EpsE	VC2732	503	ATPase	TcpT (VC0835, 503)	MshE (VC0405, 575)	PilB (VC2424, 562); PilT VC0462, 345); PilT (VC0463, 368)
EpsF	VC2731	406	–	TcpE (VC0836, 340)	MshG (VC0406, 407)	PilC (VC2425, 408)
EpsG	VC2730	146	Major pseudopilin	TcpB (VC0829, 430)		PilE (VC0857,147) FimT (VC0858, 184) PilV (VC0861,137)
EpsH	VC2729	194	Minor pseudopilin	–	–	–
EpsI	VC2728	117	Minor pseudopilin	–	–	PilV (VC0861,137)
EpsJ	VC2727	207	Minor pseudopilin	–	–	PilE (VC0857,147) FimT (VC0858,184)
EpsK	VC2726	336	Minor pseudopilin	–	–	–
EpsL	VC2725	407	Inner membrane complex	–	–	–
EpsM	VC2724	156	Inner membrane complex	–	–	–
EpsN	VC2723	252	–	–	–	–
PilD	VC2426	291	Prepilin peptidase	TcpJ (VC0839, 253)	–	–

[a] Type II secretion system (T2SS) or extracellular protein secretion (Eps) proteins
[b] Locus numbers correspond to the ORFs in the complete genome of *V. cholerae* El Tor N16961 (Heidelberg et al. 2000)
[c] The locus number and size in amino acids of each homolog are presented in parentheses. Abbreviations are: *TCP*, toxin-coregulated pilus; *MSHA*, mannose-sensitive hemagglutination; *Fim*, fimbrial assembly protein

Pseudopili are thought to function as a retractable "plug" for the outer membrane pore, EpsD, and/or as a "piston" that actively pushes secreted proteins through the EpsD pore. In the piston model, the pseudopilins are proposed to act as a piston that pushes secreted proteins into the large compartment in the lumen of the secretin channel, allowing the release of the proteins into the medium and subsequent pilus retraction. The piston model of T2SS was proposed by Hobbs and Mattick (1993) following their studies of *P. aeruginosa*, and is based on the well-documented phenomenon of type IV pilus elongation and retraction that produces twitching motility and related patterns of bacterial movement across solid surfaces. Although the mechanisms remain unclear, it seems to be accepted that pilus elongation is promoted by an ATPase (PilB), and that pilus retraction results from disassembly, as promoted by another, related ATPase (PilT) that operates in reverse. The factors controlling pilus length and the switch from elongation to retraction are unknown (Vignon et al. 2003).

10.6.5 The Outer Membrane Complex

The only T2SS protein that is integrated into the *V. cholerae* outer membrane is EpsD, which is a member of the so-called "secretin" family of proteins and forms large multimeric structures (Johnson et al. 2006). The secretins adopt a ring-shaped structure and form a pore in the membrane through which the secreted proteins pass. EpsC has been attributed the role of a regulator that links the two membranes.

EpsC of *V. cholerae* is a 305-residue bitopic inner-membrane protein, and its structure has been solved (Korotkov et al. 2006). EpsC is anchored in the inner membrane (residues 27–50), while the largest part is periplasmic, consisting of a homology region (HR) domain (residues 75–177) followed by a C-terminal PDZ domain (residues 201–305) and a short N-terminal extension (residues 1–26), predicted to be located in the cytosol. The linker regions between the transmembrane domain and the HR domain, and between the HR and PDZ domains, are rich in Pro residues and are presumed to be disordered.

PDZ domains are small globular modules of about 100 amino acid residues named after the first three proteins that were found to contain them: the post-synaptic density protein PSD, the Drosophila septate junction protein discs large, and the epithelial tight junction protein ZO-1 (Hung and Sheng 2002). It is ubiquitously present in all kingdoms of life and plays an important role in cellular signaling. In contrast to the PDZ domain, little is known about the structural properties of the HR domain. Secondary structure predictions show that this domain is likely to contain predominantly β-structural elements. Structural studies of two variants of the PDZ domain of EpsC from *V. cholerae* revealed that the PDZ domain of EpsC could adopt a more open form than previously reported structures of other PDZ domains (Korotkov et al. 2006). The HR domain of EpsC is primarily responsible for the interaction with the secretin EpsD, while the PDZ is not, or is much less so.

This finding, together with studies of others, leads to the suggestion that the PDZ domain of EpsC may interact with exoproteins to be secreted, while the HR domain plays a key role in linking the inner-membrane subcomplex of the T2SS in *V. cholerae* to the outer-membrane secretin (Korotkov et al. 2006).

EpsD, the integral outer-membrane protein, comprises a large multimeric complex of 12 monomers that form a large pore allowing the transport of macromolecules such as S-layers, pili, mature filamentous phages, and extracellular proteins such as pectinase, hydrolase, CT, chitinase, and protease. Tn*phoA* insertion into the *epsD* gene resulted in severe defects in the secretion of CT and hemolysin. Reversion of *epsD*::Tn*phoA* in this mutant to the wild type restored the phenotypes. EpsD was further shown to be involved in forming rugose phenotypes and motility (Ali et al. 2000).

Loss of motility in *epsD* or *epsE* mutants suggests a direct or indirect role for the *eps* system in flagellar biosynthesis. It is surprising that the *eps* system, which is involved in transporting proteins across the outer membrane, can also affect polysaccharide secretion (Ali et al. 2000). It may be that the rugose polysaccharide is coated with proteins during translocation. Alternatively, the *eps* system may affect polysaccharide synthesis or processing rather than secretion. It is conceivable that *eps* mutants affect the production of rugose polysaccharide indirectly, by affecting the secretion of an outer membrane protein that positively regulates polysaccharide synthesis. The positive regulation may occur through a two-component sensor–transducer regulatory cascade that relays an environmental signal. It has been shown that the *eps* mutants have a modified cell envelope due to defects in the production of certain outer membrane proteins. This membrane perturbation may affect the signaling pathway that regulates the transcription of polysaccharide biosynthesis genes.

10.6.6 The T2SS System: A Summary

The different protein components that take part in the extracellular protein secretion system (T2SS) of *V. cholerae* are shown in Fig. 10.5. How these proteins function in the secretion process is summarized in the following. Briefly, the CT assembled in the periplasmic compartment is targeted to the Eps machinery and transported through the outer membrane to the extracellular milieu. The cytoplasmic protein EpsE, proposed to be a hexameric ring, becomes associated with the inner membrane through its interaction with the β-sheet-rich cytoplamic domain of the inner-membrane protein EpsL. EpsM, a proposed dimer, interacts with EpsL. EpsG are proposed to generate a structure similar to type IV pilins, and the monomers are presumed to form a pilus-like structure which might act as a piston to push CT through the secretin EpsD (with ~12 subunits) that forms a large pore-like structure at the outer membrane. These are the core proteins, and other proteins act to stabilize the Eps structure to facilitate secretion. The minor pseudopilins EpsH, EpsI, EpsJ, EpsK likely interact transiently with the major pseudopilin EpsG. EpsC interacts with EpsD in the outer membrane and with EpsL and EpsM in the inner membrane. The proteins EpsA, EpsB, and EpsN have not been studied so far; however, these have been predicted to be involved in the stabilization of the complex structure.

10.6.7 Targeting Signals for the Translocation of CT Through the OM

The *eps* system of *V. cholerae* is one of the more promiscuous T2SS systems. Unlike in other organisms, it has the capacity to secrete at least six proteins, including CT, heat-labile enterotoxins of *E. coli* (LT-I, LT-IIa, LT-IIb), one or more proteases, and endochitinase (Connell et al. 1995,1998a, b, Sandkvist et al. 1993, Overbye et al. 1993). In addition, polysaccharide transport is believed to require EpsE, and the release of the phage CTXφ requires EpsD, which encodes the outer membrane protein of the T2SS. While numerous proteins are present in the periplasmic space of *V. cholerae*, only a few are transported across the outer membrane to the extracellular milieu. Like other bacteria, *V. cholerae* is able to discriminate between secreted and nonsecreted proteins (Sauvonnet et al. 1995, Lu and Lory 1996). It is proposed that the prospective extracellular proteins of *V. cholerae* contain an extracellular transport signal (ETS), which is recognized by *eps*-encoded transport machinery. The signals are likely to be present in secreted proteins and absent in nonsecreted proteins.

The putative extracellular transport signal for CT secretion across the OM is thought to be located in CTB (Hirst and Holmgren 1987a, b). CTB pentamers and holotoxin are secreted in *V. cholerae* with equal efficiency. CTA polypeptides are not secreted across the outer membrane by *V. cholerae* unless they are assembled into holotoxin. The proteins secreted by *V. cholerae* are divergent in sequence and structure. Comparisons of these structurally different proteins have not revealed a common amino acid sequence that might function as the translocation signal. The transport signal may be a patch signal, a structure formed by discontinuous amino acid sequences brought together by the folding of the proteins (Pugsley 1993). Genetic experiments demonstrate that B polypeptides of LT-IIa and LT-IIb are transported across the outer membrane by the same secretory apparatus that transports CT (Hirst et al. 1984a, b). The lack of significant amino acid homology between the B polypeptides of CT and LT-IIa and the B polypeptides of CT and LT-IIb precludes the use of comparative analysis of the primary amino acid sequences to locate the extracellular transport signals in each of the toxins. To locate amino acid residues in CTB that are important for extracellular secretion, a panel of plasmids encoding mutant CTB polypeptides with single-point amino acid substitutions was analyzed in *V. cholerae* (Connell et al. 1995). E11K, a mutant CTB polypeptide in which lysine was substituted for glutamate at amino acid position 11, showed a significant reduction in secretion efficiency when expressed in *V. cholerae*, but the lysine substitution did not affect several other properties of

CTB that require proper folding of the CTB polypeptide. The probability is high, therefore, that the decrease in extracellular transport of the E11K CTB polypeptide is due to a specific and localized change in a domain that interacts with the secretory apparatus of *V. cholerae*.

In another study, the heat-labile enterotoxin B chain (LTB) was used as a reporter gene to study the mechanisms of protein exocytosis, the assembly of multimeric protein complexes, and the secretion of proteins into the extracellular milieu of *V. cholerae* (Ali et al. 2000). Deletion, addition, or mutation of random nucleotides at the 3′ end of the LTB gene has been thoroughly investigated. LTB with seven extra amino acids at the C-terminus was secreted from the *V. cholerae* cells as the authentic B subunit, while deletions or mutations of amino acids at the C-terminus of LTB have a severe effect on its secretion, oligomerization, and its interaction with the A subunit. In another study, modification of LTB at its C-terminus reduced its secretion into the extracellular milieu to some extent. On the other hand, the extension of LTB at its N-terminus hindered its translocation from the periplasm of *V. cholerae* cells into the extracellular milieu. Neither monomer nor pentamer of LTB could be detected under these conditions. Thus, the N-terminus of LTB plays an important role in the secretion into the extracellular milieu by *V. cholerae*. The ETS of *V. cholerae* endochitinase has been identified using a genetic approach that delimited the regions of ChiA that were essential for secretion (Folster and Connell 2002). In this study, a region located between amino acids 75 and 555 was identified as being necessary and sufficient to promote secretion by the *eps* machinery of *V. cholerae*.

Chapter 11
Cholera Toxin (CT): Entry and Retrograde Trafficking into the Epithelial Cell

Abstract The cholera toxin secreted extracellularly by the organism *Vibrio cholerae* within the human intestine is attached to the GM1 ganglioside molecules of the epithelial cell membranes through its B pentamers. This attachment takes place preferentially in the lipid raft regions of the cell membrane, which aids in its endocytosis via a clathrin- or caveolin-dependent pathway. The CT holotoxin then undergoes retrograde trafficking within the epithelial cell via a glycolipid-dependent pathway through the trans-Golgi network to the lumen of the endoplasmic reticulum. Here the A1 fragment of the toxin CT is dissociated from the holotoxin, unfolded by the action of the endoplasmic reticulum (ER) chaperon, protein disulfide isomerase (PDI), translocated across the Sec61 channel of the ER membrane, and released in the cytosol of the epithelial cell. The A1 chain is then quickly reformed. This reformed structure is then activated by reaction with a cytosolic protein, adenosine diphosphate ribosylation factor 6 (ARF6), complexed with guanosine triphosphate (GTP). The activated CTA1 peptide (CTA1*) catalyzes the transfer of ADP-ribose from nicotinamide adenine dinucleotide (NAD^+) to an arginine residue on a G protein. The ADP-ribosylation of the α-subunit of G protein complexed with GTP activates the membrane-bound adenylate cyclase, which in turn catalyzes the conversion of ATP into cyclic AMP (cAMP). Elevated levels of cAMP in crypt cells activate the cAMP-dependent chloride channel (CFTR), leading to a massive secretion of water and electrolytes into the lumen that is manifested as severe diarrhea—a characteristic of the disease cholera. The details of this intracellular voyage of cholera toxin and the reactions that take place during it are presented in this chapter.

11.1 Introduction

In order that the cholera toxin can function and produce the heavy outpouring of intracellular fluid containing different electrolytes that leads to the diarrhea and severe dehydration that is typical of the disease cholera, it must enter the intestinal epithelial cell and reach the cytosol. It can then activate the adenylate cyclase over

K. Chaudhuri and S.N. Chatterjee, *Cholera Toxins*.

a long period of time in order to produce an intracellular accumulation of cyclic AMP, leading to the outpouring of chloride and other electrolytes. This has to take place in several stages: (i) the attachment of the toxin molecule to the plasma membrane of the epithelial cells; (ii) the internalization of the toxin by one or other of the various methods of endocytosis; (iii) retrograde trafficking of the toxin within the cell to reach the lumen of the endoplasmic reticulum; (iv) unfolding of the A1 fragment of the toxin and its disassembly from the holotoxin; (v) translocation of the unfolded A1 fragment to the cytosol, and; (vi) reactions of the A1 fragment in the cytosol leading to the activation of the adenylate cyclase, the opening of the chloride channel and the outpouring of the fluid. This chapter presents a concise account of these reactions at different stages with a view to explaining how the cholera toxin (CT) molecule produces the disease cholera.

11.2 Attachment to the Cell Membrane

The B subunits of CT bind to the GM1 gangliosides on the plasma membrane of the epithelial cells and serve as a vehicle for delivering CT to the endoplasmic reticulum (ER). CTB has a high affinity for GM1 ganglioside (Galβ1-3GalNAcβ1-4(NeuAcα2-3) Galβ1-4Glcβ1-1ceramide), which is found in the outer leaflet of the plasma membranes of virtually all cell types, including enterocytes and lymphocytes (Holmgren et al. 1973, van Heyningen et al. 1971, Angstrom et al. 1994). The X-ray structure of the CTB pentamer complexed with GM1 pentasaccharide revealed that each B subunit has a GM1 binding pocket (Fig. 7.4), with the B subunits interacting mainly with the terminal galactose, and to a lesser extent, with the sialic acid and N-acetylgalactosamine of GM1 (Merritt et al. 1994a, b). Upon receptor binding, a flexible loop comprising amino acids 51–58 becomes more ordered due to hydrogen bond interactions with the GM1 pentasaccharide and water molecules that occupy the binding site, and it is thought that these interactions stabilize the toxin–GM1 complex (Merritt et al. 1994a).

It is now generally recognized that lipids are organized into membrane microdomains or "lipid rafts," specialized membrane microdomains rich in cholesterol and glycosphingolipids (GSL) (Brown and London 2000, Pike 2003) (Fig. 11.1). These lipid rafts are formed through the segregation of lipids in the lateral plane of the membrane. They are thought to play roles in signaling and membrane trafficking (Simons and Ikonen 1997, Munro 2003). The best-characterized form of raft is composed of GSL, cholesterol, and phospholipids with long and saturated acyl chains, as found in caveolae. Since complex GSL can be sequestered into rafts, it is likely that toxins interacting with GSL receptors are also localized to such microdomains (Wolf et al. 1998, Katagiri et al. 1999). Interfering with cholesterol has been shown to inhibit endocytosis and intracellular transport of CT (Kovbasnjuk et al. 2001, Wolf et al. 2002). It is now confirmed that association with rafts is required for efficient uptake at the plasma membrane (Rodighiero et al. 2001). The CTB subunit tethers the toxin to the membrane, leading to the association of CT

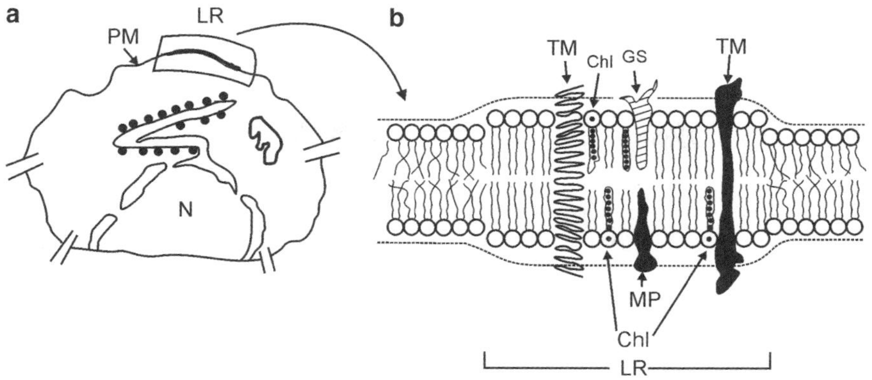

Fig. 11.1 a–b The presence of lipid raft (*LR*) in the plasma membrane (*PM*) of the epithelial cell. **a** Part of a cell showing the plasma membrane (*PM*), nucleus (*N*), portion of PM containing lipid raft (*LR*), etc. **b** Enlarged view of the portion of the plasma membrane within the *box in a* showing the organization of the PM containing lipid raft (*LR*). The LR is a specialized membrane domain containing high concentrations of cholesterol (Chl), sphingomyelin, gangliosides (GS), etc. This region is also enriched in phospholipids that contain saturated fatty acyl chains (*straight lines in lipid tails*). A variety of proteins—membrane proteins (MP) and transmembrane proteins (TM)—partition into lipid rafts. This composition of lipid raft results in lateral phase separation and generation of a liquid-ordered domain

with lipid rafts, which are required for toxin function (Wolf et al. 1998, Lencer and Tsai 2003).

11.3 Endocytosis

Endocytosis is the mechanism by which an object gains entry into the cell, and this mechanism may vary depending on the object (Fig. 11.2). CT is internalized from the eukaryotic plasma membrane by an endocytic mechanism that utilizes lipid rafts (Lord and Roberts 1998a, Lencer and Tsai 2003) (Figs. 11.2 and 11.3). Endocytosis of CT can take place by different pathways, and these different pathways can function simultaneously (Lu et al. 2005). Different workers (Wolf et al. 1998, 2002, Shogomori and Futerman 2001, Pang et al. 2004) have demonstrated that attachment to GM1 partitions CT into detergent-insoluble glycosphingolipid (DIG)-rich membranes that are critical to the toxin's function. In a number of cell types, endocytosis of CT can occur in dynamin- and caveolae-mediated pathways (Le and Nabi 2003, Nichols 2003). CT can also undergo endocytosis by a clathrin-dependent pathway (Henning et al. 1994). There is also strong evidence that in some cell types CT can be internalized via clathrin- and caveolae-independent pathways (Sandvig and van Deurs 1994, Lamaze and Schmid 1995, Lamaze et al. 2001, Massol et al. 2004), and both dynamin-dependent and dynamin-independent

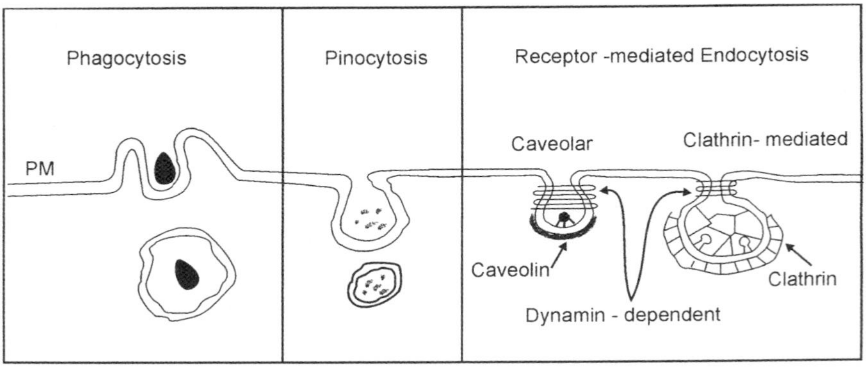

Fig. 11.2 Different forms of endocytosis. Pinocytosis is concerned with the uptake of solutes and single molecules of proteins, etc. It literally means drinking by the cell. Phagocytosis is the process by which a cell ingests large objects, such as cells, bacteria, etc. Receptor-mediated endocytosis specifically involves the attachment of ligands to the specific receptors on the cell surface, which folds inward to form coated pits. This coat may be of the protein, caveolin or clathrin, and accordingly endocytosis is either a caveolin-mediated or clathrin-mediated one. Both these modes of endocytosis are dependent on the function of another protein, dynamin. Dynamin is actually a large GTPase involved in the synthesis of nascent vesicles from parent membranes. It forms a helix around the neck of a nascent vesicle (as shown in the figure), and is involved in the lengthwise extension of the helix and breaking of the vesicle's neck, thereby releasing the vesicle into the cell cytoplasm

processes have been described. In a recent study, CTB-bound GM1 labeled with a fluorophore was internalized in a clathrin-independent manner in cells that expressed high levels of caveolin, and in a clathrin-dependent manner in the presence of low levels of caveolin (Singh et al. 2003). Further, Lu et al. (2005) presented evidence that the enterocyte endocytosis of CT is developmentally regulated. It was shown that internalization of CT in adult enterocytes is less and occurs via a clathrin-dependent pathway. Badizadegan et al. (2004) showed, on the other hand, that trafficking of the CT-GM1 complex into Golgi and induction of toxicity depend on actin cytoskeleton. The authors proposed that the CT-GM1 complex is associated with the actin cytoskeleton via the lipid rafts, and that the actin cytoskeleton plays a role in the trafficking of CT from the plasma membrane to the Golgi/ER and the subsequent activation of adenylate cyclase.

There are at least two general transport pathways from the cell surface to ER. One requires binding to the KDEL-receptor ERD2 in the Golgi apparatus, and the other requires binding to a glycolipid associated with the lipid rafts at the cell surface. The glycolipid acts as the transport vehicle from plasma membrane to ER; CT is transported via this glycolipid-dependent pathway (Lencer and Tsai 2003). The toxins can eventually be found in the early and recycling endosome (EE/RE) (Nichols et al. 2001, Richards et al. 2002). The CTs must undergo retrograde transport to a location in the cell where protein translocation channels already exist: the endoplasmic reticulum (ER). The reason that the glycolipid receptor ganglioside

GM1, but not ganglioside GD1a, has the ability to transport cholera toxin (CT) backwards into the Golgi and ER in intestinal cells is probably associated with the structure of the lipid moiety. Gangliosides have two domains, an extracellular oligosaccharide domain and an intramembrane ceramide lipid. The gangliosides are classified based on the different structures of their oligosaccharides, but the ceramide domain can also show structural heterogeneity. The structure of the ceramide domain probably defines the affinity of the molecule for association with lipid rafts. However, raft association might not be sufficient for retrograde transport into the secretory pathway. It is possible that only ceramide-based raft glycolipids move in a retrograde manner from plasma membrane to ER. Some of the molecules that enter cells by endocytosis are either recycled back to the plasma membrane or targeted to the late endocytic pathway for degradation. Although some recycling occurs, GSL-bound toxins eventually escape from the endosomal compartment by following a third pathway, the retrograde route that allows them to reach other compartments of the cell: trans-Golgi network (TGN), Golgi cisternae, and the endoplasmic reticulum (ER).

11.4 Retrograde Trafficking to the Golgi Complex and Endoplasmic Reticulum

The toxin, after being internalized, moves to the endoplasmic reticulum by a retrograde vesicular transport mechanism (Fig. 11.3). Recent studies show that the lipid raft ganglioside, GM1, is responsible for transport into the Golgi and endoplasmic reticulum (ER) compartments. The pathway backwards from cell surface to ER for CT is dependent on lipid transport. Toxin binding to glycolipids with a strong affinity for lipid microdomains (detergent-resistant membranes or lipid rafts) appears to be critical for sorting into this pathway, for a specific mechanism of internalization, and also for a subsequent sorting step at the level of early endosomes/recycling endosomes (EE/RE). It was found that the drug brefeldin A completely inhibited the action of ricin toxin (heat stable toxin, ST) and CT (Donta et al. 1993, Sandvig et al. 2002, Lencer and Tsai 2003). Brefeldin A is known to disrupt the Golgi apparatus and ER membrane transport. Furthermore, it was found that certain ER luminal proteins were localized to the ER by displaying a Lys/Arg-Asp-Glu-Leu [(K/R)DEL] motif at their C-terminus (Lewis et al. 1990, Orlandi et al. 1993, Lencer and Tsai 2003). The K/RDEL motif binds the KDEL receptor, ERD2, which cycles between the ER and Golgi, thus retrieving itinerant ER proteins from the Golgi apparatus. Cholera toxin and all *Escherichia coli* heat-labile toxins have [K/R]DEL motifs located in a position on the molecule that can bind to ERD2. Subsequently, mutagenesis of the motif revealed that these toxins interact with ERD2 to affect toxicity (Chaudhary et al. 1990, Lencer et al. 1995).

Although CT does not need to bind the KDEL receptor to move from plasma membrane to ER, it does interact with the KDEL receptor to make its action more

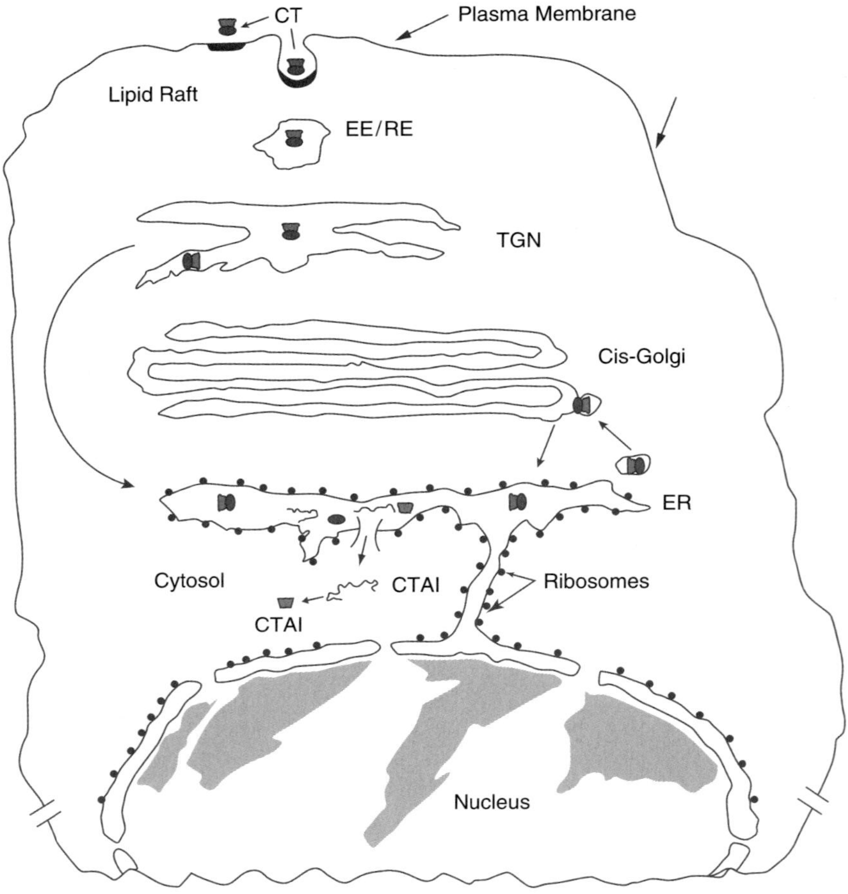

Fig. 11.3 Schematic diagram illustrating the endocytosis of cholera toxin, *CT*, and its retrograde trafficking via the trans-Golgi network (*TGN*) to the lumen of the endoplasmic reticulum (*ER*), bypassing the cis-Golgi complex (*Cis-Golgi*). Some of the CT molecules in the ER lumen may travel backward to the cis-Golgi complex and again return to the lumen of ER (shown by *arrows*) in a KDEL-mediated retrieval step. In the ER lumen, CT undergoes disassembly and the dissociated CTA1 fragment gets unfolded and translocated across the Sec61 channel of the ER membrane to the cytosol of the cell, where it gets quickly refolded to its native structure (shown by *arrow*)

efficient. The lipid- and KDEL-dependent pathways might work together in membrane transport through the Golgi (Nambiar et al. 1993).

It has now been recognized that at least two routes are available for toxins to progress from the TGN to ER. In one, the toxin, carried by the KDEL receptor or some other cycling protein receptor, moves via the Golgi complex in a coatamer protein 1 (COP1)-dependent manner. This appears to be the route followed by *Pseodomonas* exotoxin PETx and not CT. In the second, the toxin is transported directly from the trans-Golgi network (TGN) to ER, bypassing the Golgi complex altogether, perhaps by virtue of an interaction with a raft-associated glycolipid receptor. This appears to be the pathway taken by CT and ST (Lord et al. 2005). The lipid pathway may move directly from the TGN to ER without passing through the Golgi cisternae, and independently of COP1 vesicles that typify retrograde transport in the classic secretory pathway. Once in the ER, it is known—at least for CT—that the toxins bound to their lipid receptors can move anterograde from the ER to the cisGolgi and recycle back to the ER in a KDEL-mediated retrieval step. This recycling for CT by binding to the KDEL receptor is proposed to retain CT in the ER and maximize the retro-translocation of the A1-chain to the cytosol (Lord et al. 2005). Mutation of the KDEL motif in CT interrupts the recycling pathway and causes a loss in the efficiency of toxin action.

11.5 Unfolding of CTA1 and Its Translocation to the Cytosol

Cholera toxin enters the lumen of the ER as a fully assembled protein complex. Here, the A subunit is reduced, and the resulting A1 chain is unfolded, disassembled from the rest of the toxin, and transported to the cytosol (Fig. 11.3). Since the AB_5 toxins generally require proteolytic cleavage for activation, the CT is cleaved after secretion from the vibrio into the intestinal lumen. It is also potentially true for CT that the toxins are activated during entry into target cells by endogenous proteases. The A and B subunits of CT are stably folded even after proteolytic cleavage and reduction of the A subunit; the A1 chain remains tightly bound to the B subunit. Thus, disassembly of the A1 chain from the rest of the toxin requires unfolding of the peptide. This is accomplished by the ER chaperone, protein disulfide isomerase (PDI) (Tsai et al. 2001, Lord et al. 2005). The unfolding reaction catalyzed by PDI is driven by a redox cycle rather than by an ATP cycle (Orlandi 1997, Lencer and Tsai 2003); in its reduced state, PDI releases the toxin. It is probable that only the A1 chain is unfolded and dissociated from the B subunit (Tsai et al. 2001, Lencer and Tsai 2003, Lord et al. 2005) (Fig. 11.3).

In the ER, the toxin hijacks the machinery that monitors and degrades misfolded proteins. This protein quality control system, termed ER-associated degradation (ERAD), normally ensures that terminally misfolded proteins in the secretory pathway are recognized, unfolded and translocated to the cytosol for degradation by the proteasome (Schmitz et al. 2000, Teter and Holmes 2002, Teter et al. 2002, 2003, 2006). In the final step of this process, almost all known endogenous substrates for ERAD are polyubiquitinated in the cytosol for targeting to the proteasome (Tsai et al. 2002, Kostova and Wolf 2003). However, CT, and perhaps other toxins, do not share this last step (Hazes and Read 1997, Lord and Roberts 1998a, b, Rodighiero et al. 2002).

The mammalian ER is a protein-folding environment functionally similar to both the bacterial periplasm and the plant cell ER where the toxins are initially produced. In one instance, the folded toxins and specific SH groups are oxidized to form disulfide bonds, and in the other instance, the toxins are unfolded and the disulfide bonds reduced. In the case of CT, the explanation for this problem is known. The structure of CT contains a molecular switch that allows for toxin folding and assembly in the bacterial periplasm, but signal entry into the ERAD pathway for unfolding and subunit dissociation in the ER of target mammalian cells. This molecular switch is the critical protease site in the loop connecting the A1 and A2 chains that allows for toxin activation. Proteolytic cleavage of this loop converts CT into a substrate for protein disulfide isomerase (PDI). This ER chaperone unfolds and dissociates the A1 chain from the rest of the toxin in preparation for retro-translocation to the cytosol (Tsai et al. 2001, Lord et al. 2005). The exact motif that allows PDI to act on CT is not known, but it is likely associated with the exposure of hydrophobic domains in the A1 chain that are buried in the uncleaved pro-form of the toxin.

In CT, a subdomain of the A1 chain near its C-terminus termed the $A1_3$, domain contains an extended hydrophobic sequence, which is proposed to serve as an essential motif for retro-translocation (Hazes and Read 1997). However, this motif is partially buried inside the fold of the native toxin, although it might be exposed after proteolytic cleavage of the A subunit. The structure of the cleaved and reduced A subunit is not known. Presumably, a conformational change occurs that renders the A1 chain a substrate for unfolding by PDI, and activates its enzymatic activity. However, the actual recognition motif remains to be discovered.

The next step in retro-translocation is targeting of the unfolded A1 chain to the translocation channel. It has been found that the PDI–toxin complex is targeted to the luminal surface of the ER membrane. Here, the membrane-associated ER oxidase, Ero1, oxidizes PDI to release the unfolded A1 chain (Tsai et al. 2001, Tsai and Rapoport 2002). Because the A1 chain refolds spontaneously once it is released from PDI, it is proposed that the release of the toxin from PDI is mechanistically coupled to the retro-translocation channel, although how this occurs remains a mystery.

The protein-conducting channel that mediates retro-translocation of CT is not known with certainty, but there is strong evidence to suggest that this channel might be the same protein-conducting pore that functions in the biosynthetic pathway, i.e., the Sec61 channel (Hazes and Read 1997, Matlack et al. 1998). Experimental evidence in support of this idea was obtained by reconstituting the retro-translocation reaction for the CTA1-chain in rabbit reticulocyte lysates (Schmitz et al. 2000). The A1 chain was co-immunoprecipitated with components of Sec61, suggesting the capture of a translocating intermediate (Schmitz et al. 2000). Other toxins that undergo retro-translocation, such as ricin and *Pseudomonas* ExoA, have also been found to interact with the Sec61 channel. Thus, it appears that the Sec61 complex might be the common channel used by the toxins to exit the ER. The driving force for retro-translocation of CT might instead be based on the ability of the A1 chain to refold rapidly and spontaneously upon emerging from the channel (Rodighiero et al. 2002). Thus, the toxin might ratchet itself into the cytosol, or the rapidly refolding A1 chain might generate an active pulling force (Lencer and Tsai 2003).

11.6 Reactions in the Cytosol Leading to the Activation of Adenylate Cyclase, Chloride Channel Outpouring, and Diarrhea

The A1 chain enters the cytosol and causes disease. Unlike misfolded endogenous proteins, the A1 chain of CT must avoid degradation by the proteasome for at least as long as it takes for the toxin to reach its substrate in the cytosol and induce toxicity. There are two mechanisms by which this might occur. As in almost all other bacterial toxins, the A1 chain of CT has a paucity of lysines, which are the sites for polyubiquitination (London and Luongo 1989, Hazes and Read 1997). Polyubiquitination is the primary motif for targeting proteins to the proteasomes, but the proteasome can also recognize and degrade unfolded proteins in the absence of polyubiquitination. Thus, it is probable that the paucity of lysines in the A1 chain, together with its ability to refold rapidly, protects the A1 chain from degradation in the cytosol (Rodighiero et al. 2002, Lencer and Tsai 2003).

Once inside, CTA1 interacts with a protein known as adenosine diphosphate (ADP)-ribosylation factor 6 (ARF6) (Moss and Vaughan 1991), which enhances the activity of the toxin (Welsh et al. 1994, Teter et al. 2006). Almost 30 years ago, ARFs were originally discovered and defined as activators of CTA1. They have since been revealed to also play essential roles in the trafficking of vesicles within cells during normal physiological conditions. While this multispecific binding strategy gives ARF proteins broad flexibility, it also leaves them vulnerable to interaction with unintended partners. The $CTA1_3$ may promote an interaction with the cytosolic ADP ribosylating factors (ARFs), which serve as allosteric activators of CTA1. Teter et al. (2006) found that the $A1_3$ subdomain of CTA1 is important for both interaction with ARF6 and full expression of enzyme activity in vivo.

It was not clear how the two proteins—cholera toxin and ARFs—interact, and more interestingly, how ARF interaction enhances the cholera toxin's activity. To investigate, O'Neal et al. (2005) determined the 1.8-Å-resolution crystal structure of a CTA1 variant bound to a complex of human ARF6 and guanosine triphosphate (GTP), a mediator of the interaction. The structural basis of this activation of CTA1 by ARF6, as revealed by O'Neal et al. (2005), has been discussed previously in Chap. 7 of this book. The biochemical part of this story is given below.

Once in the cytosol, the activated CTA1 (CTA1*) stimulates the membrane-bound adenylate cyclase, which in turn raises the level of cyclic AMP in these cells, resulting in a very large efflux of Na^+, other electrolytes and water into the gut. But how does this stimulation of adenylate cyclase take place? It is known that adenylate cyclase is not directly activated by the binding of any hormone or any similar molecule to a specific receptor. Instead, it is stimulated or activated by another protein, the G protein, which binds guanyl nucleotides (GTP or GDP). Only the GTP complex of the G protein activates adenylate cyclase. Its GDP complex cannot do so. This G protein has to be converted from the inactive

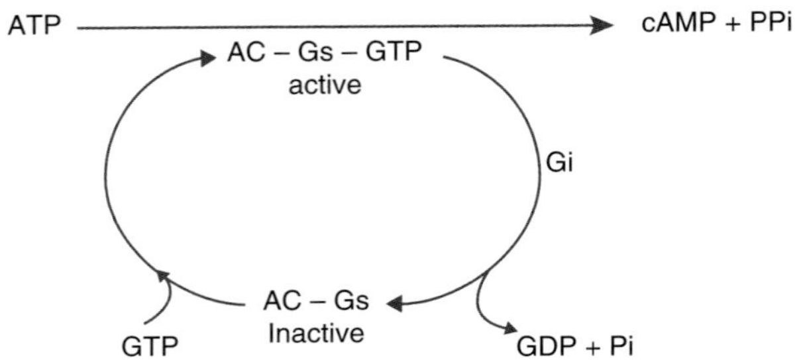

Fig. 11.4 Activation of adenylate cyclase (AC) by the G protein GTP complex. Under normal conditions, the GTPase component of the G protein (G_i) hydrolyses the bound GTP to GDP + Pi, thereby deactivating the adenylate cyclase bound G protein (AC-Gs). During its activated state (which is normally very short) the adenylate cyclase (AC) catalyzes the conversion of ATP into cyclic AMP (cAMP) + PPi

GDP complex into the active GTP complex by the exchange of GTP for bound GDP. The activation of adenylate cyclase is, on the other hand, switched off when the G protein is deactivated. The G protein is also a GTPase, and through this GTPase activity the bound GTP is hydrolyzed to GDP. The proportion of G protein in the GTP-complexed state depends on the rate of exchange of GTP for GDP compared with the rate of hydrolysis of bound GTP. The normal situation is described in Fig. 11.4.

The activated CTA1 peptide catalyzes the transfer of ADP-ribose from NAD+ to an arginine residue (no.187) on the G protein (Moss and Vaughan 1977, Kaper et al. 1995), as shown in Fig. 11.5a, b. The α-subunit of G_s contains a GTP-binding site and an intrinsic GTPase activity (Hepler and Gilman 1992). Binding of GTP to the α-subunit leads to dissociation of the α- and β-γ subunits and subsequent increased affinity of α for adenylate cyclase. The resulting activation of adenylate cyclase continues until the intrinsic GTPase activity hydrolyzes GTP to GDP, thereby inactivating the G protein and the adenylate cyclase. ADP-ribosylation of the α-subunit by the CTA1 peptide inhibits the hydrolysis of GTP to GDP, thus leaving adenylate cyclase constitutively activated (Cassel and Selinger 1977, Kahn and Gilman 1984). The activated adenylate cyclase catalyzes the conversion of ATP to 3',5'-cAMP and PP_i (Fig. 11.6).

Adenylate cyclase is a transmembrane protein (Fig. 11.7). It passes through the plasma membrane 12 times. The important parts for its functions are located in the cytoplasm, and the cytoplasmic part can be subdivided into two regions or domains, C1 and C2. The C1 region exists between transmembrane helices 6 and 7, and the C2 region follows transmembrane helix 12. These two domains form a catalytic dimer where ATP binds and is converted to cAMP (Fig. 11.7). Thus, the net effect of the toxin is to cause cAMP to be produced at an abnormally high rate, which stimulates mucosal cells to pump large amounts of Cl^- into the intestinal contents.

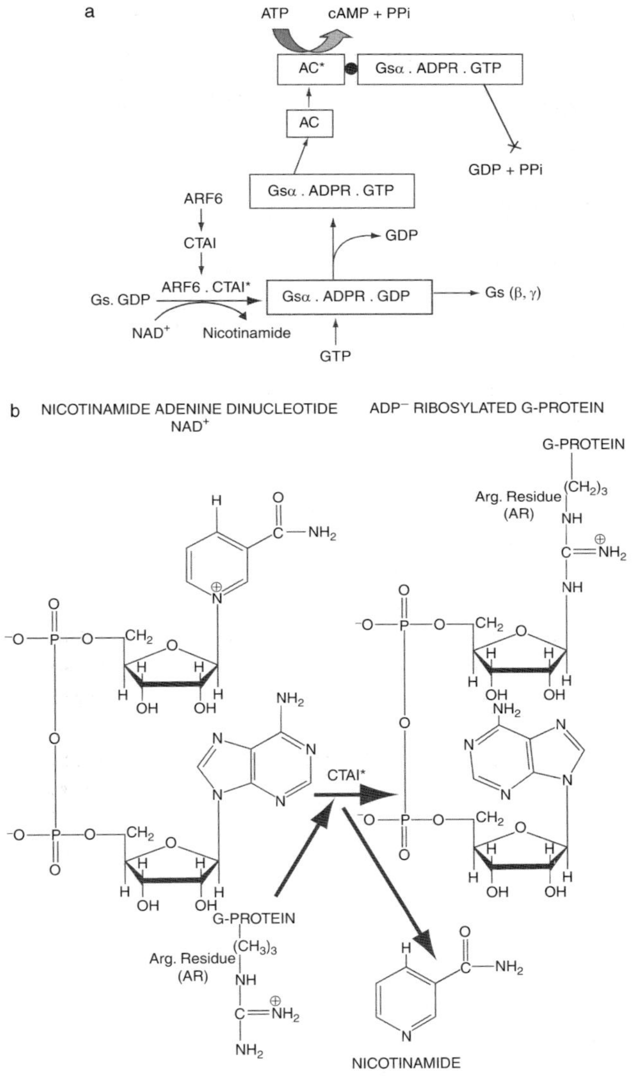

Fig. 11.5 **(a)** Role of the A1 fragment of cholera toxin (CTA1) in prolonged activation of the adenylate cyclase (*AC*). In the cytosol, the ARF6 molecule binds with CTA1 to form the ARF6. CTA1 complex, where the CTA1 is in an activated state (CTA1*). This activated complex catalyzes the transfer of ADP-ribose from NAD+ to an arginine residue (*AR*) on the G protein and the formation of the complex $G_{s\alpha}$.ADPR.GDP. Binding of GTP to $G_{s\alpha}$ leads to dissociation of the α- and β-γ subunits and subsequent increased affinity of $G_{s\alpha}$ to adenylate cyclase (*AC*). The resulting activation of adenylate cyclase (AC*) continues for a long period of time, particularly since ADP-ribosylation of $G_{s\alpha}$ ($G_{s\alpha}$.ADPR) inhibits the hydrolysis of GTP to GDP **(b)** Chemical reactions showing how the activated CTA1 (CTA1*) catalyzes the transfer of ADP-ribose to an arginine residue (AR) of the G protein with the conversion of NAD^+ to nicotinamide

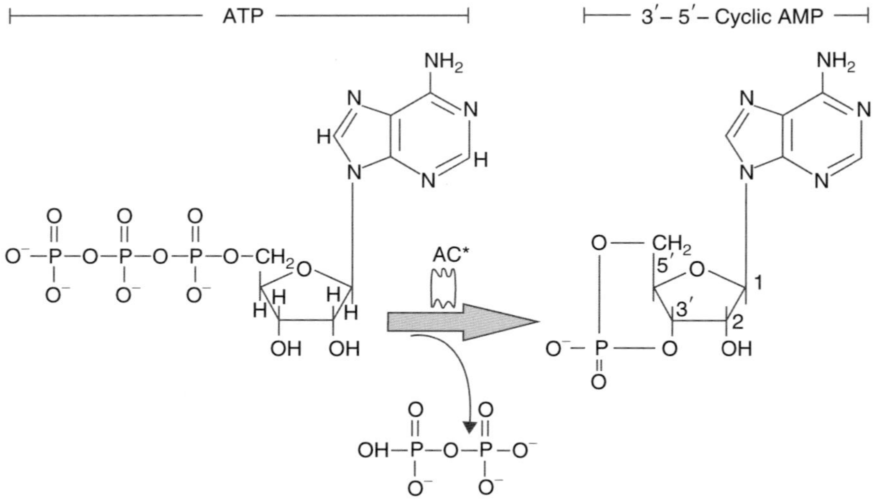

Fig. 11.6 Chemical reaction showing how the activated adenylate cyclase (AC*) catalyzes the conversion of ATP to 3'-5'-cyclic AMP

This change results in prolonged opening of the chloride channels that are instrumental in uncontrolled secretion of water from the crypts.

The apical or luminal membranes of crypt epithelial cells contain a ion channel (Field et al. 1971) of immense medical significance: a cyclic AMP-dependent chloride channel known also as the cystic fibrosis transmembrane conductance regulator, or CFTR (Bear et al. 1992). Mutations in the gene for this ion channel result in the disease cystic fibrosis (Berschneider et al. 1988). This channel is responsible for secretion of water by the following steps (Fig. 11.8):

1. Chloride ions enter the crypt epithelial cell by co-transport with sodium and potassium; sodium is pumped back out via sodium pumps, and potassium is exported via a number of channels
2. Activation of adenylate cyclase leads to generation of cyclic AMP
3. Elevated intracellular concentrations of cAMP in crypt cells activate the CFTR, resulting in secretion of chloride ions into the lumen
4. Accumulation of negatively charged chloride anions in the crypt creates an electric potential that attracts sodium, pulling it into the lumen, apparently across tight junctions; the net result is secretion of NaCl
5. Secretion of NaCl into the crypt creates an osmotic gradient across the tight junction and water is drawn into the lumen

Abnormal activation of the cAMP-dependent chloride channel (CFTR) in crypt cells has resulted in the deaths of millions of people. Activation of adenylate cyclase—often for a long period of time—by cholera toxin (CT), as described

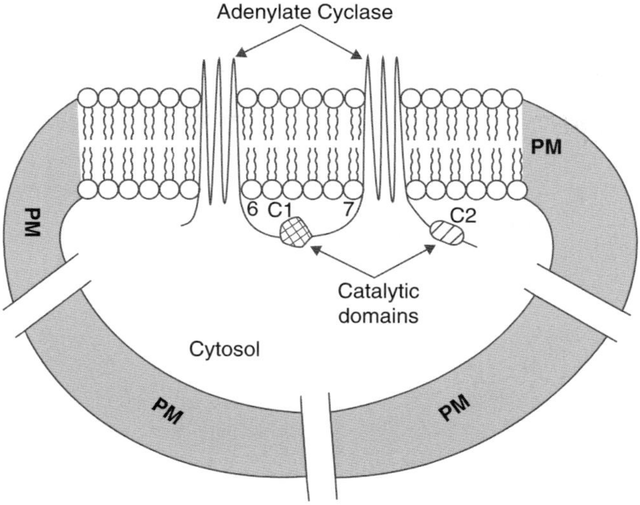

Fig. 11.7 Adenylate cyclase, the integral membrane protein, consists of six transmembrane segments and the two catalytic domains, *C1* and *C2*, extending as loops into the cytoplasm

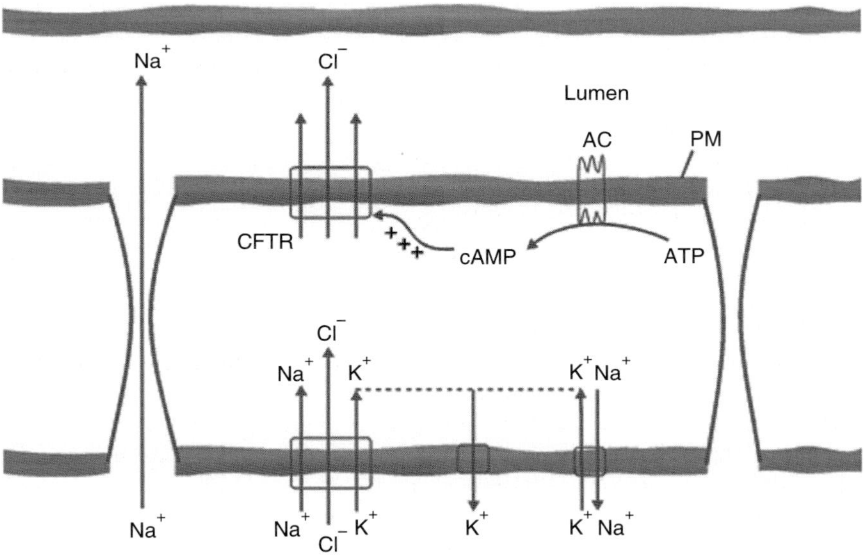

Fig. 11.8 Different steps in the activation of the cAMP-dependent chloride channel (CFTR) in the apical or luminal membranes of crypt epithelial cells, leading to a massive secretion of water and electrolytes. For details, see text

above, thus leads to elevated level of cyclic AMP and causes the chloride channels to essentially become stuck in the "open position," resulting in the massive secretion of water that is manifested as severe diarrhea, a characteristic of the disease cholera.

Chapter 12
Cholera Toxin (CT): Immune Response of the Host and Vaccine Production

Abstract Cholera toxin (CT) is also a potent immunomodulator, and it exhibits both mucosal and systemic adjuvant activities. The mechanism by which CT exhibits its adjuvant effect has still remained largely unknown. CT or its component (particularly CTB) also plays a significant role in the preparation of different types of cholera vaccines. While summarizing the important observations that are likely to act as guiding factors in the preparation of vaccines, this chapter presents a brief account of the construction and actions of different killed or live oral cholera vaccines with a short concluding note on the current status of cholera vaccine.

12.1 Introduction

Natural infection by *Vibrio cholerae* is known to provide significant protection to the subject against subsequent further infections, although the immunity may not last long. Antibacterial immunity is found to provide better protection than antitoxin immunity. The cholera toxin (CT) has, on the other hand, been found to exhibit strong adjuvant activities, and this property of CT has been utilized for different purposes. Production of cholera vaccine has focused on the use of killed or live bacteria. In order to ensure that the vaccine using live bacteria cannot produce, by itself, any significant diarrhea in the vaccinees, the vaccine strains are being constructed by deleting genes producing CT and other toxins of *V. cholerae*. Keeping in mind the better efficacy of local immunity, live oral vaccines are being constructed utilizing *V. cholerae* of different biotypes, serotypes, serogroups, and particularly serogroups O1 and O139. A brief account of these immunological aspects of *V. cholerae* strains is presented in this chapter

12.2 Protective Immunity by Anti-CT

It is well established that clinical cholera caused by *V. cholerae* O1 strains confers a substantial degree of protection against subsequent attacks (Levine et al. 1981, 1983), although the mechanism of this protection has not been clearly defined.

K. Chaudhuri and S. N. Chatterjee, *Cholera Toxins.*

Despite the discovery of other toxins in *V. cholerae*, the cholera toxin (CT) has been held to provide the major contribution to the clinical disease and some contribution to the protective immune response (Levine et al. 1983). Thus, some attention has been focused on the importance of antitoxic immunity for protection against cholera. Immunization with toxin or toxoid preparations was found to induce protection against experimental cholera in dogs, rabbits, mice, and rats (Curlin et al. 1970, Finkelstein 1970, Fujita and Finkelstein 1972, Holmgren et al. 1972, Pierce et al. 1988). Toxoids consisted of formaldehyde-treated CT or glutaraldehyde-treated CT or procholeragenoid (heat-treated CT, with or without formaldehyde treatment). The protective immunity could be achieved to a varying degree by parenteral or oral routes of immunization. The protection appeared to be mediated by locally synthesized antibodies (IgA) after oral immunization, but after parenteral immunization, both serum-derived IgG and locally produced IgA antibodies seemed to be effective in rabbits (Holmgren et al. 1975). In mice, oral administration of CT induced an IgA anti-CT antibody response associated with elevated IgG anti-CT antibody in serum (Xu-Amano et al. 1994). However, upon repeated oral administration of CT, a large number of anti-CT IgA antibody secreting cells were observed in lamina propria, providing concomitant protection against CT challenges in the intestinal loops (Lycke et al. 1989). Similar observations were found in rats as well (Yoshiyama and Brown 1987, Boesman-Finkelstein et al. 1989). Several studies by Holmgren suggested that per-oral administration of a few micrograms of CT was sufficient to induce a strong immune response (Lange and Holmgren 1978, Svennerholm and Holmgren 1976). The high immunogenicity of orally administered CT was probably due to two different properties of this antigen: its high-affinity binding to the intestinal mucosa, and its ability to stimulate adenylate cyclase activity. Purified B subunits, used orally or parenterally in volunteer studies in areas of Bangladesh (endemic) and Sweden (nonendemic), stimulated the appearance of secretory IgA (sIgA) antitoxin in intestinal fluid (Svennerholm et al. 1982). The duration of intestinal sIgA was much greater after oral than after parenteral administration. Its safety and capacity to adhere to the enterotoxin receptors were considered advantages of using B subunits, but animal studies suggested that it was less potent than native holotoxin in stimulating antitoxin response (Pierce et al. 1982).

On the other hand, studies with both human volunteers and animals infected with live *V. cholerae* have provided evidence of strong antibacterial immunity that inhibited and often completely prevented bacterial colonization upon rechallenge (Jertborn et al. 1994). Multiple antigens like lipopolysaccharide, toxin-coregulated pili, outer-membrane proteins, cell-associated hemagglutinins, etc. elaborated by *V. cholerae* served as mediators of the potent immunity. A synergistic protection against experimental cholera was observed upon immunization with toxin and bacterial antigens, i.e., the protective effect of combined immunization was better than the sum of the effects induced by each antigen component alone (Jertborn et al. 1994).

12.3 Immunomodulation by Cholera Toxin (CT)

Cholera toxin, aside from its pathogenic functions, is also a potent immunomodulator, and exhibits both mucosal and systemic adjuvant activities. Adjuvants are substances that are coadministered with an antigen in a vaccine in order to enhance the desired immune response. It is generally accepted that initiation of a specific immune response requires the activation of a specific defense mechanisms, resulting in a proinflammatory response where cytokines and chemokines are released which assist in activating and directing the adaptive immune response. Most antigens—whether food and other environmental antigens or commensal microorganisms—are not immunogenic enough, and need help to induce a good immune response. Whereas there is a long list of systemic adjuvants, the number of mucosal adjuvants is far more limited. CT does not fit to the classical definition of an adjuvant, as it stimulates an immune response to itself. However, CT can enhance the immune response to other antigens and has been used as a carrier for several parenterally as well as orally administered experimental vaccines (Lycke and Holmgren 1986a, Liang et al. 1988, Xu-Amano et al. 1993).

12.3.1 Adjuvant Properties of CT

Initially, Northrup and Fauci (1972) reported that CT, when delivered by the intravenous route with a foreign antigen, behaved as an adjuvant. Subsequently, low doses of CT were shown to potentiate the mucosal immune response to a number of unrelated antigens, which would induce little or no response by themselves (Elson and Ealding 1984, Lycke and Holmgren 1986b, Liang et al. 1988). The above studies revealed that the adjuvant effects were observed upon coadministration of CT and the antigen through an oral route. For the past few years, CT has been used in a number of studies as the mucosal adjuvant. These studies have used intranasal, transcutaneous, or parenteral routes of administration, and have shown substantial increases in the antigen-specific mucosal IgA and serum IgG responses (Table 12.1). In addition to enhancing humoral immunity, CT also augmented cellular immune responses to coadministered antigens. In most cases, CT induced a T-helper cell Th2 bias response based on the finding that interleukins IL-4, IL-5, and IL-10 were produced with little IFN-γ (Connell 2007) (Table 12.1). The impact of the above findings became all the more important when subsequent studies showed that the ability of CT to augment immune responses was not limited to short-term effects. Reactivity to antigen and CT elicited a long-term memory response and thus was detectable long after the initial immune response (Vajdy and Lycke 1992, 1993).

Table 12.1 Adjuvant effect of CT and its derivatives

Adjuvant (CT/derivative)[a]	Route[a]	Species	Adjuvant dose (μg)	Antigen (dose)[a]	Measured antigen-specific response[a]		References
					Cytokines	Antibody	
CT	Oral	Mice	10	TT (250 μg) 3 doses, 1 per week	IL-4, IL-5	sIgA in GI tract; IgG, IgA in serum	Xu-Amano et al. (1993)
CT	IP	Mice	10	TT (250 μg) 3 doses, 1 per week	IL-4, IL-5	IgG, IgM in serum	Xu-Amano et al. (1993)
CT	IN	Mice	0.5	OVA (100 μg)	IL-4, IL-5, IL-6, IL-10	IgG, IgA, IgM in serum, mucosal IgA	Yamamoto et al. (1997)
CT	SC	Mice	1	KLH (20 μg)	IL-4, IL-5, IL-10	IgG1, IgG2a	Lavelle et al. (2003)
mCT (S61F)	IN	Mice	5	OVA (100 μg)	IL-4, IL-5, IL-6, IL-10	IgG, IgA, IgM in serum, mucosal IgA	Yamamoto et al. (1997)
mCT (E112K)	IN	Mice	0.5	OVA (100 μg), 3 doses, 1 per week		Detectable OVA-specific IgA	Yamamoto et al. (1997)
mCT (E29H)	IN	Mice	1	Natural fusion (F) protein of RSV (0.1 μg)	ND	IgA in nasal washes and sera	Tebbey et al. (2000)
rCT-B	IN	Mice	5	OVA (100 μg)	Detectable but low IL-4, IL-5, IL-6, IL-10	Detectable IgM, IgA in serum, very low IgG	Yamamoto et al. (1997)
CTA2/B	Oral	Mice	100	Saliva-binding region (SBR) of *Streptococcus mutans* AgI/II	IL-4	IgG, salivary IgA	Toida et al. (1997)
CTA1-DD	IP	Mice	1	OVA (100 μg), 2 × 10-day interval	ND	IgG1, IgG2a, IgG2b	Agren et al. (1997)

[a]Abbreviations used; *CT*, cholera toxin; *mCT*, mutant CT; *S61F*, S to F substitution at position 61 of A1 subunit of CT; *E29H*, E to H substitution at amino acid 29 of the CT-A1 subunit; *E112K*, E to K substitution at position 112 of CT-A1, *rCT-B*, recombinant B-subunit of CT; *CTA2/B*, cholera toxin A2 and B subunits; *CTA1-DD*, a fusion protein of the A1 subunit of CT with a dimer of the Ig-binding domain (designated D) of the staphylococcal protein A; *IP*, intraperitoneal; *IN*, intranasal; *SC*, subcutaneous; *TT*, tetanus toxoid; *OVA*, ovalbumin; *KLH*, keyhole limpet hemocyanin; *IL*, interleukin; *ND*, not done

The inherent toxicity of CT has limited the widespread use of CT as vaccine components and adjuvants. Murine models demonstrated that intranasal sensitization with CT as adjuvant led to increased lung inflammation with a massive recruitment of macrophages (Fischer et al. 2005). Studies on nasal immunization in mice have further revealed that nasally administered CT accumulates in the olfactory nerves and epithelium and the olfactory bulbs of mice after binding to GM1 gangliosides. Furthermore, CT redirects the co-adjuvant antigen into neuronal tissues and ultimately to CNS (van Ginkel et al. 2000). Transcriptional activation of the proinflammatory mediators interleukin 1β (IL-1β), tumor necrosis factor TNF-α, cyclooxygenase enzyme COX2, monocyte chemotactic protein-1 (MCP-1), and macrophage inflammatory protein 1α (MIP-1α) were observed in murine brain following intranasal administration of CT (Armstrong et al. 2005). These limitations have led to mucosal strategies involving nontoxic mutants and purified B subunits.

12.3.2 Adjuvant Properties of B Subunits of CT

Early reports suggested that the B subunit by itself lacked the adjuvant activity when coadministered with the protein antigen (Van der Heijden et al. 1991, Lycke et al. 1991, 1992). Other studies used commercial preparations of cholera toxin B subunit purified from holotoxin, and these preparations showed strong adjuvant activity; it was later found that much of the activity was due to contamination with the A subunit (Tamura et al. 1994). Recombinant cholera toxin B subunits were subsequently available, and these showed less adjuvant activity compared to CT (Table 12.1). However, conjugating CTB subunit to the protein antigen does produce adjuvant activity, and others have detected adjuvant activity upon coadministration of protein antigen with purified CTB or recombinant CTB subunits. Some have suggested, however, that this activity could be due to the presence of small amounts of holotoxin as impurity in the CTB preparation (Table 12.1). It has been related that binding to GM1 is critically important to the immunostimulatory properties of CT or CTB (Smith et al. 2001).

12.3.3 Adjuvant Properties of Nontoxic Derivatives of CT

Nontoxic mutant derivatives were generated by the substitution of single amino acids such as CT (E29H), CT (S61F) and CT (E112K), where the mutation is denoted by the wild-type amino acid and its position in the particular wild-type sequence (CT, A subunit) followed by the substituted amino acid, with amino acids being denoted by a single letter code. All three mutants retained adjuvant activity. The mutants CT (S61F) and CT (E112K) did not induce ADP-ribosylation and cAMP formation (as the substitution was in the ADP-ribosyltransferase active center of the A subunit); CT (E29H), however, retained residual enzymatic activity. A double mutant that was substituted at the ADP-ribosyltransferase active center E112K and at the COOH-terminal KDEL (E112K/KDEV or E112K/KDGL) retained strong biological adjuvant activity without CNS toxicity (Hagiwara et al. 2006). All of these mutants showed considerable adjuvant activity (Table 12.1). Recently, a nontoxic molecule has been constructed which combines the enzymatic activity of CT with a B cell targeting moiety. The gene encoding cholera toxin A1 subunit was fused at its C-terminal end with a dimer of a DNA segment coding for a synthetic Ig-binding domain (designated D) of the staphylococcal protein A produced in *Escherichia coli*. The highly purified, monomeric fusion protein, which has a molecular mass of 37 kDa and is known as CTA1-DD, was found to exhibit strong ADP-ribosyltransferase activity and to bind to all IgG subclasses via the DD moiety (Agren et al. 1997). This fusion protein has been shown to possess adjuvant activity like CT (Table 12.1) (Akhiani et al. 2006).

12.3.4 Mechanism of Adjuvant Activity of CT

The precise mechanism by which CT exerts its adjuvant effect is still unclear. Several propositions exist to explain the potent immunogenicity and adjuvant action of CT and its derivatives. Enhanced uptake of antigens by mucosal surfaces is one mechanism by which CT may enhance the immunogenicity of coupled or coadministered antigens. When rabbit nasal mucosal membrane was exposed to CTB and CT in Ussing chambers, the short-circuit current and conductance of the mucosa increased with increasing concentration of toxin, and these changes could be blocked by the addition of ganglioside GM1 (Lycke et al. 1991). Following oral delivery of keyhole limpet hemocyanin (KLH), dextran and CT, CT strongly increased the gut permeability for dextran concomitantly with a strong enhancing effect on the KLH-specific immune response in the lamina propria (Gizurarson et al. 1991). By contrast, CTB failed to increase gut permeability and KLH-specific immune responses. It was therefore proposed that the adenylate cyclase/cAMP system plays a regulatory role in gut permeability, facilitating the access of luminal antigens to the gut mucosal immune system, and thereby enhancing mucosal immune responses (Gizurarson et al. 1991).

The activation of the cells of the immune system leading to the potentiation of antigen-presenting function and T-cell responses are likely to be a key feature in the adjuvant action. CT enhances alloantigen presentation by cultured intestinal epithelial cells (Bromander et al. 1993). Pretreatment of antigen-presenting cells (APCs) with CT or CT-B can enhance IL-4 production by a T-cell clone, through direct interactions of CTB-associated APCs and non-GM1 receptors on T-cells (Li and Fox 1996). AB_5 toxins can also modulate the production of regulatory cytokines and chemokines by cells of the innate immune system. Studies have correlated the adjuvant activity of CT with a Th2-type polarization, a model that is supported by measurements of cytokine production and antibody isotype expression (Xu-Amano et al. 1993, Gagliardi et al. 2000). CT selectively inhibits IL-2 and interferon IFN-γ production from Th1 cells and increases IL-4 and IL-5 production from Th2 cells.

It has been difficult to evaluate the correlation between CT and GM1 binding, adjuvanticity and immunomodulation due to the ubiquitous presence of the GM1 gangliosides on most of the cell types, including lymphocytes. Cross-linking of gangliosides has been shown to trigger signal transduction pathways that promote the regulation of genes involved in immunomodulation. Studies with CTA-DD fusion proteins (Agren et al. 1997, Lycke and Schon 2001) suggested that potent adjuvant activity could be mediated by a GM1-independent pathway. Therefore, the modulation of immune response by CT may be dependent on distinct effects of enzymatic activity, B-subunit interactions with GM1, and other nonenzymatic activities of the A subunit on biochemical signaling pathways in cells of the innate immune system and perhaps on T cells.

12.4 Role of Cholera Toxin (CT) in the Preparation of Vaccines

In the context of the previous discussion on the protective immunity afforded by anti-CT and the adjuvant activities of CT, the practical use of the cholera toxin or toxin components in the preparation of vaccine strains and their modes of administration are described in the following.

12.4.1 Background Knowledge

The following observations of different investigators (Levine et al. 1988a, Kaper 1989, Levine and Pierce 1992, Levine and Tacket 1994, Tacket et al. 1995b, Kabir 2005) are important as guiding factors in the preparation of vaccines against cholera:

1. A single clinical infection due to wild-type *V. cholerae* O1 confers significant protection against cholera upon subsequent exposure to wild-type *V. cholerae* O1.
2. Although cholera is a toxin-mediated disease, the predominant protective immune mechanism appears to be antibacterial rather than antitoxic.
3. Parenteral whole-cell inactivated vaccines that elicit serum vibriocidal antibody but not antitoxin confer significant protection, albeit only for short periods.
4. Parenteral toxoids that stimulate high levels of serum antitoxin do not confer credible protection, even in the short term.
5. It has become apparent from animal studies that protective cholera immunity does not depend on the type of serum antibodies that are mainly stimulated by parenteral vaccination but rather on the mucosal secretory immunoglobulin A (IgA) antibodies produced locally in the gut, and that these latter antibodies are only inefficiently stimulated by parenteral antigen administration.

6. It is now almost axiomatic that the stimulation of the local gut mucosal immune system is achieved much more effectively by administering the vaccine orally rather than parenterally.
7. Therefore, since the late 1970s, attention has turned to the development of oral vaccines that can stimulate intestinal immunity more efficiently.
8. Another important observation guiding the design of the new cholera vaccines relates to the synergistic cooperation between the antitoxic and antibacterial immune mechanisms in cholera. Two main protective antibodies have been identified; one directed against the *V. cholerae* O1 cell wall lipopolysaccharide (LPS), and the other directed against the cholera toxin B subunits. Either of these two types of antibody can confer protection against disease by inhibiting bacterial colonization and toxin binding, respectively, and when present together in the gut, they can have a strongly synergistic protective effect.
9. Administration of minute quantities of purified cholera enterotoxin (CT) to healthy adult North American volunteers induced severe purging, and a syndrome that resembled cholera gravis. These data emphasize the fundamental importance of eliminating cholera toxin expression from candidate live vaccine strains.
10. Studies showed that the purified B-subunit portion of cholera toxin was entirely devoid of toxicity, but that it selectively contained the protective toxin epitopes against which neutralizing antibodies were directed. The B subunit is particularly well suited to oral immunization, not only because it is stable in the intestinal milieu, but also because it retains the ability to bind to the intestinal epithelium (including the M cells of the Peyer's patches). Both of these properties have been shown in animals to be important for stimulating mucosal immunity, including local immunologic memory.
11. An ideal attenuated oral live vaccine should evoke a strong immune response similar to that of natural cholera infection through sustained presentation of antigens. This would be an economic approach to mass immunization, in contrast to the requirement for multiple doses of the expensive WC-BS vaccine used in field trials. However, there is a note of caution associated with the development of attenuated strains. As the cholera toxin genes are carried on a lysogenic phage, there is the possibility that an attenuated vaccine strain could reacquire the cholera toxin gene because of phage infection and become virulent. However, this problem can be resolved by removing the *att* site for the filamentous phage (CTXφ), or by introducing mutations into the *recA* gene.

12.4.2 The Vaccine Strains

12.4.2.1 Whole-Cell (WC) Vaccine with B Subunit (BS) of Cholera Toxin (WC-BS)

This vaccine usually comprises a mixture of heat- or formalin-killed *V. cholerae* of classical biotype Inaba, classical biotype Ogawa, El Tor biotype Inaba, and El Tor biotype Ogawa supplemented with CTB or recombinant CTB (rCTB) (Silva et al. 2008). Usually 1 mg of the nontoxic B subunit (BS) and 10^{11} killed organisms are present in the vaccine, which is administered in two doses separated by 7–42 days.

The results obtained with the oral B-subunit whole-cell vaccine represent marked improvements over those achieved previously with parenteral cholera vaccines. In contrast to the parenteral vaccines, which cause frequent local side effects and occasional systemic adverse reactions (mainly fever), the oral B-subunit whole-cell vaccine has not been associated with any side effects that would reduce the acceptability of the vaccine in areas where cholera is endemic. Of even greater importance, though, is the fact that the protection achieved by the B-subunit whole-cell vaccine has been superior to that obtained with the parenteral vaccines in terms of both the level and duration of protection.

To facilitate the production and use of the oral B-subunit whole-cell cholera vaccine, the recombinant DNA technology was used to produce the B-subunit and whole-cell (WC-rBS) components in a single step. This could greatly simplify the large-scale production of the B-subunit whole-cell vaccine, reduce costs, and allow local production of this vaccine in developing countries. In field trials, the WC-rBS vaccine has been reported to provide 80–85% protection from cholera caused by the *V. cholerae* serogroup O1 for at least six months.

Overexpression systems in which the gene encoding the B subunit of cholera toxin has been placed under the control of *tacP* or other strong foreign promoters in a wide-host-range multicopy plasmid have recently been constructed. Recombinant nontoxinogenic classical and El Tor *V. cholerae* strains of different serotypes harboring such plasmids have been made to excrete more than 500 mg of B subunit per liter, and may therefore be useful killed oral vaccine strains. Thus, a future B-subunit whole-cell vaccine might be based on two strains, one of classical and the other of El Tor biotype, and representing the Inaba and Ogawa serotypes, that (through genetic manipulation) lack the ability to produce cholera toxin but instead produce large amounts of cholera B subunits.

Recently, a double-blinded randomized phase III trial of the oral killed bivalent cholera vaccine was carried out in an urban slum site in Kolkata by the National Institute of Cholera and Enteric Diseases (NICED), in collaboration with the International Vaccine Institute, Seoul, Korea. Each dose of the vaccine contains the following ingredients:

Cell number	Inactivation	Strain/serotype	Biotype	Others
2.5×10^{10}	Heat	O1/Inaba	Classical	Cairo 48
5.0×10^{10}	Formaldehyde	O1/Inaba	El Tor	Phil 6973
2.5×10^{10}	Heat	O1/Ogawa	Classical	Cairo 50
2.5×10^{10}	Formaldehyde	O1/Ogawa	Classical	Cairo 50
5.0×10^{10}	Formaldehyde	O139	–	4260B

Total number of O1 cells: 1.25×10^{11}; total number of O139 cells: 5.0×10^{10}

E. coli K12 strain was used as placebo: each dose of placebo contains sufficient heat-killed *E. coli* K12 to give the same optical turbidity as that of the cholera vaccine. The vaccination was administered to a population of 110,000 in September 2006, and post-vaccination surveillance is ongoing. The authors believe that the results of this study will pave the way for the use of this vaccine in India and other cholera-endemic areas (G.B. Nair, personal communication).

12.4.2.2 CVD 103-HgR Vaccine Strain

The vaccine CVD103-HgR (Levine et al. 1988a, b Ketley et al. 1993, Levine and Tacket 1994) was constructed from the classical biotype strain Inaba 569B by deleting the *ctxA* and *ctxB* genes from the chromosome and reinserting *ctxB* encoding the nontoxic CTB subunit into the *hlyA* (hemolysin) locus. WHO and some relevant American authorities recommended that a unique marker be introduced into CVD103 so as to allow the vaccine strain to be readily differentiated from wild-type strains. Accordingly, a gene encoding resistance to Hg^{2+} was introduced into the chromosome of CVD103. The wild-type *V. cholerae* O1 strains do not generally contain this Hg^{2+} resistance gene.

This vaccine has consistently been shown to be safe, immunogenic, and protective in healthy volunteers. Adverse diarrheal reactions were observed less often in vaccines than in placebo recipients in controlled trials. The protective efficacy (PE) of this strain in an endemic area could not be conclusively demonstrated due to the low incidence of cholera during a large field trial conducted in Indonesia.

In challenge studies on American volunteers, a single dose of the vaccine containing 5×10^8 CFU (colony-forming units) demonstrated a one-month PE of around 85% against challenge with *V. cholerae* O1 of either serotype and 67% against El Tor Inaba. However, this vaccine dose was poorly immunogenic among the less-privileged people of Peru, Indonesia, and Thailand, with the impaired immunity being attributed to the presence of intestinal parasites among these vaccines.

A single dose of CVD103-HgR was administered to participants in a double-blinded, randomized and placebo-controlled field trial in Indonesia in 1992. Very few cholera cases were observed within six months of vaccination, making assessment of any short-term efficacy of the vaccine difficult. The serum vibriocidal response among the vaccines was modest and observed in about two-thirds of the recipients. The trial was monitored over a period of 4.5 years and produced a PE of around 13% for all age groups, demonstrating the vaccine's failure to confer protection in a cholera-endemic area. However, as protection was observed one week after a single dose in volunteers in the USA, it is currently sold as a tourist vaccine in some countries under the trade names of Mutachol (Canada) and Orochol-E (Switzerland). However, there are certain restrictions on the use of this vaccine. It should not be used by phenylketonurics and should not be administered concurrently with antibiotics.

12.4.2.3 Peru 15 Vaccine Strain

A new oral live cholera vaccine candidate, Peru 15, derived from a *V. cholerae* O1, Inaba, El Tor strain, is being tested in North America and Bangladesh (Kenner et al. 1995, Qadri et al. 2002). The strain is a spontaneous nonmotile mutant containing a deletion of CTXφ and cholera phage *attRS1* attachment sites. In addition, the *ctxB* gene was reinserted within the *recA* locus. Deletion of the *attRS1* site and interruption of *recA* prevents reacquisition of the phage-encoded CT genes and homologous recombination. This strain has been attenuated such that reversion to a virulent phenotype is considered to be extremely unlikely.

The strain has been demonstrated to be clinically safe, immunogenic, and protective in phase I and II clinical trials. Although the vaccine protected naïve North American volunteers against cholera in the USA, studies in developing countries are needed to validate its usefulness in an endemic setting, especially its capacity to protect infants and small children.

12.4.2.4 Live Bivalent (Ogawa and Inaba) Attenuated Strains

CVD111 is an attenuated strain derived from a wild-type Ogawa El Tor strain by removing the virulence cassette containing the toxin genes (*ctx, zot, cep*, and *ace*) and inserting the binding unit of cholera toxin (*ctxB*) and that for mercury resistance into the hemolysin locus (*hlyA*). A mixture of CVD111 and CVD103-HgR representing both the biotypes and serotypes of *V. cholerae* O1 was administered orally to a number of US military personnel who showed elevated levels of serum vibriocidal responses against both Ogawa and Inaba serotypes. Although CVD111 is a good intestinal colonizer, it caused diarrhea in a significant number of volunteers, thus limiting the use of this combination vaccine without further attenuation of CVD111.

12.4.2.5 The Indian Vaccine Strains VA1.3 and VA1.4

Several scientists from India have constructed an oral cholera vaccine strain starting from a nontoxinogenic clinical El Tor Inaba isolate which possessed *toxR* and *tcpA* genes but was devoid of the genes for CT and other virulence determinants (Thungapathra et al. 1999, Kabir 2005). Using recombinant DNA technology, the *ctxB* gene of *V. cholerae* was introduced into the cryptic hemolysin locus of the strain. The resulting strain, termed VA1.3 (where VA denotes "vaccine attempt"), did not induce any diarrhea when administered orally to rabbits, and offered protection against challenges with homologous and heterologous *V. cholerae* O1 strains performed three weeks after immunization. The phase I and II clinical trials were performed with 238 human subjects consisting of 91 placebos and 147 volunteers. The vaccine group received 5×10^9 CFU of the candidate vaccine VA1.3. Fourfold or more increase in the vibriocidal titer was recorded among 48.3% of the volunteers who received the vaccine and in the case of placebo the value remained 5.5%. In this study, none of the volunteers excreted the candidate strain as detected by the stool enrichment culture.

The strain VA1.3 had an ampicillin marker for easy detection. A new strain VA1.4 was constructed by deleting the ampicillin marker from the VA1.3 strain. A bridging study has been planned on a small number of volunteers to assure that VA1.4 (though identical to VA1.3) is as safe and immunogenic as VA1.3 in human volunteers. This will be followed, as planned, by a large scale phase III trial (G.B. Nair, personal communication).

12.4.2.6 The Cuban Vaccine Strain, *V. cholerae* 638

V. cholerae 638 (El Tor, Ogawa) was developed in Cuba by deleting the CTXφ prophage from the chromosome of a toxinogenic clinical isolate (Robert et al. 1996). The hemagglutinin/protease (*hap*) gene was subsequently inactivated by the insertion of *celA* encoding *Clostridium thermocellum* endoglucanase A (Robert et al. 1996, Benitez et al. 1999, Kabir 2005). Strain 638 induced strong local and systemic immune responses against cholera antigens in animal models and humans (Robert et al. 1996, Benitez et al. 1997) without inducing significant side effects (Benitez et al. 1999). Although the strain is a good colonizer of the human bowel, as observed in a placebo-controlled volunteer study in Cuba, it induced adverse effects such as diarrhea (10%), abdominal cramps (30%), and gurgling (33%) among the vaccinees (Benitez et al. 1999, Kabir 2005). HA/protease was found to be a reactogenicity factor. A subsequent volunteer challenge study demonstrated the protective efficacy of strain 638 (Garcia et al. 2005, Silva et al. 2008).

12.4.2.7 Live Oral Whole-Cell Vaccine Strains Against *V. cholerae* O139

A number of live attenuated *V. cholerae* O139 strains have been constructed for use as vaccines, and a brief outline of them is presented below.

Bengal-3 and Bengal-15 Strains

Bengal-3 is a derivative of the wild-type epidemic strain M010 of *V. cholerae* O139 (Coster et al. 1995, Ramamurthy et al. 2003). A spontaneous nonmotile derivative of Bengal-3 was isolated and designated Bengal-15. The Bengal-15 strain carries deletions in four specific virulence determinants (*ctxA, ace, zot*, and *cep*) and in other factors involved in site-specific and homologous recombination (RS1, *attRS1*, and *recA*). A few volunteers in the USA immunized orally with this strain did not develop diarrhea. The one-month PE of this vaccine after challenge with the live homologous epidemic strain MO10 was 83%.

CVD112 Strain

CVD112 (Tacket et al. 1995a, Ramamurthy et al. 2003) was constructed from the wild-type strain *V. cholerae* O139 A11837 by recombinant DNA techniques similar to those used in the construction of the strain CVD111. A few volunteers were immunized orally with two different doses (10^6 and 10^8 CFU) of this strain, with the second dose inducing diarrhea in some of them. A short-term (five-week) PE for the vaccine against live oral challenge with the parent wild-type strain was 84%.

VR1-16 Strain

VR1-16 is a derivative of the wild-type strain A14456, both of which are nonmotile (Coster et al. 1995, Ramamurthy et al. 2003). However, VR1-16 was shown by electron microscopy to produce flagella. This strain was derived by deleting the core region of CTXφ (containing *cep, zot, ace*, and *ctx AB*). It was assumed that this approach provided a high degree of recombinational stability to this vaccine strain. The VR1-16 vaccine strain did not exhibit any reactogenicity in volunteers, but its protective efficacy remains to be determined.

Other Vaccine Strains

Among the other O139 vaccine strains under construction or in trials, mention may be made of the CH 25 strain (Favre et al. 1996, Ramamurthy et al. 2003) and oral bivalent B-subunit O1/O139 cholera vaccine. The strain CH 25 is derived from the strain CH 19, where the genes encoding for both O139 short core-linked O-polysaccharide (SOPS) and CPS were cloned and inserted into the chromosome. CH 19 is a *wbe* deletion mutant of CVD103-HgR deficient in the production of the homologous Inaba O antigen. Rabbits immunized with CH 25 elicited high titers of anti-O139 SOPS and CPS-specific serum antibodies.

In order to produce a cholera vaccine providing protection against cholera caused by either of the two serogroups O1 and O139, an oral bivalent B-subunit O1/O139 cholera vaccine was constructed which showed significant IgA and IgG anti toxin and vibriocidal antibody responses in volunteers. Results of further trials with these vaccines are awaited.

12.5 Where Do We Stand Now?

A very brief overview is given above of the design and use of the vaccine strains that retain the cholera toxoid or the cholera toxin B subunit gene and are devoid of all or some of the toxin genes from their chromosomes. Among the other vaccines used, mention may be made of the killed whole cell vaccine with adjuvants that is used parenterally. A classical bivalent (Ogawa + Inaba) WC vaccine adsorbed onto aluminum phosphate (adjuvant) was subjected to a large-scale, double-blinded trial in India. It was claimed that the vaccine offered 100% protection to children under five years of age for six months, 89% for 12 months, etc. Some adverse reactions were detected and were claimed to be nonsignificant. However, WHO observed that killed parenteral vaccines do not prevent transmission of cholera, and that their efficacies are modest and of short duration, and it recommended that such vaccines should not be used in endemic settings. WHO has instead recognized the potential use of oral vaccines in some endemic and epidemic situations to complement existing control strategies. There are claims and counterclaims about the use and efficacy of different vaccines, including the live oral vaccine candidates, and no unequivocal solution has yet been achieved. The problem is very complex, particularly since the ideal cholera vaccine should take care of the different biotypes, serotypes and serogroups of *V. cholerae*. For a long time, attention focused on the serogroup O1, until the serogroup O139 made its appearance in 1992, with devastating effects. Now, over 200 serogroups of *V. cholerae* have been identified, and no one knows which of them will acquire toxin genes sometime in the future and produce havoc, leaving us totally unprepared. Thus, the production of a cholera vaccine that confers reliable and long-term protection on all age groups without exhibiting any adverse reactions has remained elusive.

Chapter 13
Other Toxins of *Vibrio cholerae*

Abstract The other toxins produced by *Vibrio cholerae* include (i) zonula occludens toxin (Zot), (ii) accessory cholera enterotoxin (Ace), (iii) hemolysin, (iv) repeats in toxin (RTX), (v) Chinese hamster ovary (CHO) cell elongation factor (Cef), (vi) new cholera toxin (NCT), (vii) Shiga-like toxin (SLT), (viii) thermostable direct hemolysin (TDH), (ix) heat-stable enterotoxin of nonagglutinable vibrios (NAG-ST), and (x) the toxin WO-7. This chapter presents brief accounts, to the extent that they are available, of the discovery, structures, modes of action, genetics, etc., of these other toxins produced by *V. cholerae* under different conditions.

13.1 Introduction

The endotoxin (LPS) and the cholera enterotoxin (CT) are the two principal toxins produced by *V. cholerae* cells. These two toxins play important and primary roles in the pathogenesis of the disease cholera and have been investigated thoroughly by numerous workers, and they were discussed extensively in the preceding chapters of this book. With a view to developing an effective and long-lasting cholera vaccine, and particularly the live oral vaccines that will act locally at the intestinal mucosal level, different vaccine strains—from which the *ctx* gene in particular has been removed—have been developed, with the expectation that these strains will not themselves cause any cholera-like diarrhea in vaccinated individuals. Investigations of several vaccine strains led to the discovery, at different times and by different investigators, that *V. cholerae* produces several other enterotoxins besides CT that play significant roles (although not primary) in the causation of milder diarrhea and other symptoms in infected patients. This chapter presents an account of these other enterotoxins produced by *V. cholerae*.

K. Chaudhuri and S.N. Chatterjee, *Cholera Toxins*.

13.2 Zonula Occludens Toxin (Zot)

13.2.1 Zonula Occludens (ZO)

The intestinal epithelium represents the largest interface between the external environment and the internal host milieu, and constitutes the major barrier through which molecules can be either absorbed or secreted. There is now substantial evidence that zonula occludens or tight junctions play a major role in regulating epithelial permeability by influencing the paracellular flow of fluid and solutes (Gumbiner 1987, Anderson et al. 1993). The zonula occludens (ZO) are specialized intercellular junctions in which the two plasma membranes are separated by only 1–2 nm, are found near the apical surface of cells in simple epithelia (Fig. 13.1), and form a sealing gasket around the cell (Salama et al. 2006). Tight junctions consist of a branching network of sealing strands, each strand acting independently of the others. Therefore, the efficiency of the junction at preventing ion passage increases exponentially with the number of strands. Each strand is formed from a row of transmembrane proteins embedded in both plasma membranes, with extracellular domains joining one another directly. Although more proteins are present, the major types are the claudins and the occludins. These associate with different peripheral membrane proteins located on the intracellular side of plasma membrane, which anchor the strands to the actin cytoskeleton. Thus, tight junctions join the cytoskeletons of adjacent cells together. They prevent fluid moving through the intercellular gaps and the lateral diffusion of intrinsic membrane proteins between apical and basolateral domains of the plasma membrane. The intestinal ZO act to: (i) hold cells together; (ii) block the movement of integral membrane proteins between the apical and basolateral surfaces of the cell,

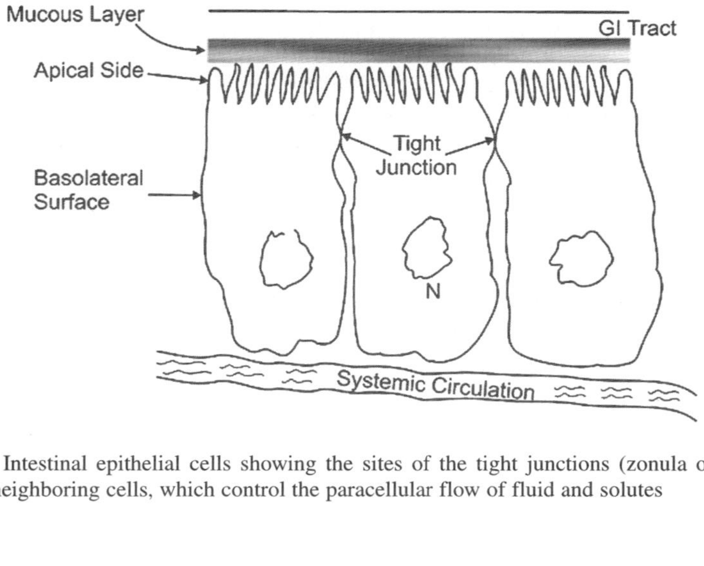

Fig. 13.1 Intestinal epithelial cells showing the sites of the tight junctions (zonula occludens) between neighboring cells, which control the paracellular flow of fluid and solutes

allowing the specialized functions of each surface (e.g., receptor-mediated endocytosis at the apical surface and exocytosis at the basolateral surface) to be preserved; this is done in order to preserve the transcellular transport, and; (iii) prevent the passage of molecules and ions through the space between cells. Therefore, materials must actually enter the cells (by diffusion or active transport) in order to pass through the tissue. This pathway provides control over the substances that are allowed through. The ZO thus restricts or prevents the diffusion of water-soluble molecules through the intercellular space (the paracellular pathway) back into the lumen. The diffusion is driven by the concentration gradients created by the active transepithelial transport processes. As a consequence of altering the paracellular pathway, the intestinal mucosa becomes more permeable, and water and electrolytes (driven by hydrostatic pressure) leak into the lumen, resulting in diarrhea. Moreover, structural features of occluding junctions such as strand number often correlate inversely with the permeability of epithelia, as measured electrophysiologically. There is now abundant evidence that tight junctions are dynamic structures that readily adapt to a variety of developmental, physiological, and pathological circumstances. In recent years, much has been discovered about the structure, function, and regulation of tight junctions (Fasano 1999a, b). However, the precise mechanism(s) through which they operate are still incompletely understood.

13.2.2 Discovery of Zot

Attenuated *V. cholerae* vaccine strains mutated in genes encoding cholera toxin (CT) were still found to cause mild to moderate diarrhea. Despite the absence of CT, these strains were still capable of inducing an unacceptable amount of diarrhea. It was then hypothesized by Fasano et al. (1991) that *V. cholerae* produced a second toxin, which was still present in strains with the *ctxA* sequence deleted. The activity was discovered by testing culture supernatants of *V. cholerae*, both wild-type and Δ*ctx*, in Ussing chambers, a classical technique for measuring the transepithelial transport of electrolytes across intestinal tissue. When culture supernatants of *V. cholerae* CVD101 (Δ*ctx*) were added to rabbit ileal tissue mounted in Ussing chambers, an immediate increase in tissue conductivity (i.e., a decrease in tissue resistance) was observed. Unlike the increase in potential difference observed in response to CT, which reflects ion transport across the membrane (i.e., the transcellular pathway), variations in transepithelial conductance primarily (although not exclusively) reflect the modification of tissue permeability through the intercellular space (i.e., the paracellular pathway). Electron microscopy of the epithelial tight junctions, the major barrier in this paracellular pathway, revealed that exposure of ileal tissue to culture supernatants of CVD101 resulted in the "loosening" of the tight junction so that an electron-dense marker could permeate the paracellular space. In contrast, tissue treated with uninoculated control broth was not permeable to this marker. Freeze-fracture electron microscopy showed that the anastomosing network of strands made up of tight junctions suffered a striking decrease in strand complexity in tissue treated with supernatants. Strands lying perpendicular to the long axis of the ZO appeared to be preferentially lost, resulting in a decreased number of strand intersections. Detailed analysis showed that there was a significant overall decrease in the number of ZO strands, reflecting an increased number of altered ZO, in tissues treated with culture supernatants of CVD101 compared with tissues treated with uninoculated broth or culture supernatant of wild-type *V. cholerae* O395N1. Because the toxin had a striking effect on ZO, it was designated the zonula occludens toxin or Zot (in short). Fasano et al. (1991) speculated that, by increasing intestinal permeability, Zot might cause diarrhea due to the leakage of water and electrolytes into the lumen under the force of hydrostatic pressure.

13.2.3 Properties of Zot

Zot is a single polypeptide chain 44.8 kDa in size and 399 amino acids (AA) in length, with a predicted pI of 8.5. It has a bacteriophagal origin, and is present in toxigenic strains of *V. cholerae* with the ability to reversibly alter intestinal epithelial tight junctions (TJs), allowing the passage of macromolecules through mucosal barriers (Salama et al. 2006). Zot possesses multiple domains that lend it dual functionality: it can act as a morphogenetic phage protein and as an enterotoxin. After cleavage at AA residue 287, a carboxy-terminal fragment of 12 kDa is excreted that is probably responsible for the biological effect of the toxin. The ~33 kDa N-terminal left after cleavage is probably involved in CTXφ assembly and remains associated with the bacterial membrane. Studies in Ussing chambers have shown that the activity of Zot is reversible, heat-labile, sensitive to protease digestion, and found in culture supernatant fractions containing molecules between 10 and 30 kDa in size. To identify the Zot domains directly involved in the protein permeating effect, several Zot gene deletion mutants were constructed and tested for their biological activity in Ussing chamber assays and for their ability to bind to the target receptor on intestinal epithelial cell cultures. In these studies, the biologically active domain of Zot was localized toward the carboxy-terminus of the protein and coincided with the predicted cleavage product generated by *V. cholerae*. It was hypothesized that Zot may mimic the effect of a functionally and immunologically related endogenous modulator of epithelial TJs. The combination of affinity-purified anti-Zot antibodies and the Ussing chamber assay revealed an intestinal Zot analog that was named zonulin. It has been demonstrated that Zot induces modifications of the cytoskeletal organization that lead to the opening of TJ related to the PKCα-dependent polymerization of actin monomers into actin filaments that are strategically localized in order to regulate the paracellular pathway. Further, this modification is known to be reversible and time- and dose-dependent, and is confined to the small intestine, since Zot does not affect the colon permeability. The selective effect of the toxin on the small intestine seems related to the regional distribution of the Zot receptors that are present in the jejunum and ileum, but not in the colon. The aforementioned properties make this moiety a potential tool for modulating the TJ permeability. These data show that Zot regulates tight junctions in a rapid, reversible, and reproducible fashion and probably activates intracellular signals that are operative during the physiological modulation of the paracellular pathway. Figure 13.2 shows a schematic view of Zot trafficking and processing in *V. cholerae* cell.

Zot showed a selective effect among the various cell lines tested, suggesting that it may interact with a specific receptor whose surface expression varies depending on the cell. When tested in rat intestinal epithelial cell IEC6 monolayers, Zot-containing supernatants induced a redistribution of the F-actin cytoskeleton. Similar results were obtained with rabbit ileal mucosa, where the reorganization of F-actin paralleled the increase in tissue permeability. Pretreatment with a protein kinase C inhibitor CGP41251 completely abolished the effects of Zot on both tissue permeability and actin polymerization. In IEC6 cells, Zot induced a peak increase in the PKC-α isoform after 3 min of incubation. Taken together, these results suggest that Zot activates a complex intracellular cascade of events that regulate tight junction

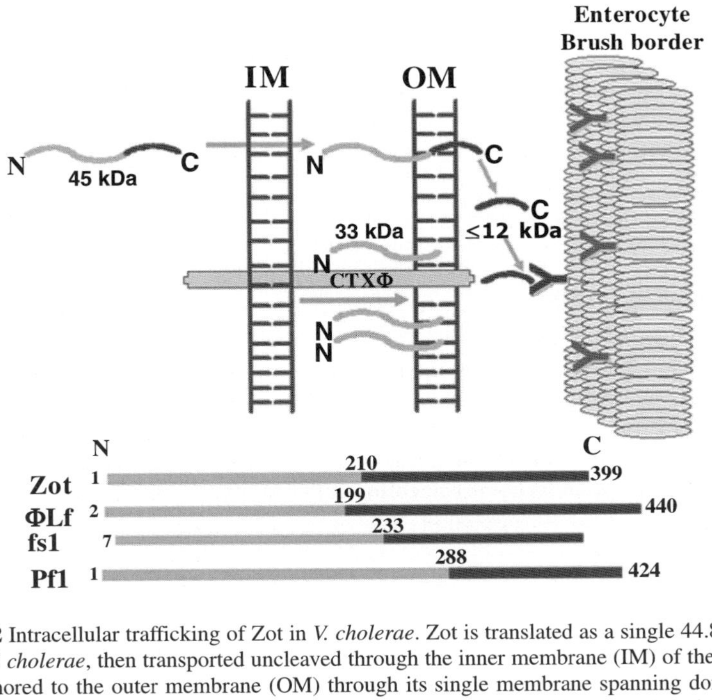

Fig. 13.2 Intracellular trafficking of Zot in *V. cholerae*. Zot is translated as a single 44.8 kDa protein in *V. cholerae*, then transported uncleaved through the inner membrane (IM) of the organism and anchored to the outer membrane (OM) through its single membrane spanning domain. The Zot N-terminal part remains intracellularly located, while its 12 kDa C-terminal domain is outside and is cleaved by a vibrio-specific protease. Once cleaved, the C-terminus binds to its enterocyte target receptor, triggering the intracellular signaling that leads to the opening of the tight junctions of intercellular enterocytes. The *lower panel* shows a comparison of Zot with other p-proteins of several filamentous phages. (Figure obtained through the kind courtesy of A. Fasano, University of Maryland School of Medicine, Baltimore, USA)

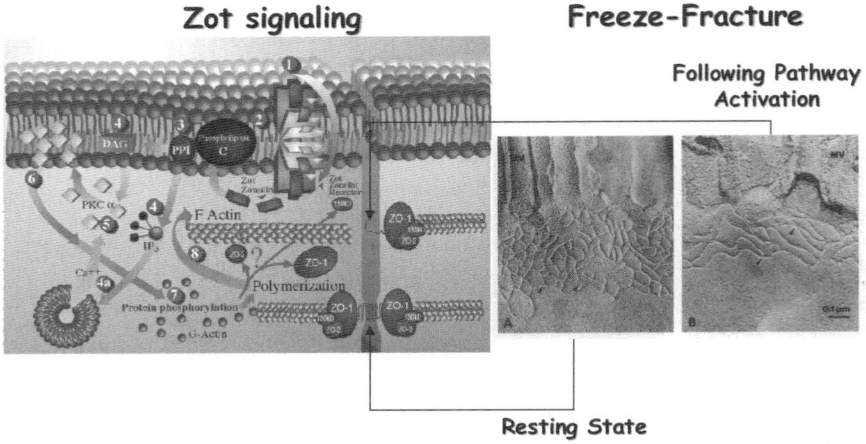

Fig. 13.3 Proposed zonulin intracellular signaling leading to the opening of intestinal TJ. Zonulin interacts with a specific surface receptor (1) whose distribution within the intestine varies. The protein then activates phospholipase C (2) that hydrolyzes phosphatidyl inositol (3) to release inositol 1,4,5-triphosphate (PPI-3) and diacylglycerol (DAG) (4). PKCα is then activated (5) either directly (via DAG) or through the release of intracellular Ca^{2+} (via PPI-3) (4a). Membrane-associated, activated PKCα (6) catalyzes the phosphorylation of target protein(s), with consequent polymerization of soluble G-actin in F-actin (7). This polymerization causes the rearrangement of the filaments of actin and the subsequent displacement of proteins (including ZO-1) from the junctional complex (8). As a result, intestinal tight junction (TJ) becomes looser (see the freeze-fracture electron microscopic pictures: *A*, control; *B*, Zot-treated). Once the zonulin signaling is over, the TJ resume their baseline steadiness. (Figure obtained through the kind courtesy of A. Fasano, University of Maryland School of Medicine, Baltimore, USA)

permeability, probably mimicking the effect of physiologic modulator(s) of epithelial barrier function. Figure 13.3 presents a summarized account of the Zot signaling leading to tight junction modulation.

The observations that, in rabbit small intestine (Fasano et al. 1997), Zot (a) does not affect Na^+-glucose coupled active transport; (b) is not cytotoxic; (c) fails to completely abolish the transepithelial resistance; and, most importantly; (d) induces a reversible increase in tissue permeability, suggest that this moiety a potential tool for studying intestinal TJ regulation.

13.2.4 Genetics

The gene encoding Zot was cloned and found to be located immediately upstream of the *ctx* locus (Baudry et al. 1992). The genes encoding Ace, Zot, and CT are located in a genetic element that has been recently shown to be the chromosomally integrated genome of a temperate filamentous bacteriophage (CTXφ) (Waldor and Mekalanos 1996). The CTX element also contains the genes encoding the core-encoded pilus (*cep*), and an open reading frame (ORF) (*orfU*) whose function remains unknown (Pearson et al. 1993). Waldor and Mekalanos have described that CTXφ gives rise to replicative forms (RF) and produces phage particles without affecting the cell's ability to grow and divide. These authors also demonstrated that mutagenesis of *zot* and *orfU* impairs the ability of the CTX element to be self-transmissible under appropriate conditions. Therefore, it has been suggested that Zot and the product of *orfU* are involved in CTXφ morphogenesis (Waldor and Mekalanos 1996). The *zot* gene sequences are present in both *V. cholerae* O1 and non-O1 strains, and strains that contain *ctx* sequences almost always contain *zot* sequences and vice versa. The *zot* gene consists of a 1.3 kb open reading frame which could potentially encode a 44.8 kDa polypeptide (Baudry et al. 1992). The predicted amino acid sequence of the Zot protein shows no homology to any other bacterial toxin, including the toxin A of *Clostridium difficile*, a large 300 kDa protein which also alters tight junctions. Some homology to ATPase protein sequences has been reported, but the significance of this observation is unknown. Zot sequence analysis previously revealed that the toxin is related to the product of the class 1 filamentous bacteriophage gene 1, a putative ATPase noncapsid protein that is required for bacteriophage assembly at the cell envelope (Koonin 1992). Phage gene 1 protein appears to be involved in the formation of adhesion zones between the inner and the outer membranes where morphogenesis of the filamentous bacteriophage f1 occurs. Sequence analysis revealed that Zot, similar to the gene 1 product, may possess a transmembrane topology (Koonin 1992).

13.2.5 The Zot Receptors

Zot binding to the surface of rabbit intestinal epithelium has been shown to vary along the different regions of the intestine. Zot increases tissue permeability in the ileum but not in the colon, where the presence of colonic microflora and/or their bioproducts could be harmful if the mucosal barrier is compromised. This binding distribution coincides with the regional effect of Zot on the intestinal permeability and with the preferential F-actin redistribution induced by Zot in the mature cells of the villi.

The identification of Zot receptor was initiated by binding studies using maltose-binding protein (Mbp)-Zot in different epithelial cell lines, including IEC6 (rat, intestine), Caco-2 (human, intestine), T84 (human, intestine), MDCK (canine, kidney), and BPA endothelial cells (bovine, pulmonary artery) (Salama et al. 2006). The Zot receptor was purified by ligand-affinity chromatography. Only IEC6 cells (derived from rat crypt cells) and Caco-2 cells (that resemble mature absorptive enteric cells of the villi), not T84 (crypt-like cells) or MDCK, express receptors on their surface. Zot action is tissue specific; it is active on epithelial cells of the small intestine (jejunum and distal ileum) and on endothelial cells, but not on colonocytes or renal epithelial cells. Moreover, Zot is active only on the mucosal side. Zot showed differential sensitivity among the enterocytes, with the mature cells of the tip of the villi being more sensitive than the less mature crypt cells. The number of Zot receptors may decrease along the villus crypt axis, and this explains why the effect of Zot on IEC6 cells requires a longer exposure time than for whole tissue. Fluorescence staining of Zot binding to the intestinal epithelium is maximal on the surface of the mature absorptive enterocytes at the tip of the villi, and completely disappears along the surface of the immature crypt cells, suggesting that the expression of Zot/zonulin receptor is upregulated during enterocyte differentiation. It was proposed that the 66 kDa Zot/zonulin putative receptor identified in intestinal cell lines is an intracellular modulator of the actin cytoskeleton and the tight junctional complex (Uzzau et al. 2001), whose activation is triggered by ligand engagement and subsequent internalization of the ligand/receptor complex. Earlier studies discussed the partial characterization of the zonulin receptor in the brain and revealed that it is a 45 kDa glycoprotein containing multiple sialic acid residues that has structural similarities to myeloid-related protein, a member of the calcium-binding protein family possibly linked to cytoskeletal rearrangements. The functional significance of the glycosylation of the zonulin/Zot receptor remains unknown. However, the sialic acid residues may contribute to the stability (protection from proteolysis) or assist in translocation to the cell surface during synthesis. Affinity column purification revealed another ~55 kDa Zot-binding protein in the human brain plasma membrane preparation (Wang et al. 2000), which was identified as tubulin. Pretreatment with a specific protein kinase C inhibitor CGP41251 completely abolished the effects of Zot on both tissue permeability and actin polymerization. Zot decreases the soluble G-actin pool, thus exerting an effect on actin filaments (polymerization). Thus, Zot triggers a cascade of intracellular events that lead to a protein kinase C (PKC) α-dependent polymerization of actin microfilaments strategically localized in order to regulate the paracellular pathway (Fasano et al. 1995), and consequently leading to the opening of the TJs at toxin concentrations as low as 1.1×10^{-13} M. So far, the receptor has been identified in the small intestine, brain, nasal epithelium, and possibly the heart (since zonulin was identified there).

13.2.6 Applications of Zot

The low bioavailability of drugs remains an active area of research (Salama et al. 2006). Novel approaches that aim at improving drug delivery have been introduced lately. Among these, the modulation of TJs to improve paracellular drug transport appears to be a very appealing and attractive solution (Salama et al. 2006). The transient opening of TJs would be beneficial to the therapeutic effect (Fasano and Uzzau 1997), because it avoids entry of metabolic waste as well as the leakage of important proteins and nutrients. Zot was identified as a potential modulator of the TJs that initiate a cascade of intracellular events upon binding to its receptor that lead to the opening of the TJs. Extensive in vivo and in vitro studies (Salama et al. 2006) have identified Zot receptors in the small intestine, the nasal epithelium, the heart, and the brain endothelium. The discovery of Zot triggered studies aimed at identifying possible eukaryotic analogs. Comparison of the amino termini of the secreted Zot fragment (AA 288–399) and its eukaryotic analog, zonulin, which governs the permeability of intercellular TJs, revealed an octapeptide amino acid shared motif corresponding to the 291–298 AA region (Di Pierro et al. 2001). The physiological role of the zonulin system remains to be established, but it is likely that this system is involved in several functions, including TJ regulation during developmental, physiological and pathological processes, such as tissue morphogenesis, protection against microorganism's colonization, as well as the movement of fluid, macromolecules and leucocytes between the blood stream and the intestinal lumen and vice versa (Fasano 2001). In vitro and in vivo studies across the blood brain barrier have shown that Zot increases the transport and tissue accumulation of therapeutic agents when administered concurrently (Salama et al. 2006). In addition, Zot and its biologically active fragment ΔG displayed significant potential to enhance the oral bioavailability of anticancer, immunosuppressant and antiretroviral drugs in Caco-2 cells (Cox et al. 2001, 2002) and Sprague Dawley rats. Moreover, toxicity studies have shown that Zot and ΔG do not compromise cell viability or cause membrane toxicity, unlike other absorption enhancers. Collectively, this report indicates that modulation of the TJs provides a promising route for enhanced drug delivery approaches, as evidenced by the novel modulator, Zot (Salama et al. 2006). Further, Zot was shown to act as adjuvant through different mucosal routes and to induce protective immune responses (Marinaro et al. 1999, 2003).

13.3 Accessory Cholera Enterotoxin (Ace)

Along with CT and Zot, accessory cholera enterotoxin (Ace) has been identified by Trucksis et al. (1993) as being the third potential enterotoxin of *V. cholerae*. Crude toxin extracts in an animal model indicated that Ace increases transcellular ion transport, which is proposed to contribute to diarrhea in cholera. Ace increases short-circuit current (I_{sc}) in rabbit ileal tissue mounted in Ussing chambers and in an intestinal epithelial cell T84 model, and it causes fluid secretion in ligated rabbit ileal loops. Like CT, and in contrast to Zot, Ace increases potential difference (PD) rather than tissue conductivity.

13.3.1 Discovery of Ace

Derivative CVD110 of a virulent *V. cholerae* (El Tor Ogawa strain E7946) containing a deletion of a 4.5 kb "core region" with *ctx* and *zot* was nonreactive in the Ussing chamber when it was tested in order to measure net ion transport across isolated intestinal mucosa. CVD110 also contained a mutation of the *hlyA* gene.

The deletion of the core region was mediated by recombination between the flanking RS1 elements with site-specific transposase activity. However, the genes *zot* and *ctx* comprise only 55% of the core region and the function of the remaining part of the core was unknown. When Trucksis et al. (1993) introduced a 2.9 kb *EcoR*V fragment containing a sequence upstream of *zot* but lacking intact *ctx* and *zot* genes into CVD110 on a plasmid, giving CVD110 (pCVD630), a significant increase in I_{sc} was observed in the Ussing chamber. The increase in I_{sc} was secondary to an increase in PD, with tissue resistance remaining stable. The strain showed 15% greater fluid accumulation compared to CVD110 in an in vivo sealed infant mouse model. These observations suggested a potential pathogenic role for this 2.9 kb region.

The DNA sequence analysis of the 2.9 kb region revealed the presence of two open reading frames; the functions of these were unknown at that time. The introduction (through a plasmid) of a fragment containing a deletion of 309 bp from one of these ORFs, *orfU* (signifying an unknown ORF), gave CVD110(pCVD630A), which retained full Ussing chamber activity and caused significant fluid accumulation when tested in an in vivo ligated rabbit ileal loop model. The other ORF presumed to be responsible for Ussing chamber activity was denoted Ace or accessory cholera enterotoxin. Cultures of CVD110 containing the cloned *ace* gene showed an increase in I_{sc} of a size that was equal to that produced by the positive control CVD110 (pCVD630), and significant fluid accumulation compared to CVD110, supporting the role of the *ace* gene in secretory action (Trucksis et al. 1993).

13.3.2 The ace *Gene and the Encoded Protein*

The *ace* open reading frame is located upstream of *zot* and *ctx*, the locus ID being VC1459 in the complete genome of El Tor N16961 (Heidelberg et al. 2000); two copies are present in the *V. cholerae* O395 genome, with the loci being denoted VC0395_A0510 and VCO395_A1062 in chromosomes I and II, respectively (NCBI database). It comprises a 289 bp gene coding for 96 amino acid protein with a predicted molecular mass of about 11.3 kDa. The termination codon of *ace* overlaps with the initiation codon of *zot*, an arrangement involving translational coupling between *ace* and *zot*, like the one found between the *ctxA* and *ctxB* genes (Betley et al. 1986).

The analysis of the predicted protein sequence revealed the presence of an alpha-helical C-terminal region of high hydrophobic moment. The amphipathic nature of this 20-residue region in which nearly all of the hydrophobic residues are on one side of the helix suggested that multimers of Ace possibly insert into the eukaryotic membrane, with the hydrophobic surfaces facing the lipid bilayer and hydrophilic sides facing the interior of the transmembrane pore (Trucksis et al. 1993).

The predicted amino acid sequence of Ace shows similarity to a family of eukaryotic ion-transporting ATPases, including the human plasma membrane calcium pump (CaPM) (Verma et al. 1988), the calcium-transporting ATPase from rat

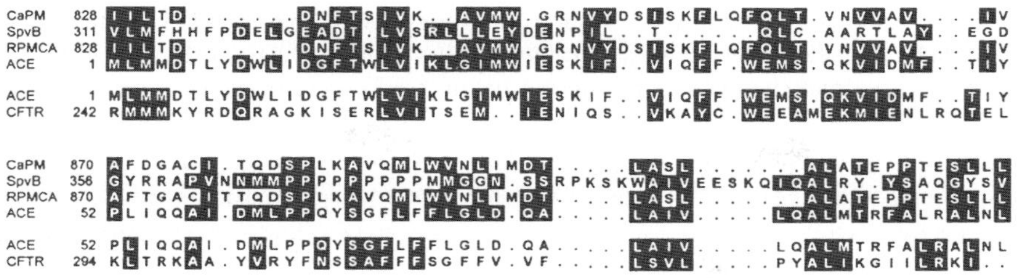

Fig. 13.4 Amino acid sequence comparison of the predicted *ace* gene product and the predicted sequences of two ion-transporting ATPases, the human plasma membrane Ca^{+2} pump (CaPM) (Verma et al. 1988) and Ca^{+2}-transporting ATPase from the rat brain (RPMCA) (Schull and Greeb 1988); a virulence protein of Salmonella Dublin, SpvB (Krause et al. 1991); and the cystic fibrosis transmembrane conductance regulator (CFTR) (Riordan et al. 1989). *Boxed residues* indicate an identical or similar amino acid residue between Ace and other sequences. Similar residues are those belonging to the groups P/A/G/S/T, H/K/R, Q/N/E/D, I/L/V/M and F/W/Y (single letter codes). Sequence alignment was performed with the BESTFIT program. Ace was aligned separately with CFTR to maximize sequence similarity. (Reproduced from Trucksis et al. 1993 with permission from the Proceedings of the National Academy of Sciences USA, and J.B. Kaper)

brain (RPMCA) (Shull and Greeb 1988), and the cystic fibrosis transmembrane conductance regulator (CFTR) (Riordan et al. 1989), all of which are involved in ion transport across membranes. In addition, Ace shows sequence similarity with SpvB, a virulence protein of Salmonella Dublin (Krause et al. 1991). The C-terminal region of Ace further shows homology with delta-toxin of *Staphylococcus aureus* (Fig. 13.4).

Moreover, the gene coding for Ace is part of the filamentous phage CTXφ (Waldor and Mekalanos 1996). It exhibits homology with gene VI of the M13 filamentous phage, and ORFs of *V. parahaemolyticus* phages f237 (ORF 6, Accession No. ZP_02194770), VfO3K6 (Vp114, Accession No. NP_059536), and VfO4K68 (Vp114, Accession No. NP_059546). Gene VI encodes a small hydrophobic protein of unknown function in the M13 phage (van Wezenbeek et al. 1980). Two of the open reading frames, *zot* and *orfU*, located downstream of *ace*, are required for the production of CTXφ phage particles. It is expected that Ace may play some role in phage function, although this has not been demonstrated experimentally. Thus, Ace appears to be bifunctional, having both a phage function and toxin activity.

13.3.3 Purification of Ace

The Ace toxin expression in *V. cholerae* is quite low, which hampered the purification and further characterization of this toxin. Attempts to express Ace in a prokaryotic expression system did not work due to the toxicity of Ace towards the host *Escherichia coli*. Later on, Ace toxin was successfully expressed in the methylotrophic yeast *Pichia pastoris* (Trucksis et al. 1997). *Pichia ace* integrant clones containing the *ace* gene integrated at the alcohol oxidase I (AOX1) gene were tested for the expression of Ace protein in comparison with the Ace^- control strain. Analysis of the crude supernatant by SDS-PAGE revealed two new protein bands in the Ace^+ clone following double staining with Coomassie brilliant blue and silver stain. The sizes of the protein bands were consistent with the presumed monomer and dimer forms of Ace, with molecular masses of 9 and 18 kDa, respectively. The 18 kDa form was the predominant one. Furthermore, when analyzed by SDS-PAGE and silver staining, concentrated *V. cholerae* culture supernatants of Ace^+ [CVD110(pCVD630)] and Ace^- [CVD110(pCVD315)] clones revealed the presence of monomer and dimer forms of Ace in the Ace^+ clone (Trucksis et al. 1997). The Ace protein showed the unusual property of staining with silver but not with Coomassie blue stain—a property usually observed for sialoglycoproteins or lipoproteins. The Ace protein did not stain with PAS, suggesting the absence of glycosylation. It is unlikely to be a lipoprotein, as it lacks the consensus motif (Leu-Leu-Ala-Gly-Cys) of a bacterial lipoprotein and the Cys residue, which is universally conserved at the +1 residue of mature lipoprotein (Allaoui et al. 1992). Ace protein was purified from concentrated yeast supernatant from the Ace^+ clone as well as from concentrated and size-fractionated culture supernatant of the *V. cholerae* (Ace^+) clone. The quantity of Ace from *V. cholerae* was estimated to be about 0.6 µg/l, 10,000-fold less than that produced in the yeast expression system. The yeast recombinant Ace protein showed characteristic biological activity in the Ussing chamber (Trucksis et al. 1997).

13.3.4 Mode of Action of Ace

The mechanism of action of Ace has been delineated by Trucksis et al. (2000) in T84 cells, a human colonic tumor cell line that resembles crypt cells morphologically and maintains vectorial electrolyte transport. This cell line secretes Cl^- in response to secretagogues, the effects of which are mediated via cAMP, cGMP, or Ca^{+2}-related mechanisms. T84 monolayers mounted in modified Ussing chambers served as an in vitro model of a Cl^- secretory cell. Partially purified Ace^+ and Ace^- supernatants obtained from *V. cholerae* CVD113 (pCVD630), CT^-, Zot^-, Ace^+, and CVD113 (CT^-, Zot^-, Ace^-), respectively, were used. The supernatant of the Ace^+ strain, when added to the apical surface, stimulated a rapid increase in I_{sc} and PD, which was concentration dependent and reversed as soon as the toxin was removed. Basolateral addition alone had no effect. When purified Ace protein containing both monomeric and dimeric forms of Ace (with the dimeric form being predominant) were tested on T84 cells, the monomeric form of Ace produced a concentration-dependent increase in I_{sc} compared with the negative control, while the dimer demonstrated less activity. Ion replacement studies established that the current was dependent on Cl_2 and HCO_3^-, suggesting stimulation of anion secretion in T84 cells. The Ace-stimulated secretion was dependent on an influx of extracellular Ca^{2+} across the apical membrane as well as intracellular Ca^{2+} stores, but was not associated with an increase in intracellular cyclic nucleotides. Thus, the mechanism of secretion by Ace involves Ca^{2+} as a second messenger, and this toxin stimulates a novel Ca^{2+}-dependent synergy.

Ace, like CT, is secreted by *V. cholerae* independent of phage production, and exhibits physiological activity (Trucksis et al. 2000). Its role in the pathogenesis of cholera has not been determined with the use of isogenic mutants in human volunteers. It has been suggested that Ace may contribute to an early phase of intestinal secretion in *V. cholerae* infection before the onset of the secretion stimulated by the more slowly acting cholera toxin (Trucksis et al. 2000).

13.4 Hemolysin

Hemolysin is an exotoxin that attacks blood cell membranes and causes cell rupture with the liberation of hemoglobin. It is the most widely distributed toxin among the pathogenic vibrios. The production and secretion of hemolysin into the culture medium was traditionally used as an important criterion to distinguish the El Tor biotype (Hly+) from the classical biotype (Hly−) of *V. cholerae* O1 (Politzer 1959), although in the mid-1960s and 1970s, many of the El Tor isolates were found to be nonhemolytic (Gallut 1971). This toxin, often called *V. cholerae* cytolysin (VCC), is produced by most O1 biotype El Tor and non-O1 non-O139 isolates, including many non-CT strains (Honda and Finkelstein 1979, Goldberg and Murphy 1984, Yamamoto et al. 1984). It has been proposed that VCC contributes to the pathogenesis of gastroenteritis caused by *V. cholerae*, especially with non-CT producing strains (Pichel et al. 2003), although the significance of VCC in the disease in humans has previously been questioned (Levine et al. 1988a). This exotoxin is a pore-forming toxin (PFT) that lyses rabbit erythrocytes (Zitzer et al. 1995) and human intestinal cells in culture (Honda and Finkelstein 1979, Zitzer et al. 1997a). It increases vascular permeability of rabbit skin and is lethal to mice (Yamamoto et al. 1984). Cell-vacuolating activity in Vero cells was linked to VCC (Figueroa-Arredondo et al. 2001), and an apoptosis-inducing property of VCC has recently been demonstrated (Saka et al. 2008).

13.4.1 The Gene and the Encoded Protein

The structural gene for hemolysin *hlyA* from *V. cholerae* El Tor has been cloned and sequenced independently by two different laboratories (Goldberg and Murphy 1984, Manning et al. 1984). The nucleotide sequence revealed an open reading frame (ORF) of 2,223 bp, which is predicted to encode a precursor protein with a deduced molecular mass of 81,977 Da. The mature form is predicted to be a 691 amino acid polypeptide with a molecular mass of 79,350 Da because of its cleavage of 25 amino acids at the N-terminal. In classical 569B, an 11 bp deletion has been observed in *hlyA*, which interrupts the ORF, introducing a termination codon at 733 bp (Rader and Murphy 1988). Thus, *hlyA* of 569B is predicted to encode, in precursor form, a 244 amino acid polypeptide with a molecular mass of about 27 kDa, resulting in the production of a nonfunctional truncated protein. The putative promoter and signal sequences for the El Tor and 569B were found to be identical.

VCC is transcribed from the *hlyA* gene as a 82 kDa prepro-HlyA, which is processed into a 79 kDa pro-HlyA following cleavage of a 25 amino acid signal peptide and is then secreted into the culture supernatant through the outer membrane (Yamamoto et al. 1990). The 79 kDa pro-HlyA is then further processed to 65 kDa mature HlyA following cleavage of the 15 kDa domain at the N-terminus. The cleavage occurs through proteolysis at one bond in the vicinity of Leu-146 to Asn-158 by exogenous or endogenous proteases (Nagamune et al. 1996). The 15 kDa pro-region acts as a chaperone for VCC, shares sequence identity with the Hsp90 family of heat shock proteins, and are required for the functional expression of the above toxin.

Two other gene products, HlyB and HlyC, are also essential for the efficient production of El Tor hemolysin (Alm and Manning 1990). The genes *hlyA, hlyB*, and *hlyC* form an operon. The gene *hlyB* encodes a product that is highly homologous to methyl-accepting chemotactic proteins (Mcps) involved in motility and chemotaxis, and *hlyC* encodes a product homologous to lipases of other bacteria. The downstream of *hlyC* (later denoted *lipA*) contains an ORF (termed *lipB*) that codes for lipase-specific foldase or accessory to lipase. A regulatory locus *hlyU* has also been identified. Downstream of *lipB* lies *prtV*, which is transcribed in the opposite orientation and codes for a metalloprotease (Ogierman et al. 1997). A lecithinase-phospholipase gene is located upstream of *hlyA* and is transcribed divergent to *hlyA*.

13.4.2 Protein Purification and the Crystal Structure

The purified protein elicits fluid accumulation in rabbit ileal loops. The crystal structure of the pro-toxin has been solved at 2.3 A by X-ray diffraction (Olson and Gouaux 2005). The VCC water-soluble monomer is a cross-shaped molecule with four distinct structural domains. At the amino terminus is the chaperone-like pro-domain with sequence homology to the Hsp90 family of heat shock proteins. The pro-domain is connected to an ~325 amino acid residue cytolysin domain, which exhibits sequence homology and shares a conserved residue cluster with alpha-hemolysin and LukF, a *Staphylococcus* pore-forming toxin. The structural study revealed that the VCC cytolysin domain shares a common fold with LukF and alpha-hemolysin. Following the cytolysin core are two domains with lectin-like folds. The first one encompasses a region of ~116 amino acid residues exhibiting sequence homology with as well as structural similarity to the carbohydrate-binding B domain of the plant toxin ricin. Ricin-like lectin domains are found in many

AB-type plant toxins and participate in the binding of a variety of sugar molecules to the surfaces of cells. The fourth domain (about 15 kDa), which does not exhibit significant sequence identity with any known protein, has a beta-prism topology and is structurally similar to the plant lectin jacalin. There is a possible receptor-binding site in this lectin domain, and removal of this domain leads to a tenfold decrease in lytic activity towards rabbit erythrocytes. The three smaller domains only form contacts with the cytolysin domain and do not interact with each other.

13.4.3 Biological Activity

VCC belongs to a family of pore-forming toxins and permeabilizes a wide spectrum of eukaryotic cells, including human and rabbit erythrocytes, by forming transmembrane oligomeric diffusion channels (Yamamoto et al. 1984, Zitzer et al. 1999). Biochemical studies indicate that VCC is most active against membranes rich in cholesterol and sphingolipids, abundant in the mucosal epithelium of the small intestine (Zitzer et al. 1999, Ikigai et al. 1996). Mature VCC forms heptameric oligomers, which insert into lipid bilayers to generate noncylindrical anion-selective membrane pores about 1–2 nm in internal diameter (Olson and Gouaux 2005). The process of pore formation by VCC may involve three discrete biochemical events, as illustrated in Fig. 13.5. The toxin monomer binds to the membrane surface, self-assembles into a pre-pore oligomer, which finally leads to transmembrane pore formation. VCC presumably interacts only with lipids and not with

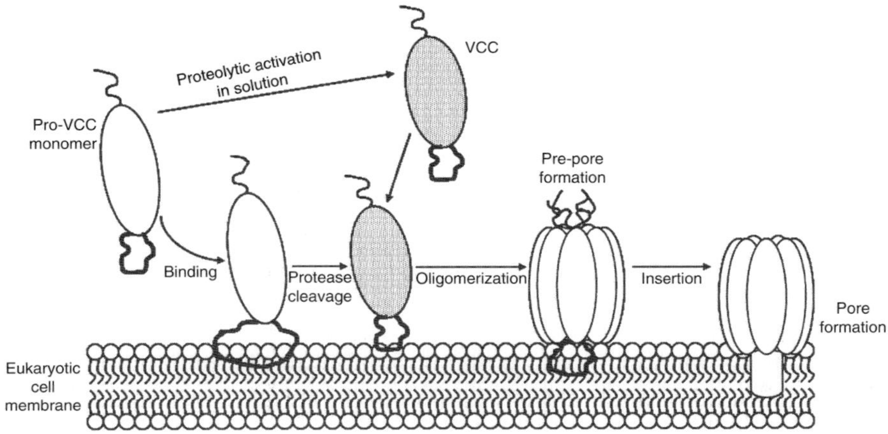

Fig. 13.5 Schematic representation of events leading to pore formation by *V. cholerae* hemolysin. Proteolytic cleavage of pro-VCC is apparently required for oligomerization to take place. Monomers diffuse laterally and assemble into a transient heptameric structure. See text for further details

membrane proteins (Olson and Gouaux 2005). The toxin binds to liposomes with different compositions. In phosphatidylcholine liposomes, the binding occurs non-specifically. VCC appears to bind reversibly to those membranes: the toxin's hemolytic potency was preserved in the presence of pure phosphatidylcholine liposomes (Zitzer et al. 1997b, 1999).

In erythrocytes, removal of the sialoglycoprotein glycophorin B results in a tenfold decrease in sensitivity to VCC and is therefore likely to be a receptor. Also, the monoclonal antibody raised by immunizing with the membrane of human erythrocyte and inhibiting VCC-induced lysis recognized glycophorin B (Zhang et al. 1999), thus suggesting its involvement in VCC-induced hemolysis. It has been observed that fewer than 10,000 toxin molecules are lethal to intestinal cells in culture (Zitzer et al. 1997a). Electron microscopic studies have shown similar membrane orientations, sizes (Zitzer et al. 1997b), and heptameric stoichiometries (Harris et al. 2002) of the alpha-hemolysin and VCC pores, supporting a shared mode of lysis among these two proteins. A cytocidal action of VCC on human intestinal cells was attributed to VCC oligomerization and pore formation on cell membranes.

In addition to membrane permeabilization, extensive cell vacuolation of HeLa and Vero cells induced by *V. cholerae* filtrates was linked to VCC. It seems likely that the vacuolating activity may be provoked by VCC-mediated pore formation. Studies with DIDS (4,4′-disothiocyanatostilbene-2,2′-disulfonic acid) and STS (4-acetamido-4′-isothiocyanatostilbene-2,2′-disulfonic acid), inhibitors of VCC anion channels, revealed that the formation of the anion channel is necessary for the development of the vacuoles and for cell death to be induced by this toxin (Moschioni et al. 2002). Using markers of cell organelles, it has been shown that vacuoles derive from different intracellular compartments and late endosomes, and the trans-Golgi network contributes to vacuole biogenesis (Moschioni et al. 2002).

Recently, a study by Saka et al. (2008) reported that VCC could cause apoptotic cell death in the intestinal epithelium. In this work, the authors studied a CT-negative, TCP-negative non-O1 non-O139 *V. cholerae* strain causing a cholera-like syndrome. The wild-type strain but not its isogenic VCC-null mutant induced extensive fluid accumulation in a rabbit ileal loop, providing evidence of the involvement of VCC in virulence. Experiments performed in rabbit ileal loops using TUNEL in situ indicated that the wild-type strain but not its isogenic VCC-null mutant induced increased apoptosis in the intestinal epithelium. In addition, internucleosomal DNA fragmentation and caspase activation, hallmarks of apoptosis, were also observed when Caco-2 cells were exposed to wild-type supernatant, while no DNA cleavage or caspase activation was observed in *hlyA*-null mutant. In an independent study, Zitzer et al. (1997a) did not detect DNA fragmentation of the human intestinal epithelial cell line Int407 after treatment with VCC. It is important to mention that previous studies on the effects of purified VCC in vivo reported that this toxin had the ability to induce fluid accumulation in rabbit ileal loops, increased vascular permeability of rabbit and mice skin, and killed mice rapidly. However, a detailed description of the intestinal pathology was not provided in those reports, and the investigation of apoptosis induction was not addressed. Furthermore, only purified VCC was tested, with the consequent limitation that an arbitrary concentration of toxin was used, which may not reflect the real situation produced during infection.

13.5 Repeats in Toxin (RTX)

RTX toxins are a family of secreted bacterial cytotoxins produced by a diverse group of Gram-negative pathogens. The nomenclature is derived from the presence of a C-terminal calcium-binding domain of acidic glycine-rich nonapeptide repeats in the members of the RTX toxin family. These toxins also share some common features: a common gene organization, post-translational maturation, and export out of the bacterial cell through a type I secretion system (T1SS)(Welch 2001). These RTX toxins generally fall into two categories: the hemolysins, which affect a variety of cell types, and the leukotoxins, which are cell type and species specific. Examples include *E. coli* alpha-hemolysin (HlyA), *Bordetella pertussis* adenylate cyclase hemolysin toxin, *Actinobacillus actinomycetemcomitans* leukotoxin (LtxA), and *Pasteurella haemolytica* leukotoxin (LktA) (Lally et al. 1999).

13.5.1 Discovery of RTX

In 1999, RTX toxins were discovered in *V. cholerae* while searching for undiscovered potential virulence determinants in the *V. cholerae* genome based on sequence similarities to virulence genes of other pathogenic microorganisms (Lin et al. 1999). The RTX gene cluster was uncovered using genome sequence analysis followed by representational difference analysis. The investigators analyzed the genome sequence data of *V. cholerae* El Tor N16961 (publicly available at TIGR, The Institute of Genome Research, at http://www.tigr.org) for contigs located in the vicinity of the CTX genetic element. Sequence analysis revealed the presence of a cluster of four genes that display a high degree of similarity to the genes involved in the biogenesis of RTX toxins in several Gram-negative organisms. This cluster was located 693 nucleotides downstream of the *rstB* gene of the CTX element. Functional analysis revealed that RtxA (the toxin) is responsible for the rounding and detachment of human laryngeal carcinoma cells (HEp-2) in culture (Lin et al. 1999), mediated by actin crosslinking (Fullner and Mekalanos 2000) or inactivation of Rho GTPases (Sheahan and Satchell 2007). Since then, further studies have been carried out on this toxin in order to understand its genetic organization, mode of action, and its role in the pathogenesis of the disease cholera.

13.5.2 Organization of the RTX Locus

The RTX operon structure in *V. cholerae* consists of six genes organized into two divergently transcribed operons (Fig. 13.6) (Lin et al. 1999, Heidelberg et al. 2000). The *rtxA* operon in El Tor N16961 consists of VC1449-*rtxC*-*rtxA*, where *rtxA* codes for the RTX toxin, *rtxC* is the putative RTX-activating acyltransferase gene, and the ORF VC1449 encodes a conserved hypothetical protein. The 13,635 bp *rtxA* encodes a protein 4545 amino acids in length with a predicted molecular mass of

484 kDa, making it the largest ORF in the genome sequence at the time of its discovery (Lin et al. 1999). The *rtxBDE* operon encodes transport ATPases (*rtxB* and *rtxE*) and the transmembrane linker (*rtxD*).

The RTX gene cluster in classical *V. cholerae* strains differs from that present in the El Tor biotype. Mapping of the RTX gene cluster in O395 identified a 7,869 bp deletion that removes the 5′ end of *rtxA*, all of *rtxC*, and the 5′ end of *rtxB*, as well as any regulatory element that lies between these divergently transcribed genes. PCR analysis detected deletion in other classical strains, like 569B, CA401 and NIH41, but not in El Tor strains P27459, E7946 and C6709 (Lin et al. 1999). The gene cluster was found to be intact in the O139 strain MO10. A large number of *V. cholerae* isolates belonging to O1 El Tor, O139, and non-O1 non-O139 serogroups, including strains causing current cholera pandemics, exhibited intact gene clusters (Chow et al. 2001, Dalsgaard et al. 2001, Cordero et al. 2007). Furthermore, several nonpathogenic environmental isolates lacking the CTX element showed intact RTX gene clusters; the RTX gene cluster may thus predate the acquisition of the neighboring CTX element (Lin et al. 1999). The genetic structures of the RTX elements extracted from genome sequences of *V. cholerae* El Tor N16961 and classical *V. cholerae* O395 are presented in Fig. 13.6.

13.5.3 Structural Features of RtxA

Rtx toxins have certain domain structures in common: an N-terminal hydrophobic domain required for pore formation; central prototoxin activation sites; C-terminal GD-rich calcium binding repeats involved in target-cell binding; and a C-terminal signal for RtxB/RtxD-dependent secretion (Lally et al. 1999).

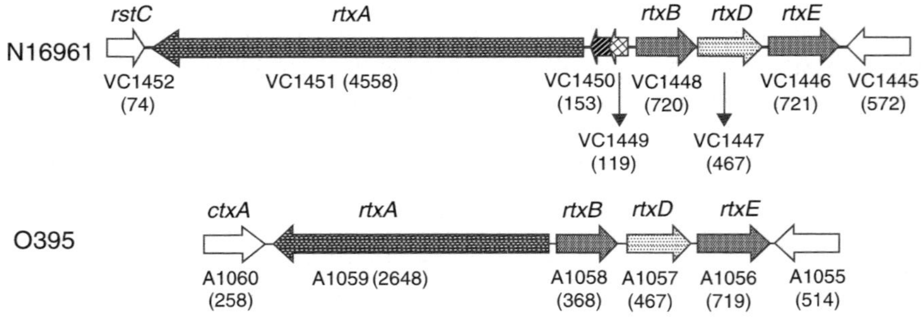

Fig. 13.6 Genomic organization of the RTX elements in *V. cholerae* El Tor N16961 and classical O395 chromosome. The *block arrows* represent open reading frames and indicate the direction of transcription. The *length of the arrow* approximately reflects the relative lengths of the ORFs. The ORFs were named according to the annotations in the complete genome sequence. The locus number and length of amino acid of each ORF is shown as the VC number (no. of amino acids) for N16961 and the A number (no. of amino acids) for O395

V. cholerae RtxA shows some unique structural features. First, a pore-forming domain is notably absent from the N-terminus, although a more internal hydrophobic region (amino acids 600–1200) may provide the pore-forming functionality. Second, the repeat region (amino acids 700–1350) overlaps with this hydrophobic region and is remarkably different in sequence from the nine-residue Gly-Asp (GD)-rich repeats commonly found in RTX toxins. The repeats in RtxA consist of 19 amino acids with a consensus motif of GXAN(I/V)XT(K/H)VGDGXTVAVMX, which is repeated 29 times consecutively. However, the C-terminal repeat region of RtxA shows the greatest sequence similarity to other RTX toxins, although, instead of the usual nine residue repeats, an 18-residue consensus motif of X(V/I)XXGXXNX(V/I)XXGDGXDX is observed.

Although related to the RTX family of pore-forming toxins, the *V. cholerae* RTX toxin has been proposed to be the founding member of a new family of RTX toxins encoded by *V. cholerae, V. vulnificus, Photorhabdus luminescens*, and *Xenorhabdus* sp. (Fullner and Mekalanos 2000). These proteins are all larger than 3500 amino acids and share consensus 20-residue N-terminal glycine-rich repeats, in addition to 18-residue C-terminal repeats that include the nonapeptide motif GGXGXDXXX (common to all RTX exoproteins), although the central regions vary dramatically (Boardman and Satchell 2004, Sheahan et al. 2004).

13.5.4 Secretion of RTX Toxin from V. cholerae

Like other RTX toxins, *V. cholerae* RTX is also secreted by a T1SS secretion apparatus (Welch 2001, Boardman and Satchell 2004). T1SS commonly consists of three components: an inner-membrane transport ATPase, a periplasmic linker protein, and an outer-membrane porin, which is exemplified by *E. coli* hemolysin T1SS, consisting of HlyD, HlyB and TolC, respectively (Holland et al. 2005). *V. cholerae* RTX toxin is secreted by an atypical four-component TISS encoded by *rtxB, rtxD, rtxE* and *tolC*. The *rtxBDE* operon is located adjacent to the *rtxA* operon and is divergently transcribed, while the *tolC* homolog is unlinked and located at VC2436 in the N16961 genome. The putative transport ATPase gene, *rtxB*, overlaps by 46 nt with the linker gene *rtxD* in different reading frames and encodes proteins that are 63 and 50% similar to HlyB and HlyD, respectively. The ORF *rtxE* encodes a protein that is 63% similar to HlyB and 60% similar to RtxB of *V. cholerae*. Thus, there are two potential TISS transport ATPases located within this operon. The deduced amino acid sequences for both RtxB and RtxE include conserved Walker box A mononucleotide-binding motifs, indicating that both are potentially capable of binding and hydrolyzing ATP (Boardman and Satchell 2004).

The putative TISS components of *V. cholerae*, as described above, were characterized by a genetic approach in which insertion/deletion mutants were created and the supernatant fluids/cell lysates were assayed for the RTX toxin-associated activities of cell rounding and actin crosslinking. A polar mutation in *rtxB* which inactivated the entire operon failed to secrete RTX toxin, which was found to accumulate intracellularly (as detected by immunoblotting) and remained active. Nonpolar deletion inactivating *rtxB* or *rtxE* resulted in supernatant fluid that lacked RTX toxin but bacterial cell lysate that retained cell-rounding and actin crosslinking activities, indicating that both of the ATPases are required for RTX secretion.

Hydrolysis of the transport ATPases is a prerequisite for activity and requires the presence of an intact nucleotide-binding site (NBS). ATP hydrolysis at both RtxB and RtxE is necessary for the secretion of RtxA; this was shown by generating point mutations K496A in RtxB and K522A in RtxE, thus changing the lysine residue of the Walker box A motif to alanine in each case. Both of the mutants failed to round HEp-2 cells and cause actin crosslinking, demonstrating that both RtxB and RtxE along with an intact NBS are required for RTX toxin secretion (Boardman and Satchell 2004). It has been proposed that, unlike in other TISS, the inner membrane transport ATPase RtxB and RtxE could form a heterodimer, as both the proteins contain the highly conserved ATP-binding cassette (ABC) family signature sequence (LSGGQ) motif, which is important for the formation of ABC dimer interfaces. These ATPases are presumed to provide energy for the secretion/efflux.

Deletion of the last four amino acids (LRER) of RtxD stops the secretion of the RTX toxin, although RTX activity could be detected in cell lysate. The involvement of TolC in RTX secretion was suggested by evidence that the absence of *tolC* abrogated cell rounding by *V. cholerae* (Bina and Mekalanos 2001). Later studies showed that the supernatant of a Δ*tolC* strain did not exhibit cell rounding or actin crosslinking activity, whereas the RTX toxin accumulated intracellularly in this mutant. TolC is believed to form a trimer of identical subunits comprising a β-barrel that spans the outer membrane, along with a series of α-helical coils that hang below the barrel like tentacles and dip into the periplasm (Koronakis et al. 2000). The periplasmic linker component (homolog of RtxD) is hypothesized to serve as a bridge between the cytoplasmic membrane ATPase and TolC in the outer membrane. The extent of bridge formation will depend on the size of the periplasm and the degree to which TolC as well as inner-membrane ATPases protrude into the periplasmic space (Wandersman and Delepelaire 1990). In *V. cholerae*, the last four amino acids, LRER, are proposed to interact with the outer membrane porin TolC (Boardman and Satchell 2004).

A model has been proposed by the group of Satchell for the type I secretion of RTX toxin in *V. cholerae* based on the above findings and our present knowledge, and is depicted in Fig. 13.7. According to this model, the RTX toxin comes into contact with RtxB and/or RtxE located in the inner membrane. RtxD then recruits TolC to the TISS complex, followed by a conformational change in all four components, thus opening the cavity to the extracellular milieu and allowing the RTX toxin to exit the cell. The two-cylinder engine model proposed for the multidrug-resistant transporters LmrA and MDR1 would drive the translocation process. The large size of the *V. cholerae* RTX toxin may require multiple contact points within the structurally diverse regions of RtxB and RtxE to enhance specificity, thereby necessitating two distinct transport ATPases (Boardman and Satchell 2004).

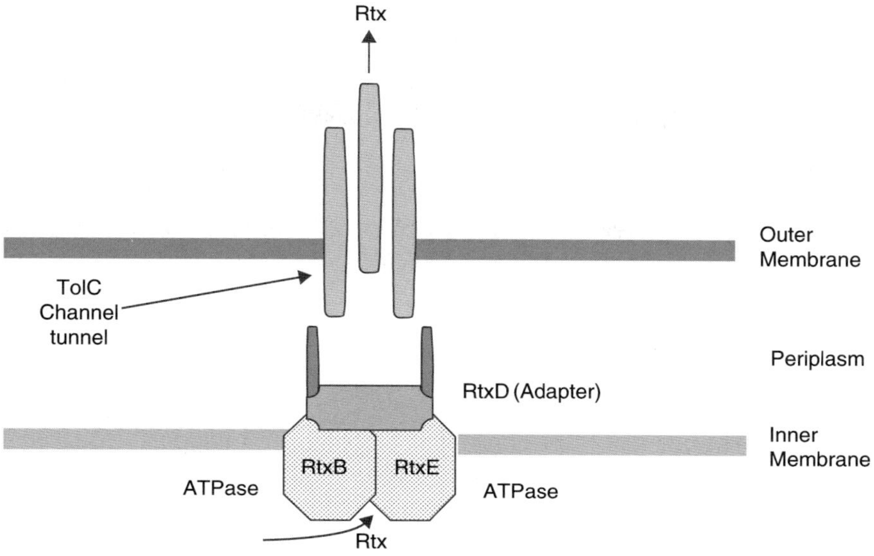

Fig. 13.7 Model of RTX toxin secretion by TISS. See text for further details

13.5.5 Mode of Action of RTX Toxin

Unlike the RTX family of toxins, *V. cholerae* RTX did not disrupt membrane integrity and did not show necrosis or leakage of cytoplasmic components from eukaryotic cells. The intact membrane integrity was assessed by releasing lactate dehydrogenase, a cytoplasmic enzyme, into the culture medium, or by excluding the fluorophore DEAD-RED as well as other membrane-impermeant dyes, including trypan blue and propidium iodide (Fullner and Mekalanos 2000). Instead, *V. cholerae* RTX causes the rounding and detachment of human laryngeal epithelial HEp-2 cells in culture. The cell-rounding effect was further observed for various other cell types, including A549 lung epithelial carcinoma cells, Henle 407 intestinal epithelial cells, L6 rat fibroblastoma cells, Chinese hamster ovary cells, Raw 264.7 and J774 macrophages, and Ptk2 kidney epithelial cells (Fullner and Mekalanos 2000).

The common toxic mechanism that leads to the rounding of a broad range of cell types without directly affecting the cell viability is the alteration of the polymerization state of the actin cytoskeleton (Steele-Mortimer et al. 2000). Actin is a highly conserved cytoskeletal protein that exists in dynamic equilibrium between globular monomer (G-actin) and filamentous polymer (F-actin). Inside living cells, the dynamics of actin polymerization are regulated by an array of actin-binding proteins (ABPs) (Pollard and Borisy 2003). Actin has a high-affinity binding site for a divalent cation (Mg^{+2} or Ca^{+2}) in complex with ATP or ADP (Gershman et al. 1986, Kinosian et al. 1993). As the salt concentration increases to physiological levels,

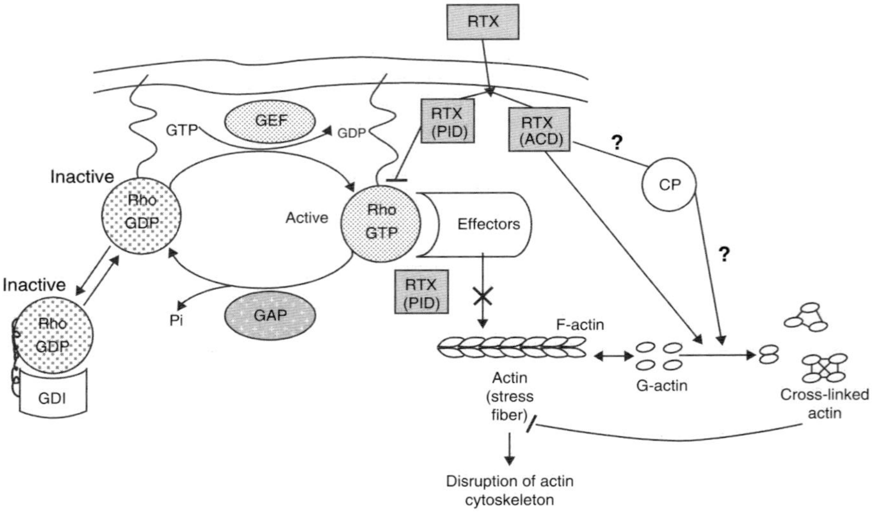

Fig. 13.8 Mechanism of disruption of the actin cytoskeleton by RTX toxin. RTX toxin targets the actin cytoskeleton by two different mechanisms: actin cross-linking and inactivation of Rho GTPases. See text for further details

G-actin spontaneously polymerizes to F-actin, and the ATP molecules bound to G-actin are hydrolyzed to ADP and Pi. It is believed that *V. cholerae* RTX toxin targets the actin cytoskeleton by two different mechanisms: either by crosslinking actin (Lin et al. 1999, Fullner and Mekalanos 2000) or by inactivating Rho GTPases (Sheahan and Satchell 2007), as illustrated in Fig. 13.8.

RTX toxin targets monomeric G-actin, resulting in the crosslinking of actin into dimers, trimers, and higher-order crosslinked products, disrupting the equilibrium between F-actin and G-actin, ultimately resulting in F-actin disassembly and an increase in paracellular permeability (Fullner and Mekalanos 2000). The actin crosslinking activity has been mapped within the RtxA protein to a 412 amino acid region located between amino acid residues 1963 and 2375, recognized to be the actin crosslinking domain (ACD). Transient expression of this domain in COS-7 (African green monkey kidney fibroblast) and HEp-2 cells leads to the formation of crosslinked actin species, demonstrating that expression of these 412 amino acids of the toxin in cytosol is sufficient to initiate actin crosslinking in the target cell (Sheahan et al. 2004). Subsequent studies demonstrated that purified ACD was sufficient to induce cell rounding and crosslinking of purified G-actin when delivered to the cell cytoplasm, and that it directly catalyzed the crosslinking of purified actin in vitro in the absence of other host cell factors (Cordero et al. 2007). RTX could act directly on G-actin or RTX could activate some endogenous crosslinking protein that carries out the crosslinking reaction. Deletion of this domain from the toxin eliminated actin crosslinking but had no effect on cell rounding, thus revealing the presence of a second mechanism for cell rounding (Sheahan et al. 2004). Further studies showed that, both in the presence and in the absence of actin crosslinking, the RTX toxin was able to induce the depolymerization of the actin cytoskeleton through the inactivation of Rho, Rac, and Cdc2. These three proteins are extensively characterized members of the Rho GTPase family that are involved in the formation of stress fibers, lamellipodia, and filopodia, respectively (Hall and Nobes 2000). Rho GTPases cycle between an active membrane-localized GTP-bound state and an inactive GDP-bound state. The inactive GDP-bound form is kept as such and sequestered in the cytosol in a complex with guanine neucleotide dissociation inhibitor (GDI). The activation state is regulated by guanine nucleotide exchange factors (GEFs), which mediate exchange of GDP for GTP. Rho GTPases are inactivated by hydrolysis of the bound GTP, a process that is largely stimulated by GTPase-activating proteins (GAPs). In their active state, Rho GTPases interact with many different downstream effector proteins, resulting in multiple effects, including the formation of actin stress fibers. RTX inactivates the Rho GTPases through its PID domain (Fig. 13.8).

Inactivation of Rho by RTX was reversed by pretreatment of cells with CNF1. CNF1 is the cytotoxic necrotizing factor produced by *E. coli*, and it modifies the GTPase proteins Rho, Rac, and Cdc2 by deaminating Gln43 (Lerm et al. 1999). This modification results in the constitutive activation of these proteins through the inhibition of GTPase activity (Aktories and Barbieri 2005). The constitutive activation of Rho GTPases can counteract the activities of most toxins that inhibit Rho GTPases, but it has no effect on those toxins that target actin directly (Fiorentini et al. 1995). Constitutive activation of Rho GTPases prevented cell rounding and actin depolymerization associated with RTX, suggesting that RTX targets Rho GTPase signaling pathways (Sheahan and Satchell 2007).

A 548 amino acid region of RtxA (called the Rho-inactivation domain or RID) located between amino acids 2552 and 3099 has been found to be associated with the toxin-induced inactivation of the Rho GTPases (Sheahan and Satchell 2007). The involvement of host factors for RID activity is expected, although no data are yet available. Whether the two cell-rounding activities of RTX act together or independent of each other is not known. Additionally, RTX toxin has been demonstrated to disrupt the paracellular tight junction of polarized T84 intestinal epithelial cells (Fullner et al. 2001).

However, the extent to which these activities of RTX influence pathogenesis in vivo is largely unknown. Expression of RTX toxin in the mouse pulmonary model results in a diffuse pneumonia characterized by proinflammatory response and tissue damage (Fullner et al. 2002). A *V. cholerae* mutant ($\Delta ctxAB\Delta hapA\Delta hlyA$) expressing only RTX contributes to lethality in mice (~31% survival at two days postinfection). During vaccine trials it was observed that volunteers who ingested $CT^{-}RTX^{+}$ strains of *V. cholerae* exhibited a higher level of lactoferrin, a physiological marker for the presence of neutrophils (Silva et al. 1996), suggestive of an inflammatory response that was probably mediated by the RTX toxin.

13.6 Chinese Hamster Ovary (CHO) Cell Elongation Factor (Cef)

Clinical studies with vaccine strains consistently reported a residual level of diarrhea and reactogenic activity. Searches for the presence of other toxins led to the discovery of toxins like Zot and Ace. However, a *V. cholerae* strain CVD110 without *ctx, zot, ace* and *hlyA* still caused mild diarrhea in volunteers (Cryz et al. 1995), indicating the presence of additional toxins. The secreted enterotoxic activities are often characterized in model systems such as Chinese hamster ovary (CHO) cell assay (Guerrant et al. 1974) or via fluid accumulation in an infant mouse model (Kothary and Richardson 1987). While testing the cytotonic activity of culture supernatants of *ctxA*-negative mutants of *V. cholerae* classical 569B (CVD103, CVD103-HgR, CVD103-HgR2), El Tor N16961 (JBK 70), and nontoxinogenic environmental isolates from Brazil on cultured CHO cells, Hall, McCardell and colleagues (Sathyamoorthy et al. 2000, McCardell et al. 2000) detected a CHO elongation factor in two of these strains, JBK 70 and CVD103-HgR. Morphological examination of CHO cells maintained in tissue culture by light microscopy showed marked elongation of CHO cells in the presence of virulence factors from several bacterial pathogens, including CT (Guerrant et al. 1974). The cytotoxic activity was measured in CHO units. One CHO unit is defined as the reciprocal of the dilution that caused the cells to elongate by >50%. The specific activity observed for CT was about 1000×10^6 CHO units per mg of protein (McCardell et al. 2000). The CHO cell elongation activity was partially purified and was found to cause fluid accumulation in a sealed infant mouse model, thus suggesting its enterotoxic nature, and was termed CHO elongation factor (Cef).

13.6.1 Purification, Stability, and Biological Activity

Cef has been partially purified from the supernatant of lysozyme-EDTA treated JBK 70 cells by anion exchange chromatography and eluting from a pH 3–9 isoelectric focusing (IEF) gel (McCardell et al. 2000). CHO cell activity was associated with a single band at 85 kDa on SDS-PAGE and a pI of 3.8, and this 85 kDa protein causes fluid accumulation in suckling mice and exhibits esterase activity for 4–10 C esters but not lipase activity. On the other hand, Cef from CVD103-HgR was purified to electrophoretic homogeneity by using $(NH_4)_2SO_4$ fractionation, anion exchange, cation exchange, hydrophobic interaction, and gel filtration chromatography (Sathyamoorthy et al. 2000). The purified Cef migrated with a molecular mass of 79 kDa with no visible copurifying subunit. The toxin was stable from pH 5 to pH 10 and at 56 °C (15 min), but the activity was abolished at 100 °C.

Although the molecular mass of Cef as measured by SDS-PAGE was similar to that of native CT, Cef was found to be distinct from CT based on the following evidence: (i) unlike CT, no subunit structure was observed for Cef in SDS-PAGE;

(ii) the CHO cell activity of Cef was not inhibited by ganglioside GM1 or mixed gangliosides, as is the case with CT or LT (Schengrund and Ringler 1989); (iii) Cef activity on CHO cells did not decrease after pre-incubating with the antiserum to CT; (iv) neither cAMP or prostaglandin E2 (PGE2) levels were elevated by Cef in CHO cell assays, in contrast to CT, for which a greatly elevated cAMP level and a slightly elevated level of PGE2 are observed (Kaper et al. 1995).

Cef, however, does have enterotoxic properties, as partly purified toxin caused fluid accumulation in the infant mouse model. The unique molecular mass, the N-terminal sequence, and the activity on CHO cells indicate that this factor is not zonula occludens toxin (Zot), accessory cholera enterotoxin (Ace), or HlyA hemolysin. There is a possibility that either the esterase activity of Cef or as-yet unidentified lipase or phospholipase activity associated with it may play a role in the fluid accumulation and elongation of CHO cells.

13.6.2 Cloning of the cef *Gene and the Encoded Protein*

The *cef* gene was further cloned in *E. coli* using a yeast vector and subsequently expressed in the yeast *P. pastoris* (McCardell et al. 2002). The *cef* genes from *V. cholerae* candidate vaccine strains JBK 70 and CVD 103-HgR were sequenced from the *E. coli* clone and found to be nearly identical (100 and 99.9%, respectively) to an open reading frame coding for a 796 amino acid protein located in chromosome II of *V. cholerae* N16961 (VCA0863) (Heidelberg et al. 2000). This protein has been functionally designated a putative lipase.

The cloned Cef protein expressed in yeast was purified to homogeneity and yielded a size of 114 kDa on SDS-PAGE. The increased size was probably due to glycosylation by yeast, as the cloned protein reacted strongly with PAS, the glycoprotein stain. The cloned purified toxin showed CHO cell elongation activity, esterase activity for 2–14 carbon *p*-nitrophenylesters, but no suckling mouse activity. This result differs from that of Cef purified from *V. cholerae* (Sathyamoorthy et al. 2000, McCardell et al. 2000). Understanding the mode of action of Cef requires further studies with mutants lacking Cef, and also studies in tissue culture and animal models.

13.7 New Cholera Toxin (NCT)

In 1983, it was reported from India that some nontoxinogenic environmental strains of *V. cholerae* O1, which failed to hybridize with CT or LT probes, could cause fluid accumulation in rabbit ileal loop and diarrhea in infant rabbits (Sanyal et al. 1983). Culture filtrates of these strains were able to cause fluid accumulation. The filtrates also increased the capillary permeability of rabbit skin, but unlike CT, caused blueing accompanied by blanching or necrosis (Sanyal et al. 1983). The toxin in the culture filtrate could not be neutralized in rabbit skin or ileal loop assays by antisera against CT or its A and B subunits. The toxicity to loop and skin was lost when the culture filtrate was heated to 100 °C for 10 min. The toxin was designated a new cholera toxin and proposed to be the cause of diarrhea in the CT^- *V. cholerae* strain (Sanyal et al. 1983). Later studies on CT^+ and CT^- *V. cholerae* O1 classical and El Tor biotype strains of clinical and environmental sources revealed that NCT is produced by O1, O139, non-O1 non-O139 strains (Tikoo et al. 1996).

NCT has been partially purified. This partially purified NCT also showed enterotoxic activity in a rabbit ileal loop model, suggesting some similarity in secretory response between NCT and CT (Saha and Sanyal 1989). No antigenic relationship between partially purified NCT and CT could be demonstrated in gel-diffusion tests. On the other hand, similarities between the NCTs produced by all strains have been demonstrated in immunodiffusion studies (Saha and Sanyal 1990).

Production of NCT by the CT^+ strain, in addition to CT, was confirmed by in vivo neutralization tests. The activity of the 569B enterotoxin was only partially neutralized (44%) by anti-NCT, while purified anti-CT caused about 66% neutralization, suggesting that the fluid accumulation produced in the RIL by 569B enterotoxin was the combined effect of both CT and NCT (Saha and Sanyal 1990). However, additional genetic and biochemical characterization of NCT is lacking, and so the identity of the protein in the complete genome could not be established.

13.8 Shiga-Like Toxin (SLT)

Shiga-like toxins (SLTs) (or verotoxins), named after the prototype toxin produced by *Shigella dysenteriae* type 1, and also produced by certain strains of *E. coli*, are well-documented causes of bloody and nonbloody diarrhea throughout the world. SLTs are cytotoxic to certain tissue culture cells, enterotoxic in rabbit ileal loops, and neurotoxic (causing limb paralysis and death) in rabbits and mice (O'Brien and Holmes 1987).

In 1984, the presence of SLT in *V. cholerae* was reported on the basis of cytotoxic activity of cell lysates towards HeLa cells, and this activity was found to be neutralized by antibody against purified Shiga toxin (O'Brien et al. 1984). The investigators detected moderate to high levels of heat-stable (56 °C, 15 min) cell-associated cytotoxin in cell lysates of *V. cholerae* O1, non-O1, and *V. parahaemolyticus* strains.

It was thought that SLT could be responsible for residual diarrhea in vaccine strains. Supporting evidence came from the fact that the vaccine strain *V. cholerae* CVD103-HgR (Δ*ctx*), which does not produce detectable SLT, caused little or no reactogenicity (Levine et al. 1988b). Genes encoding SLT have not been cloned from *V. cholerae* despite repeated efforts (Pearson et al. 1990). *V. cholerae* genomic fragments hybridized with *slt-1* of *E. coli* under low stringency, and DNA sequence analysis of these cloned fragments did not show significant homology to *slt-1* (Pearson et al. 1990). No homologs of SLT could be found in the complete genome sequence of *V. cholerae* (Heidelberg et al. 2000).

13.9 Thermostable Direct Hemolysin (TDH)

Thermostable direct hemolysin (TDH) is a putative toxin that has been epidemiologically associated with cases of gastroenteritis in humans caused by *V. parahaemolyticus* (Nishibuchi et al. 1992). Production of TDH is routinely tested for via the β-type hemolysis of erythrocytes incorporated into a special medium called Wagatsuma agar (Miyamoto et al. 1969). This hemolytic reaction is known as the Kanagawa phenomenon (KP). This hemolysin was named thermostable direct hemolysin (TDH), as it is stable upon heating (100 °C, 10 min), and the hemolytic activity did not increase upon the addition of lecithin, indicating its direct action on erythrocytes. The biological activities of TDH include hemolysis of various species of erythrocytes, cytotoxicity, lethal toxicity towards small experimental animals, and increased vascular permeability in rabbit skin (Nishibuchi and Kaper 1995).

The TDH toxin has not been reported in *V. cholerae* O1 (Terai et al. 1991), but it is found in plasmids as well as in chromosomal locations in a few non-O1 *V. cholerae* and was named NAG-rTDH, as it was related to TDH (Honda et al. 1985, Yoh et al. 1986). NAG-rTDH has been purified; like TDH, it migrated with a molecular mass of about 18.5 kDa on SDS-PAGE and showed lytic activities on erythrocytes which were stable at 100 °C for 10 min; it also cross-reacted with Vp-TDH in both the Ouchterlony and the neutralization tests (Yoh et al. 1986). Furthermore, the gene coding for TDH was cloned and sequenced from *V. cholerae* non-O1 and was about 96–98.6% homologous with the genes *tdh1* to *tdh5* of *V. parahaemolyticus* (Terai et al. 1991). All *tdh* genes, including those from *V. cholerae* non-O1, were flanked by insertion sequence-like elements (Terai et al. 1991).

13.10 Heat-Stable Enterotoxin of Nonagglutinable Vibrios (NAG-ST)

13.10.1 Discovery

Some strains of *V. cholerae* non-O1 were found to be associated with an illness that was clinically indistinguishable from cholera, while some others caused fever and bloody diarrhea (Blake et al. 1980, McIntyre et al. 1965). Most of these non-O1 strains do not produce CT. Arita et al. (1986) first described a new type of heat-stable enterotoxin (ST) in *V. cholerae* non-O1 that exhibits 50% amino acid sequence homology with ST of enterotoxigenic *E. coli* ETEC. This was designated nonagglutinable vibrio ST or NAG-ST. Like the ST of *E. coli*, NAG-ST of *V. cholerae* caused rapid fluid accumulation in a suckling mouse assay (Giannella 1976, Arita et al. 1986). The occurrence of NAG-ST in non-O1 isolates is quite low, and it is only rarely found in *V. cholerae* O1 (Hoge et al. 1990, Takeda et al. 1991).

13.10.2 Purification of NAG-ST

NAG-ST was purified by the method developed for the purification of EC-ST, with minor differences (Arita et al. 1986). The HPLC profile of a mixture of NAG-ST and EC-ST showed distinct peaks for these two toxins. Like EC-ST, purified NAG-ST was heat stable even at 100 °C for 30 min, but was inactivated at 120 °C for 10 min (Arita et al. 1986). NAG-ST did not react with anti-CT, but immunological cross-reactivity was observed between NAG-ST and EC-ST. Thus, NAG-ST was proposed to be an enterotoxin produced by non-O1 *V. cholerae* that is similar, but not identical, to EC-ST and is distinct from CT.

13.10.3 Amino Acid Sequence of NAG-ST

The amino acid composition of purified NAG-ST showed similarities to and differences from that of EC-ST in terms of size as well as composition (Arita et al. 1986). However, the amino acid sequence of the purified NAG-ST, a 17 amino acid peptide toxin (Takao et al. 1985), showed a close resemblance to *E. coli* STa and *Yersinia enterocolitica* ST (Takao et al. 1983, 1984). The amino acid sequence of NAG-ST from TQ-C3, a non-O1 *V. cholerae*, tentatively named Hakata, possessing the C (Inaba) factor but not the B (Ogawa) and A factors of *V. cholerae* O1, also resembled EC-ST (Arita et al. 1991). Interestingly, NAG-ST has also been detected in a CT-producing *V. cholerae* O1 strain (GP156) strain which belongs to biotype El Tor and serotype Inaba and was originally isolated from Australia (Yoshino et al. 1993). Four molecular species of heat-stable enterotoxins (designated as O1-ST-1 to O1-ST-4) were isolated from the culture supernatant and their primary structures determined. They had the same amino acid sequences but varied in length at the N-terminal portions. A comparison of the amino acid sequences of ST from *V. cholerae, E coli* and *Y. enterocolitica* is presented in Fig. 13.9a. Vc-H-ST showed comparable toxic effects to that of NAG-ST in suckling mouse assay. In the case of variants of O1-ST, enterotoxins with shorter N-terminal sequences showed more potent toxicities, as measured by fluid accumulation in suckling mice (Yoshino et al. 1993).

```
a NAG-ST     ----------  -----IDCCE  ICCNPACFGC  LN
  Vc-H-ST    ----------  ----LIDCCE  ICCNPACFGC  LN
  O1-ST-1    ----------  -----IDCCE  ICCNPACFGC  LN
  O1-ST-2    ----------  ----LIDCCE  ICCNPACFGC  LN
  O1-ST-3    ----------  ---NLIDCCE  ICCNPACFGC  LN
  O1-ST-4    ----FILQVD  GNGNLIDCCE  ICCNPACFGC  LN
  EC-STh     ----------  --NSSNYCCE  LCCNPACTGC  Y-
  EC-STp     ----------  ---NTFYCCE  LCCNPACAGC  Y-
  Y-ST       QACDPPSPPA  EVSSDWDCCD  VCCNPACAGC  --
                                **:  :******  **
```

```
b MRNLFIALML LFSSIAFSQT VENNKKTVQQ PQQIESKVNI
      Signal peptide              Propeptide

  KKLSENEECP FIKQVDENGN LIDCCEICCN PACFGCLN
      Propeptide               Processed active peptide
```

Fig. 13.9 **(a)** Comparison of the amino acid sequences of heat-stable enterotoxins of non-O1 and O1 vibrios with those of STs produced by other bacteria. *NAG-ST*, ST produced by a non-O1 *V. cholerae* (Takao et al. 1985); *VC-H-ST* denotes ST produced by *V. cholerae* non-O1 serogroup possessing the C (Inaba) factor but not the B (Ogawa) and A factors of *V. cholerae* O1 (Arita et al. 1991); *O1-ST1-1 to 4* represent the four molecular species of STs elaborated by GP156, a CT-producing strain of *V. cholerae* O1 (Yoshino et al. 1993); *EC-STh and EC-STp* are the enterotoxins produced by enterotoxigenic *E. coli* isolated from human and porcine sources (Arita et al. 1986); *Y-ST* is the ST produced by *Y. enterocolitica* (Takao et al. 1984). *indicates the positions where all of the chains have the same amino acid; :indicates the presence of positively charged amino acids. **(b)** Predicted amino acid sequence of the NAG-ST gene. The processed active peptide is preceded by the propeptide. The putative signal peptide is located upstream of propeptide. The predicted amino acid data are taken from GenBank Accession No. M36061 (Ogawa et al. 1990)

13.10.4 Cloning and Sequencing of the stn Gene Encoding NAG-ST

Although the amino acid sequence of NAG-ST was very similar to those of other STs from *E. coli* and *Y. enterocolitica, E. coli* STa probe was nonreactive in a colony hybridization assay (Sommerfelt et al. 1988), which suggested a lack of nucleotide sequence similarity among these STs. The NAG-ST gene was cloned from NRT36, a non-O1 *V. cholerae* known to produce NAG-ST (Ogawa et al. 1990). Unlike *E. coli* ST, which is found in plasmids, *V. cholerae* NAG-ST is located chromosomally. Two sets of chromosomal libraries were constructed, one in plasmid vector pUC18 and another in cosmid vector pHSG262, and screened for NAG-ST activity by competitive ELISA using a monoclonal antibody raised against synthetic NAG-ST (Ogawa et al. 1990). The ELISA-positive clones were tested for the

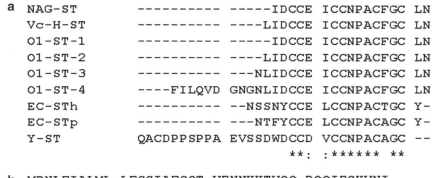

production of NAG-ST by suckling mouse assay. Three clones positive for both of the assays were found to share a common 1.6 kb *Hind*III fragment derived from NRT36 chromosomal DNA. Deletion analysis further narrowed down the NAG-ST positive region to about 0.4 kb, which was then cloned and sequenced. The ORF coding for NAG-ST was designated *stn* (Accession No. P0A4M3), and it encodes a protein product of 78 amino acids, with an estimated molecular mass of 8,814. The predicted amino acid sequence is described in Fig. 13.9b. The first 18 amino acids at the N-terminus constitute the signal peptide, amino acids 19–61 constitute the propeptide, and the last 17 amino acids at the C-terminal (62–78) correspond to the processed active peptide—identical to that of purified NAG-ST. The genes encoding ST enterotoxin in *V. cholerae* O1 and non-O1 are flanked by 123 bp direct repeats (Ogawa and Takeda 1993), which indicates that this toxin might have been introduced into the chromosome via a transposition event.

13.10.5 Mode of Action of NAG-ST

The mode of action of NAG-ST has been studied by Chaudhuri et al. (1998). Like *E. coli* STa, NAG-ST has been shown to raise the intracellular calcium level in rat enterocytes in a dose-dependent manner, and was associated with the activation of guanylyl cyclase, thereby increasing the intracellular level of cyclic guanosine-3′,5′-monophosphate (cGMP) in rat intestinal epithelial cells (Chaudhuri et al. 1998). From subsequent studies, the authors proposed that NAG-ST caused inositol trisphosphate (IP3)-mediated calcium release from intracellular calcium stores, which then stimulates nitric oxide production by activating nitric oxide synthase, and the nitric oxide activates calcium influx through cGMP. Enhanced intracellular calcium potentiates the translocation of PKCα from cytosol to membrane, which then phosphorylates some membrane proteins, one of which is guanylate cyclase. This subsequently leads to the accumulation of cyclic GMP, resulting in fluid loss and subsequent diarrhea (Hoque et al. 2004).

13.11 Toxin WO7

13.11.1 Discovery

An extracellular toxin was identified in an Inaba El Tor *V. cholerae* strain isolated from an outbreak of cholera in Warangal City in Andhra Pradesh in southern India (Walia et al. 1999). This strain, denoted *V. cholerae* W07, was found to be devoid of the *ctx, ace* or *zot* genes (as determined by hybridization studies), but still caused a disease that closely resembled cholera. The culture supernatant of this strain caused fluid accumulation in the rabbit ileal loop assay, elongation of Chinese hamster ovary (CHO) cells and rounding of Vero cells. The toxin was antigenically distinct from CT, Zot, Ace or hemolysin, and was termed WO7 toxin according to the nomenclature of the parent strain. Maximal production of WO7 was observed in AKI media (1.5% Bacto Peptone, 0.4% yeast extract, 0.5% NaCl, 0.3% filter-sterilized $NaHCO_3$ at pH 7.4) adjusted to pH 8.5 and 37 °C under shaking conditions (Walia et al. 1999).

13.11.2 Purification of WO7 Toxin

WO7 toxin was initially purified to homogeneity by sequential ammonium sulfate precipitation, affinity chromatography with a fetuin-Sepharose CL-4B column, and gel filtration chromatography (Walia et al. 1999). The purity and yield of this process was improved by employing a modified method using FPLC with a Superdex 200 HR 10/30 column (Bhattacharyya et al. 2008). On SDS-PAGE, WO7 toxin showed two subunits of molecular masses 58 and 40 kDa, and heating was essential for the dissociation of these two subunits. This is in contrast to CT, where heating as well as the presence of a reducing agent is required for the separation of A and B subunits under denaturing conditions. Like CT, the receptor for WO7 was GM1-coated erythrocytes (Walia et al. 1999). The 40 kDa subunit of WO7 serves as the carbohydrate-binding subunit, as probed by HRP-feutin in Western blot. Feutin contains the *N*-acetyl neuraminyl lactosamine unit that is structurally similar to the minimum essential component of GM1 required for binding to the WO7 toxin.

13.11.3 Biological Activity

The purified WO7 toxin produced fluid accumulation in the rabbit ileal loop as well as in sealed adult mice model (Bhattacharyya et al. 2008), but the secretion was not hemorrhagic in nature. Compared to CT, four times less WO7 was required to evoke a positive fluid accumulation. Although WO7 toxin resembled CT in some of its properties, polyclonal antibody raised against WO7 toxin did not cross-react with CT or *E. coli* LT in immunodiffusion, or in immunoblot assay.

This toxin has been proposed to induce apoptosis in HEp-2 cells based on the detection of a nucleosomal DNA ladder in response to WO7 and the presence of histone-associated DNA fragments in cytosol by cell death detection ELISA (Bhattacharyya et al. 2008). The WO7 toxin has been reported to cause alterations in intestinal transport in mice by altering the intracellular/extracellular Ca^{2+} level. It is proposed that in the cholera-like symptoms associated with *V. cholerae* WO7, the WO7 toxin probably plays a crucial role in altering the levels of the key mediators of different signal transduction pathways (Bhattacharyya et al. 2004).

Chapter 14
Concluding Notes

Abstract This concluding chapter of the book presents a glimpse of the authors' thoughts on the wealth of knowledge acquired on the nature and function of cholera toxins and the current status of the prevention and cure of the disease cholera.

14.1 An Introspection

This book has dealt extensively with the toxins, endotoxins and exotoxins, of *Vibrio cholerae*. The knowledge acquired can be reviewed briefly in the context of the ultimate goals of cholera research: (i) the development of an effective and long-lasting vaccine for preventing the disease at both the individual and population levels, and; (ii) the development of a simple, effective, and inexpensive treatment regime which will be easily available, even in the remotest corner or village of a developing country. Although neither of these two objectives falls within the direct scope of this book, they both almost automatically come to mind considering our unavoidable obligation to the society at large.

From as far back as John Snow's investigations (1847–1854), it has been clear that cholera is basically a waterborne disease, and that improvements in public health sanitation hold the major key to the prevention or even eradication of the disease. However, for many developing and less affluent countries, such improvements are difficult to achieve. Hence, there is obvious need for an effective vaccine and easily available treatment.

It is widely acknowledged that antibacterial immunity works better than anti-toxin immunity, and further that local immunity (i.e., at the intestinal mucosal level) is preferable to immunity at the serum level. Hence, the preparation of vaccines generally involves whole cells or bacteria (live attenuated or killed) with or without the enzymatically inactive B subunit of cholera toxin, and more attention is being given to the oral administration of vaccines. The lipopolysaccharides (LPS) covering the outermost layer of the *V. cholerae* organism act as the bacterial antigens against which the proper antibodies must be developed either locally or at the serum level. Although this is a highly simplified picture of the actual state of affairs, it is the basis for the development of a vaccine. However, the complexity of the problem only increases from this point.

For a long time *V. cholerae* serogroup O1 reigned the field as the causative organism of the disease cholera, and accordingly vaccines were developed against serogroup O1. Suddenly (without giving any notice!), the O139 serogroup appeared around 1992 and caused widespread havoc. The scientific world was caught totally unprepared and was inexperienced with respect to this new serogroup. The fact that vibrios belonging to any serogroup may acquire the virulence genes (genes for cholera toxin, *ctxAB, tcp*, etc.), may become virulent, and may rapidly cause a cholera epidemic or pandemic emerged. Already there are more than 200 serogroups, and more will be discovered in the near future. Each serogroup has a distinct LPS structure and hence has a distinct O-antigen. To remain prepared with vaccine or vaccines against 200 or more serogroups is a formidable task.

The structures and functions of *V. cholerae* LPS were discussed in Chaps. 4–6 of this book. The role of LPS in the preparation of cholera vaccine was also discussed in these chapters. One line of work that was discussed involved the preparation of synthetic mono- or oligosaccharide fragments that mimic the terminal mono- or oligosaccharide residues of the O-PS of *V. cholerae* LPS, and the study of their interactions with monoclonal anticholera antibodies, with a view to developing a synthetic carbohydrate-based anticholera vaccine (Liao et al. 2002). Perhaps the design and preparation of a synthetic vaccine against any serogroup can be achieved quickly using a suitable computer program, so that the delay between the outbreak of a disease and the production of the specific vaccine for it can be as short as possible. The induction of protective immunity by synthetic antigens that mimic the terminal hexasaccharide epitope of the O-PS of *V. cholerae* O1, serotype Ogawa, conjugated to bovine serum albumin has already been demonstrated (Chernyak et al. 2001, 2002). Extensive and intensive works along these lines are needed to achieve the ultimate goal. Instead of sticking to the conventional lines of work, the development of an effective and long-lasting cholera vaccine against more than one serogroup and/or serotype thus demands bold and novel approaches to the problem.

Of all the exotoxins produced by *V. cholerae*, the cholera toxin, CT, is the major one responsible for the causation of the disease. S.N. De's work (1959) established for the first time that the causative agent, cholera toxin (CT), is an exotoxin. This book has presented, in detail, the properties and the structures of the various toxins produced by *V. cholerae* cells, the structural basis for their toxicity, as well as their interactions with the cells and molecules of the human system, etc., which have been so elegantly elucidated by workers (following on from S.N. De) using sophisticated techniques. The wealth of knowledge acquired is certainly amazing, and speaks volumes about the illustrious investigators whose untiring efforts and skill have made this possible. This book on cholera toxins has, in fact, become a record of these achievements in cholera research.

In the context of all of these achievements one cannot, however, avoid the question that may be asked by the general public: does all of this information help us to overcome the cholera problem? Certainly we have learned a lot about the nature and lifestyle of the organism that causes the disease. Unfortunately, we have not yet been able to find an antidote to the cholera toxin, CT, released by the bacteria within the human system. Chapter 7 of this book presented a brief account of the initiatives undertaken by investigators towards the structure-based design and development of drugs against or inhibitors of the cholera toxin, CT (Fan et al. 2004). More research is certainly needed in this area. There are, at least theoretically, a number of stages where the toxin can be challenged and made inactive within the human system as effective therapeutic measures. However, this requires novel and ingenious approaches and a new direction of research.

The complete genome sequence of a clinical isolate of *V. cholerae* N16961, serogroup O1 and biotype El Tor, was deciphered in 2000 (Heidelberg et al. 2000); the presence of two circular chromosomes in *V. cholerae* cell was discovered (Trucksis et al. 1998), and the two chromosomes together were found to encode 3885 open reading frames (Heidelberg et al. 2000). The genomes of some other strains of *V. cholerae* have subsequently been or are now being sequenced. Since then, the idea of a new generation of vaccines has been gaining ground. Based on an increased understanding of the functional and comparative genomics of *V. cholerae*, attempts are being made to identify the common virulence factors among epidemic strains that may be the vaccine targets. It has been found that there is little sequence heterogeneity between the pathogenic strains of *V. cholerae* identified so far (LaRocque et al. 2006). This approach has already been applied to some other bacterial pathogens, including *Neisseria meningitides* (Pizza et al. 2000), but its fruitful application to *V. cholerae* is yet to be seen.

Fortunately, better therapeutic measures have already been devised practically, by applying a clinical approach. To give a simplified version of this story, the oral rehydration solution (ORS)—formulated based on the results of clinical studies (Phillips 1964, Cash et al. 1970, Sack et al. 1970), and recommended by the World Health Organization (WHO)—has been applied pretty successfully to the treatment of cholera cases. It is simple, inexpensive, and easily available in the remotest areas. Of course, a suitable antibiotic must be administered to shorten the lifetime of the disease and to achieve an effective cure. The exact molecular mechanism of action of ORS in preventing dehydration of infected individuals and its action on the infecting agent, *V. cholerae,* need to be elucidated. An in vitro study has, however, suggested that the ORS fluid has a bacteriostatic effect on *V. cholerae* cells (Bhattacharya and Chatterjee 1994).

But what is the role of the wealth of knowledge that we have acquired about the cholera toxin? Can it not help in the development of a really effective and long-lasting preventive measure against the disease? Scientists do have an obligation to society to provide a positive answer to these questions.

Appendix

Table A1 Genome information for *Vibrio cholerae* strains that are currently being sequenced*

Vibrio cholerae strain	Serogroup	Country and year of isolation	Source	Genotype	Length (nt)	GC (%)	% Coding	Protein coding	Structural RNAs	Accession number	Sequencing center and status
1587	O12	Peru	Clin	$CT^- TCP^-$	4,137,501	47.5	82	3758	97	NZ_AAUR00000000 AAUR00000000	TIGR (DA)
2740-80	O1 El Tor	Gulf Coast, 1980	Env	$CT^- TCP^+$	3,945,478	47.6	87	3771	68	NZ_AAUT00000000 AAUT00000000	TIGR (DA)
623-39					3,975,259	47.5	86	3777	73	NZ_AAWG00000000 AAWG00000000	TIGR (DA)
AM-19226	O39	Bangladesh, 2001	Clin	$CT^- TCP^-$	4,056,157	47.7	ND	ND	ND	NZ_AATY00000000 AATY00000000	TIGR (DA)
B33	O1	Mozambique, 2004		$CT^+ TCP^+$	4,026,835	47.5	83	3677	61	NZ_AAWE00000000 AAWE00000000	TIGR (DA)
MAK 757	O1 El Tor		Clin	$CT^+ TCP^+$	3,917,446	47.6	82	3501	81	NZ_AAUS00000000 AAUS00000000	TIGR (DA)
MO10	O139	India, 1992	Clin	$CT^+ TCP^+$	4,079,046	47.6	83	3441	ND	NZ_AAKF00000000 AAKF00000000	TIGR (DA)
MZO-2	O14	Bangladesh, 2001	Clin	$CT^- TCP^-$	3,862,985	47.6	83	3425	65	NZ_AAWF00000000 AAWF00000000	TIGR (DA)
MZO-3	O37	Bangladesh, 2001	Clin	$CT^- TCP^-$	4,146,039	47.6	86	3897	71	NZ_AAUU00000000 AAUU00000000	TIGR (DA)
NCTC 8457	O1 El Tor	Saudi Arabia, 1910	Clin		4,063,388	47.4	86	3975	72	NZ_AAWD00000000 AAWD00000000	TIGR (DA)

RC 385	O135	Chesapeake Bay	Env	CT^- TCP^-	3,634,985	47.5	81	3229	ND	NZ_AAKH00000000 AAKH00000000	TIGR (DA)
V51	O141	United States, 1987	Clin		3,782,275	47.5	83	3323	ND	NZ_AAKI00000000 AAKI00000000	TIGR (DA)
V52	O37	Sudan, 1968	Clin	CT^+ TCP^+	3,974,495	47.4	86	3815	87	NZ_AAKJ00000000 AAKJ00000000	TIGR (DA)

*Last accessed on Aug 11, 2008

Abbreviations used: *Clin*, clinical isolate; *Env*, environmental isolate; *TIGR*, The Institute of Genome Research; *DA*, draft assembly; *ND*, not determined; *CT*, cholera toxin; *TCP*, toxin-coregulated pili.

Table A2 CTXφ and RS1 gene cluster and their possible functions in *V. cholerae* El Tor N16961 (from the NCBI database at http://www.ncbi.nlm.nih.gov/genomes/lproks.cgi and Heidelberg et al. 2000)

Gene name	Locus no.	Gene ID	Position start	Position end	Strand	GC content	Length of amino acids	Putative function
rstC	VC1452	2613958	1564152	1564376	–	37.84	74	Antirepressor of RstR
rstB	VC1453	2613959	1564470	1564844	–	45.43	124	Provirus integration
rstA	VC1454	2613960	1564822	1565901	–	45.68	359	Viral genome replication
rstR	VC1455	2613961	1566027	1566365	+	37.80	112	Transcriptional repressor
ctxB	VC1456	2613962	1566967	1567341	–	32.80	124	Cholera enterotoxin, B subunit; pathogenesis, protein amino acid ADP-ribosylation, host cell surface binding
ctxA	VC1457	2613963	1567338	1568114	–	38.50	258	Cholera enterotoxin, A subunit; NAD+ ADP-ribosyltransferase activity, protein amino acid ADP-ribosylation
zot	VC1458	2613964	1568213	1569412	–	49.21	399	Zona occludens toxin; cytoskeleton organization and biogenesis, epithelial fluid transport; phage coat protein
ace	VC1459	2613965	1569409	1569702	–	44.67	97	Accessory cholera enterotoxin; anion transport, regulation of transport; phage coat protein
orfUpIII^{CTX}	VC1460	2613966	1569699	1570982	–	48.71	427	Hypothetical protein; phage coat protein
cep	VC1461	2613967	1570993	1571241	–	47.56	82	Colonization factor; phage coat protein
rstB	VC1462	2613968	1571377	1571760	–	44.88	127	Provirus integration
rstA	VC1463	2613969	1571738	1572817	–	45.78	359	Viral genome replication
rstR	VC1464	2613970	1572943	1573281	+	37.80	112	Transcriptional repressor

Table A3 VPI-1 genes and their possible functions in *V. cholerae* El Tor N16961 (from the NCBI database at http://www.ncbi.nlm.nih.gov/genomes/lproks.cgi and Heidelberg et al. 2000)

Gene name	Locus no.	Gene ID	Position start	Position end	Strand	GC content	Length of amino acids	Putative function
Tnp	VC0817	2615360	873242	874225	+	45.06	327	Transposase, putative
aldA	VC0819	2614486	875273	876898	–	35.80	541	Aldehyde dehydrogenase
tagA	VC0820	2614487	876974	880015	+	32.02	1013	ToxR-activated gene A protein
–	VC0821	2614488	880128	881267	+	28.58	379	Hypothetical protein
–	VC0822	2614489	881390	884728	+	32.88	1112	Inner-membrane protein, putative
–	VC0823	2614490	884725	885663	+	35.90	312	Hypothetical protein
tagD	VC0824	2614491	885773	886267	+	39.43	164	ToxR-activated gene D protein
tcpI	VC0825	2614492	886543	888405	–	38.17	620	Chemotaxis transducer
tcpP	VC0826	2614493	888846	889511	+	36.80	221	Virulence regulators
tcpH	VC0827	2614494	889496	889906	+	41.67	136	Virulence regulators
tcpA	VC0828	2614495	890449	891123	+	43.45	224	Major pilin
tcpB	VC0829	2614496	891203	892495	+	40.70	430	–
tcpQ	VC0830	2614497	892495	892947	+	36.00	150	–
tcpC	VC0831	2614498	892950	894419	+	35.72	489	–
tcpR	VC0832	2614499	894412	894867	+	28.04	151	–
tcpD	VC0833	2614500	894855	895691	+	30.46	278	–
tcpS	VC0834	2614501	895707	896165	+	31.14	152	–
tcpT	VC0835	2614502	896155	897666	+	36.18	503	–
tcpE	VC0836	2614503	897641	898663	+	35.59	340	–
tcpF	VC0837	2614504	898673	899689	+	35.01	338	–
toxT	VC0838	2614505	899896	900726	+	28.38	276	Virulence regulatory protein
tcpJ	VC0839	2614506	900726	901487	+	31.49	253	Leader peptidase
acfB	VC0840	2614507	901494	903374	+	34.35	626	Accessory colonization factor
acfC	VC0841	2614508	903380	904150	+	36.33	256	Accessory colonization factor
–	VC0842	2614509	904166	904639	+	30.36	157	Hypothetical protein
tagE	VC0843	2614510	904648	905556	+	34.88	302	ToxR-activated gene E protein
acfA	VC0844	2614511	905559	906206	–	31.32	215	Accessory colonization factor
–	VC0845	2614512	906379	910941	+	38.51	1520	Hypothetical protein
int	VC0847	2614514	912856	914124	–	46.37	422	Integrase, phage family

Table A4 VPI-2 gene loci and their possible functions in *V. cholerae* El Tor N16961 (from the NCBI database at http://www.ncbi.nlm.nih.gov/genomes/lproks.cgi and Heidelberg et al. 2000)

Locus no.	Gene ID	Position start	Position end	Strand	GC content	Length of amino acids	Putative function
VC1757	2613762	1892430	1895480	+	52.99	1016	Transporter, AcrB/D/F family
VC1758	2613764	1896092	1897327	+	44.20	411	Integrase, phage family
VC1760	2613766	1898009	1900831	–	41.13	940	Helicase, putative
VC1761	2613767	1900851	1901459	–	40.43	202	Hypothetical protein
VC1762	2613642	1901450	1902901	–	40.72	483	Hypothetical protein
VC1763	2613643	1902898	1903632	–	40.16	244	Chemotaxis protein, MotB-related protein
VC1764	2613644	1903625	1905745	–	41.31	706	Hypothetical protein
VC1765	2613645	1905790	1908858	–	43.38	1022	Type I restriction enzyme HsdR, putative
VC1766	2613646	1908886	1910877	–	43.34	663	Hypothetical protein
VC1767	2613647	1910870	1912282	–	44.96	470	Hypothetical protein
VC1768	2613648	1912283	1913671	–	40.69	462	Hypothetical protein
VC1769	2613649	1913628	1916009	–	43.93	793	DNA methylase HsdM, putative
VC1770	2613650	1916150	1918213	–	41.73	687	Hypothetical protein
VC1771	2613651	1918195	1921857	–	41.86	1220	Hypothetical protein
VC1772	2613652	1921885	1922745	+	44.87	286	Hypothetical protein
VC1773	2613653	1922810	1923880	–	36.24	356	Hypothetical protein
VC1774	2613654	1923895	1925049	–	41.75	384	Hypothetical protein
VC1775	2613655	1925231	1926067	–	43.76	278	Hypothetical protein
VC1776	2613656	1926196	1927092	–	43.29	298	*N*-acetylneuraminate lyase, putative
VC1777	2613657	1927130	1928413	–	45.36	427	Hypothetical protein
VC1778	2613658	1928421	1928942	–	41.23	173	Hypothetical protein
VC1779	2613659	1928993	1929958	–	44.76	321	C4-dicarboxylate-binding periplasmic protein
VC1780	2613660	1929989	1930282	+	34.36	97	Hypothetical protein
VC1781	2613661	1930173	1930895	+	47.22	240	*N*-acetylmannosamine-6-phosphate 2-epimerase
VC1782	2613662	1930883	1931746	+	45.18	287	*N*-acetylmannosamine kinase

VC1783	2613663	1931750	1932886	+	44.97	378	*N*-acetylglucosamine-6-phosphate deacetylase
VC1784	2613664	1933231	1935654	+	45.23	807	Neuraminidase
VC1785	2613665	1935801	1936007	–	44.61	68	Transcriptional regulator
VC1786	2613666	1936121	1936597	+	48.95	158	DNA repair protein RadC, putative
VC1787	2613667	1936594	1936731	+	45.93	45	Hypothetical protein
VC1788	2613668	1936827	1937522	–	43.15	231	Hypothetical protein
VC1789	2613669	1937519	1938391	–	44.48	290	Transposase OrfAB, subunit B
VC1790	2613670	1938388	1938732	–	43.27	114	Transposase OrfAB, subunit A
VC1791	2613671	1938739	1939779	–	44.03	346	Hypothetical protein
VC1792	2613672	1939914	1940273	–	43.14	119	Hypothetical protein
VC1793	2613673	1940322	1940699	–	39.47	125	Hypothetical protein
VC1794	2613674	1940754	1941332	–	37.85	192	Hypothetical protein
VC1795	2613675	1941351	1941671	–	39.31	106	Transcriptional regulator, putative
VC1796	2613676	1941658	1942032	–	39.78	124	Middle operon regulator-related protein
VC1797	2613677	1942842	1943303	–	40.09	153	Hypothetical protein
VC1798	2613678	1943306	1944457	–	44.56	383	Eha protein
VC1799	2613679	1944387	1946144	–	47.81	585	Hypothetical protein
VC1800	2613680	1946151	1947122	–	44.27	323	Hypothetical protein
VC1801	2613681	1947461	1947823	–	35.00	120	Hypothetical protein
VC1802	2613682	1947786	1948022	–	39.32	78	Hypothetical protein
VC1803	2613683	1948112	1948573	+	40.52	153	Hypothetical protein
VC1804	2613684	1948800	1949114	+	48.72	104	Hypothetical protein
VC1805	2613685	1949165	1949611	+	45.50	148	Hypothetical protein
VC1806	2613686	1949718	1950704	+	43.50	328	Hypothetical protein
VC1808	2613688	1951671	1952516	+	29.18	281	Hypothetical protein
VC1809	2613689	1952631	1952861	–	47.37	76	Transcriptional regulator, putative
VC1810	2613690	1953461	1953664	+	44.28	67	Hypothetical protein

Table A5 VSP-I gene loci and their possible functions in *V. cholerae* El Tor N16961

Locus no.	Gene ID	Position start	Position end	Strand	GC content	Length of amino acids	Putative function
VC0174	2614284	174143	175090	+	52.17	315	Hypothetical protein
VC0175	2614285	175343	176941	–	36.90	532	Deoxycytidylate deaminase-related protein
VC0176	2614268	177450	177758	–	39.22	102	Transcriptional regulator, putative
VC0177	2614269	177861	178424	+	37.79	187	Hypothetical protein
VC0178	2614189	179338	180405	+	42.63	355	Patatin-related protein
VC0179	2614190	180419	181729	+	41.82	436	Hypothetical protein
VC0180	2614191	181732	183486	+	43.55	584	Hypothetical protein
VC0181	2614172	183476	183946	+	45.73	156	Hypothetical protein
VC0182	2614173	183957	184388	–	37.30	143	Hypothetical protein
VC0183	2614436	184360	186471	–	37.55	703	Hypothetical protein
VC0184	2614437	186481	188169	–	39.68	562	Hypothetical protein
VC0185	2614438	188166	189380	–	36.96	404	Transposase, putative
VC0186	2614842	189457	190821	–	52.20	454	Glutathione reductase

* From the NCBI database at http://www.ncbi.nlm.nih.gov/genomes/lproks.cgi and Heidelberg et al. (2000)

Table A6 VSP-II genes and their possible functions in *V. cholerae* El Tor N16961 (from the NCBI database at http://www.ncbi.nlm.nih.gov/genomes/lproks.cgi and Heidelberg et al. 2000)

Locus no.	Gene ID	Position start	Position end	Strand	GC content	Length of amino acids	Putative function
VC0489	2615283	520634	522394	–	48.41	586	Hemolysin, putative
VC0490	2615284	523156	525117	–	37.42	653	Hypothetical protein VC0490
VC0491	2615285	525118	525654	–	34.46	178	Hypothetical protein VC0491
VC0492	2615286	525623	526789	–	37.63	388	Hypothetical protein VC0492
VC0493	2615287	527045	527920	–	39.29	291	Hypothetical protein VC0493
VC0494	2615288	528305	528949	+	43.46	214	Hypothetical protein VC0494
VC0495	2615289	529011	529685	+	45.39	224	Hypothetical protein VC0495
VC0496	2615290	529739	530338	+	41.88	199	Hypothetical protein VC0496
VC0497	2615291	530402	530602	+	45.45	66	Transcriptional regulator
VC0498	2615292	530684	531124	+	46.80	146	Ribonuclease H
VC0502	2615794	534198	534722	–	41.57	174	Type IV pilin, putative
VC0503	2615795	535418	536698	+	43.90	426	Hypothetical protein VC0503
VC0504	2615796	536876	537103	–	45.78	75	Hypothetical protein VC0504
VC0505	2615797	537151	537519	–	49.73	122	Hypothetical protein VC0505
VC0506	2615798	537550	538284	–	48.09	244	Hypothetical protein VC0506
VC0507	2615799	538423	538599	–	44.83	58	Hypothetical protein VC0507
VC0508	2615800	538603	539046	–	46.71	147	Hypothetical protein VC0508
VC0509	2615801	539097	539540	–	46.94	147	Hypothetical protein VC0509
VC0510	2615802	539531	540004	–	47.98	157	DNA repair protein, RadC-related protein
VC0511	2615803	540216	540335	+	32.48	39	Hypothetical protein VC0511
VC0512	2615804	541319	542908	+	43.67	529	Methyl-accepting chemotaxis protein
VC0513	2615805	544362	545177	+	27.43	271	Transcriptional regulator, AraC/XylS family
VC0514	2615806	545174	547054	+	35.25	626	Methyl-accepting chemotaxis protein
VC0515	2615807	547158	548390	+	33.90	410	Hypothetical protein VC0515
VC0516	2615808	548780	550021	–	40.36	413	Phage integrase
VC0517	2615810	550407	552284	–	47.47	625	RNA polymerase sigma factor RpoD

Table A7 RTX gene cluster and its possible functions in *V. cholerae* El Tor N16961*

Gene name	Locus no.	Gene ID	Position start	Position end	Strand	GC content	Length of amino acids	Putative function
rtxE	VC1446	2614078	1543231	1545396	–	49.61	721	Toxin secretion transporter, putative
rtxD	VC1447	2614079	1545399	1546802	–	50.46	467	RTX toxin transporter
rtxB	VC1448	2614080	1546757	1548919	–	45.09	720	RTX toxin transporter
–	VC1449	2614081	1549277	1549636	+	42.02	119	Hypothetical protein
–	VC1450	2613956	1549662	1550123	+	49.46	153	RTX toxin-activating protein
rtxA	VC1451	2613957	1550108	1563784	+	48.68	4558	RTX toxin RtxA

*From the NCBI database at http://www.ncbi.nlm.nih.gov/genomes/lproks.cgi and Heidelberg et al. (2000)

Table A8 Genes involved in extracellular protein secretion (Eps) in *V. cholerae* El Tor N16961 and their possible functions (from the NCBI database at http://www.ncbi.nlm.nih.gov/genomes/lproks.cgi and Heidelberg et al. 2000)

Gene name	Locus no.	Gene ID	Position start	Position end	Strand	GC content	Length of amino acids	Putative function
epsN	VC2723	2615551	2896907	2897665	–	54.23	252	T2S: GspN
epsM	VC2724	2615552	2897667	2898167	–	53.01	166	T2S: GspM, IM complex component
epsL	VC2725	2615553	2898174	2899397	–	51.27	407	T2S: GspL, IM complex component
epsK	VC2726	2614889	2899354	2900364	–	52.28	336	T2S: GspK, minor pseudopilin component
epsJ	VC2727	2614890	2900354	2901019	–	52.49	221	T2S: GspJ, minor pseudopilin component
epsI	VC2728	2614891	2901006	2901359	–	52.71	117	T2S: GspI, minor pseudopilin component
epsH	VC2729	2614892	2901349	2901933	–	51.37	194	T2S: GspH, minor pseudopilin component
epsG	VC2730	2614893	2901967	2902407	–	48.17	146	T2S: GspG, major pseudopilin component
epsF	VC2731	2614894	2902448	2903668	–	53.37	406	T2S: GspF, IM complex component
epsE	VC2732	2614895	2903668	2905179	–	52.68	503	T2S: GspE, ATPase, IM complex component
epsC	VC2734	2614897	2907245	2908162	–	49.84	305	T2S: GspC, periplasmic linker
epsB	VC2444	2622851	2623621	2612986	–	49.87	256	T2S: GspA
epsA	VC2445	2623621	2625210	2612987	–	52.11	529	T2S: GspB
pilD	VC2426	2600669	2601544	2612968	+	51.66	291	Leader peptidase

T2S, type II secretion; *Gsp*, general secretion pathway; *IM*, inner membrane

Table A9 Primers used for PCR amplification of virulence genes in *V. cholerae*

Gene	Oligont primer	Sequence of the primer	PCR conditions	Size of product (bp)	References
ctxA	Forward	5′ CGGGCAGATTCTAGACCTCCTG 3′	95°C, 5 min; 25 cycles: 95°C, 1 min, 60°C, 1 min, 72°C, 1 min; final extension 72°C, 10 min	564	Fields et al. (1992)
	Reverse	5′ CGATGATCTTGGAGCATTCCCAC 3′			
ctxB	Forward	5′ GATACACATAATAGAATTAAGGATG 3′	25 cycles: 95°C, 1 min, 55°C, 1 min, 72°C, 1 min	460	Olsvik et al. (1993)
	Reverse	5′ GGTTGCTTCTCATCATCGAACCAC 3′			
Ace	Forward	5′ TAAGGATGTGCTTATGATGGAC ACCC 3′	94°C, 1 min; 29 cycles: 94°C, 30 s, 55°C, 30 s, 72°C, 30 s; final extension 72°C, 5 min	289	Trucksis et al. (1993)
	Reverse	5′ CGTGATGAATAAAGATACTCATAG 3′			
St	Forward	5′ GAGAAACCTATTCATTGC 3′	94°C, 1 min; 29 cycles: 94°C, 30 s, 55°C, 30 s, 72°C, 30 s; final extension 72°C, 5 min	216	Vicente et al. (1997)
	Reverse	5′ GCAAGCTGGATTGCAAC 3′			
stn/sto	Forward	5′ TCGCATTTAGCCAAACAGTAGAAA3′	30 cycles: 94°C, 2 min, 55°C, 1 min, 72°C, 1 min; final extension 72°C, 10 min	172	Rivera et al. (2001)
	Reverse	5′ GCTGGATTGCAACATATTTCGC 3′			
rstR Class	Forward	5′ CTTCTCATCAGCAAAGCCTCCATC 3′	Not available	Not available	Faruque et al. (2003)
El Tor	Forward	5′-GCACCATGATTTAAGATGCTC 3′			
Calc	Forward	5′-CTGTAAATCTCTTCAATCCTAGG 3′			
	Reverse	5′-TCGAGTTGTAATTCATCAAGAGTG 3′			
rstC	Forward	5′ATGAGTTTGAAACCATACACTTT 3′	Not available	Not available	Faruque et al. (2002)
	Reverse	5′ TTACAGTGATGGAT CAGTCAAT 3′			
acfB	Forward	5′ GGACCAAGCATTATTATCTCT 3′	25 cycles: 94°C, 2 min, 50°C, 2 min, 72°C, 3 min	1.9 kb	Faruque et al. (1999)
	Reverse	5′ AATGATAAACTTACTGATTAA 3′			

Table A9 (continued)

Gene	Oligont primer	Sequence of the primer	PCR conditions	Size of product (bp)	References
ompW	Forward	5′ CACCAAGAAGGTGACTTTATTGTG-3′	94°C, 5 min; 35 cycles: 94°C, 1 min, 55°C, 1 min, 72°C, 1 min; final extension 72°C, 7 min	304	Nandi et al. (2000)
	Reverse	5′ GGTTTGTCGAATTAGCTTCACC 3′			
ompU	Forward	5′ ACGCTGACGGAATCAACCAAAG 3′	30 cycles: 94°C, 2 min, 60°C, 1 min, 72°C, 1 min; final extension 72°C, 10 min	869	Rivera et al. (2001)
	Reverse	5′ GCGGAAGTTTGGCTTGAAGTAG 3′			
toxR	Forward	5′ CCTTCGATCCCCTAAGCAATAC 3′	30 cycles: 94°C, 2 min, 60°C, 1 min, 72°C, 1 min; final extension 72°C, 10 min	779	Rivera et al. (2001)
	Reverse	5′ AGGGTTAGCAACGAT CGTAAG 3′			
tcpA	Forward	5′ CACGATAAGAAAACCGGTCAAGAG 3′	30 cycles: 94°C, 2 min, 60°C, 1 min, 72°C, 1 min; final extension 72°C, 10 min		Rivera et al. (2001)
El Tor	Reverse	5′ CGAAAGCACCTTCTTTCACGTTG 3′		451	
Classical	Reverse	5′ TTACCAAATGCAACGCCGAATG 3′		620	
tcpI	Forward	5′ TAGCCTTAGTTCTCAGCAGGCA 3′	30 cycles: 94°C, 2 min, 60°C, 3 min, 72°C, 1 min; final extension 72°C, 10 min	862	Rivera et al. (2001)
	Reverse	5′ GGCAATAGTGTCGAGCTCGTTA 3′			
hlyA (El Tor)	Forward	5′ GGCAAACAGCGAAACAAATACC 3′	30 cycles: 94°C, 2 min, 60°C, 1 min, 72°C, 1 min; final extension 72°C, 10 min	481	Rivera et al. (2001)
(ET/Class)	Forward	5′ GAGCCGGCATTCATCTGAAT 3′		738/727	
	Reverse	5′ CTCAGCGGGCTAATACGGTTTA 3′			

zot	Forward	5′ TCGCTTAACGATGGCGCGTTTT 3′	30 cycles: 94°C, 2 min, 60°C, 1 min, 72°C, 1 min; final extension 72°C, 10 min	947	Rivera et al. (2001)
	Reverse	5′ AACCCCGTTTCACTTCTACCCA 3′			
sxt	Forward	5′ TCGGGTATCGCCCAAGGGCA 3′	94°C, 2 min, 30 cycles; 94°C, 1min, 60°C, 1 min, 72°C, 1min; final extension 72°C, 10 min	946	Bhanumathi et al. (2003)
	Reverse	5′ GCGAAGATCATGCATAGACC 3′			
rfb (O139)	Forward	5′ AGCCTCTTTATTACGGGTGG 3′	94°C, 5 min; 35 cycles; 94°C, 1min,	449	Hoshino et al. (1998)
	Reverse	5′ GTCAAACCCGATCGTAAAGG 3′	55°C, 1 min, 72°C,		
rfb (O1)	Forward	5′ GTTTCACTGAACAGATGGG 3′	1min, final extension 72°C, 7 min	192	
	Reverse	5′ GGTCATCTGTAAGTACAAC 3′			

Table A10 Database and journal sources on the structure of cholera toxin and relevant proteins*

Group	Description of the protein	Submitting authors	PDB ID	References
Cholera toxin	Cholera toxin B subunit pentamer	Zhang RG, Westbrook E	1FGB	Zhang et al. (1995b)
	Cholera holotoxin, crystal form 1	O'Neal CJ, Amaya EI, Jobling MG, Holmes RK, Hol WG	1S5E	O'Neal et al. (2004)
	Cholera holotoxin, crystal form 2	O'Neal CJ, Amaya, EI, Jobling MG, Holmes RK, Hol WG	1S5F	O'Neal et al. (2004)
	Cholera toxin	Zhang RG, Westbrook E	1XTC	Zhang et al. (1995a)
	Cholera toxin A1 subunit bound to its human protein activator, ARF6	O'Neal CJ, Jobling MG, Holmes RK, Hol WGJ	2A5D	O'Neal et al. (2005)
	Cholera toxin A1 subunit bound to its substrate, NAD+, and its human protein activator, ARF6	O'Neal CJ, Jobling MG, Holmes RK, Hol WGJ	2A5F	O'Neal et al. (2005)
	Cholera toxin A1 subunit bound to ARF6(Q67L)	O'Neal CJ, Jobling MG, Holmes RK, Hol WGJ	2A5G	O'Neal et al. (2005)
	Cholera toxin B pentamer complexed with GM1 pentasaccharide	Merritt EA, Hol WGJ	2CHB	Merritt et al. (1997)
	Cholera toxin B pentamer complexed with GM1 pentasaccharide	Merritt EA, Hol WGJ	3CHB	Merritt et al. (1998)
TcpA	TcpA, the Type IV pilin subunit from the toxin-coregulated pilus of *V. cholerae*, classical biotype	Craig L, Tainer JA	1OQV	Craig et al. (2003)
HapR	*V. cholerae* HapR	Kull FJ, DeSilva RS, Kovacikova G, LinW, Taylor RK, Skorupski K.	2PBX	DeSilva et al. (2007)
Antibody S-20-4, chain L;				
S-20-4, chain H	An anti-carbohydrate antibody directed against *V. cholerae* O1 in complex with antigen	Alzari PM, Souchon H	1F4W	Villeneuve et al. (2000)
Monoclonal antibody F-22-30	A monoclonal antibody directed against an antigenic determinant common to Ogawa and Inaba serotypes of *Vibrio cholerae* O1	Ahmed F, Haouz A, Nato F, Fournier JM, Alzari PM	2UYL	Villeneuve et al. (2000)
HlyA	*V. cholerae* cytolysin (HlyA) pro-toxin with bound octylglucoside	Olson R, Gouaux E	1XEZ	Olson and Gouaux (2005)

RTX	Actin dimer crosslinked by *V. cholerae* MARTX toxin and complexed with DNAse I and Gelsolin-segment 1	Sawaya MR, Kudryashov DS, Pashkov I, Reisler E, Yeates TO	3CJC	Unpublished
TcpG	Disulfide oxidoreductase	Hu S-H, Martin JL	1BED	Hu et al. (1997)
Eps	*V. cholerae* putative NTPase EpsE	Robien MA, Krumm BE, Sandkvist M, Hol WGJ	1P9R	Robien et al. (2003)
	V. cholerae putative NTPase EpsE	Robien MA, Krumm BE, Sandkvist M, Hol, WGJ	1P9W	Robien et al. (2003)
	Periplasmic domain of EpsM from *V. cholerae*	Abendroth J, Hol WGJ	1UV7	Abendroth et al. (2004b)
	Cyto-EpsL: the cytoplasmic domain of EpsL, an inner-membrane component of the typeII secretion system of *V. cholerae*	Abendroth J, Bagdasarian M, Sansdkvist M, Hol WGJ	1W97	Abendroth et al. (2004a)
	Cyto-EpsL: the cytoplasmic domain of EpsL, an inner-membrane component of the type II secretion system of *V. cholerae*	Abendroth J, Murphy P, Mushtaq A, Sandkvist M, Bagdasarian M, Hol WGJ	1YF5	Abendroth et al. (2005)
	The general secretion pathway complex of the N-terminal domain of EpsE and the cytosolic domain of EpsL of *V. cholerae*	Abendroth J, Murphy P, Mushtaq A, Sandkvist M, Bagdasarian M, Hol WGJ	2BH1	Abendroth et al. (2005)
	PDZ domain of EpsC from *V. cholerae*, residues 204–305	Korotkov KV, Krumm B, Bagdasarian M, Hol WGJ	2I4S	Korotkov et al. (2006)
	PDZ domain of EpsC from *V. cholerae*, residues 219–305	Korotkov KV, Krumm B, Bagdasarian M, Hol WGJ	2I6V	Korotkov et al. (2006)
	Minor pseudopilin EpsH from the type II secretion system of *V. cholerae*	Yanez ME, Korotkov KV, Abendroth J, Hol WGJ	2QV8	Yanez et al. (2008a)
	A binary complex of two pseudopilins: EpsI and EpsJ from the type II secretion system of *V. vulnificus*	Yanez ME, Korotkov KV, Abendroth J, Hol WGJ	2RET	Yanez et al. (2008b)

* Data taken from the Research Collaboratory for Structural Bioinformatics Protein Data Bank (RCSB-PDB) at http://www.rcsb.org/pdb/

Table A11 Some methods for assaying cholera toxins

Toxin	Method of detection		References
	Broad classification	Method	
Cholera toxin	Bioassay based on animal model	Ligated rabbit ileal loop assay	De and Chatterjee (1953), De (1959)
		Infant rabbit infection model	Dutta et al. (1959), Finkelstein et al. (1964)
		Rabbit skin test	Craig (1965)
		Infant mouse model	Baselski et al. (1977), Ujiye and Kobari (1970)
	Tissue culture assay	Chinese hamster ovary (CHO) cell assay	Guerrant et al. (1974)
		Y-1 mouse adrenal tumor cell assay	Donta (1974), Donta et al. (1974)
	Immunoassay	Passive hemagglutination test	Finkelstein and Peterson (1970)
		Solid-phase radioimmunoassay	Ceska et al. (1978)
		GM1-ganglioside enzyme-linked immunosorbent assay (ELISA)	Sack et al. (1980)
		Bead ELISA	Oku et al. (1988)
		Monoclonal antibody-based ELISA	Svennerholm et al. (1986)
		Reversed passive latex agglutination	Ito et al. (1983)
		GM1-liposome immunoassay	Ahn-Yoon et al. (2003), Edwards and March (2007)
	Immunoassay with biosensors	GM1-ganglioside assay with biosensors	Rowe-Taitt et al. (2000), Song et al. (2000), Alfonta et al. (2001), Ewalt et al. (2001), Konry et al. (2003), Viswanathan et al. (2006)
	Colorimetric bioassay with nanoparticles	Colorimetric assay with glycol-nano particles	Schofield et al. (2007)
Zonula occludens toxin (Zot)	Bioassay based on animal model	Ussing chamber	Fasano et al. (1991)
Accessory cholera enterotoxin (Ace)	Bioassay based on animal model	Ussing chamber	Trucksis et al. (1993)
Hemolysin	Immunoassay	Reverse passive latex agglutination assay	Ichinose et al. (1987)
Repeats in toxin (RTX)	Tissue culture assay	HEp-2 cell assay	Lin et al. (1999)
Nonagglutinable vibrio ST (NAG-ST)	Bioassay based on animal model	Suckling mouse assay	Arita et al. (1986)
	Immunoassay	Monoclonal antibody-based competitive ELISA	Nair et al. (1993)

Table A12 Amino acids: one- and three-letter codes

Amino acids	Abbreviation (3-letter code)	Abbreviation (1-letter code)
Alanine	Ala	A
Arginine	Arg	R
Asparagine	Asn	N
Aspartic acid (aspartate)	Asp	D
Cysteine	Cys	C
Glutamine	Gln	Q
Glutamic acid (glutamate)	Glu	E
Glycine	Gly	G
Histidine	His	H
Isoleucine	Ile	I
Leucine	Leu	L
Lysine	Lys	K
Methionine	Met	M
Phenylalanine	Phe	F
Proline	Pro	P
Serine	Ser	S
Threonine	Thr	T
Tryptophan	Trp	W
Tyrosine	Tyr	Y
Valine	Val	V
Asparagine or aspartic acid (aspartate)	Asx	B
Glutamine or glutamic acid (glutamate)	Glx	Z

References

Abendroth J, Bagdasarian M, Sandkvist M, Hol WG (2004a) The structure of the cytoplasmic domain of EpsL, an inner membrane component of the type II secretion system of *Vibrio cholerae*: An unusual member of the actin-like ATPase superfamily. J Mol Biol 344:619–633

Abendroth J, Rice AE, McLuskey K, Bagdasarian M, Hol WG (2004b) The crystal structure of the periplasmic domain of the type II secretion system protein EpsM from *Vibrio cholerae*: The simplest version of the ferredoxin fold. J Mol Biol 338:585–596

Abendroth J, Murphy P, Sandkvist M, Bagdasarian M, Hol WG (2005) The X-ray structure of the type II secretion system complex formed by the N-terminal domain of EpsE and the cytoplasmic domain of EpsL of *Vibrio cholerae*. J Mol Biol 348:845–855

Ackermann HW (1992) Frequency of morphological phage descriptions. Arch Virol 124:201–209

Ackermann HW, DuBow MS (1987a) Vibriophages. In: Ackermann HW and DuBow MS (eds.) Viruses of Prokaryotes. CRC Press, Boca Raton, FL, pp 127–130

Ackermann HW, DuBow MS (1987b) Natural group of bacteriophages. In: Ackermann HW and DuBow MS (eds.) Viruses of Prokaryotes. CRC Press, Boca Raton, FL, pp 171–218

Ackermann HW, Eisenstark A (1974) The present state of phage taxonomy. Intervirology 3:201–219

Adeyeye J, Azurmendi HF, Stroop CJ, Sozhamannan S, Williams AL et al. (2003) Conformation of the hexasaccharide repeating subunit from the *Vibrio cholerae* O139 capsular polysaccharide. Biochemistry 42:3979–3988

Adhikari PC, Chatterjee SN (1969) Fimbriation and pellicle formation of Vibrio El Tor. Indian J Med Res 57:1897–1901

Agarwal V, Biswas M, Pathak AA, Saoji AM (1995) Rapid detection of *Vibrio cholerae* 0139 in faecal specimens by coagglutination. Indian J Med Res 101:55–56

Agren LC, Ekman L, Lowenadler B, Lycke NY (1997) Genetically engineered nontoxic vaccine adjuvant that combines B cell targeting with immunomodulation by cholera toxin A1 subunit. J Immunol 158:3936–3946

Ahn-Yoon S, DeCory TR, Baeumner AJ, Durst RA (2003) Ganglioside-liposome immunoassay for the ultrasensitive detection of cholera toxin. Anal Chem 75:2256–2261

Akhiani AA, Stensson A, Schon K, Lycke N (2006) The nontoxic CTA1-DD adjuvant enhances protective immunity against *Helicobacter pylori* infection following mucosal immunization. Scand J Immunol 63:97–105

Aktories K, Barbieri JT (2005) Bacterial cytotoxins: Targeting eukaryotic switches. Nat Rev Microbiol 3:397–410

Alam M, Miyoshi S, Tomochika K, Shinoda S (1997) Hemagglutination is a novel biological function of lipopolysaccharide (LPS), as seen with the *Vibrio cholerae* O139 LPS. Clin Diagn Lab Immunol 4:604–606

Alam M, Miyoshi S, Yamamoto S, Tomochika K, Shinoda S (1996) Expression of virulence-related properties by, and intestinal adhesiveness of, *Vibrio mimicus* strains isolated from aquatic environments. Appl Environ Microbiol 62:3871–3874

Albert MJ, Alam K, Ansaruzzaman M, Qadri F, Sack RB (1994) Lack of cross-protection against diarrhea due to *Vibrio cholerae* O139 (Bengal strain) after oral immunization of rabbits with *V. cholerae* O1 vaccine strain CVD103-HgR. J Infect Dis 169:230–231

Albert MJ, Bhuiyan NA, Rahman A, Ghosh AN, Hultenby K et al. (1996) Phage specific for *Vibrio cholerae* O139 Bengal. J Clin Microbiol 34:1843–1845

Aldova E, Laznickova K, Stepankova E, Lietava J (1968) Isolation of nonagglutinable vibrios from an enteritis outbreak in Czechoslovakia. J Infect Dis 118:25–31

Alfonta L, Willner I, Throckmorton DJ, Singh AK (2001) Electrochemical and quartz crystal microbalance detection of the cholera toxin employing horseradish peroxidase and GM1-functionalized liposomes. Anal Chem 73:5287–5295

Ali A, Johnson JA, Franco AA, Metzger DJ, Connell TD, Morris JG Jr, Sozhamannan S (2000) Mutations in the extracellular protein secretion pathway genes (eps) interfere with rugose polysaccharide production in and motility of *Vibrio cholerae*. Infect Immun 68:1967–1974

Allaoui A, Sansonetti PJ, Parsot C (1992) MxiJ, a lipoprotein involved in secretion of Shigella Ipa invasins, is homologous to YscJ, a secretion factor of the Yersinia Yop proteins. J Bacteriol 174:7661–7669

Alm RA, Manning PA (1990) Characterization of the hlyB gene and its role in the production of the El Tor haemolysin of *Vibrio cholerae* O1. Mol Microbiol 4:413–425

Anderson JM, Balda MS, Fanning AS (1993) The structure and regulation of tight junctions. Curr Opin Cell Biol 5:772–778

Angstrom J, Teneberg S, Karlsson KA (1994) Delineation and comparison of ganglioside-binding epitopes for the toxins of *Vibrio cholerae, Escherichia coli*, and *Clostridium tetani*: Evidence for overlapping epitopes. Proc Natl Acad Sci U S A 91:11859–11863

Anonymous (1979) Intestinal immunity and vaccine development: A WHO memorandum. Bull World Health Organ 57:719–734

Anonymous (2003) Research Report 1999–2003. Institut Pasteur 44

Apter FM, Michetti P, Winner LS, Mack JA, Mekalanos JJ, Neutra MR (1993) Analysis of the roles of anti-lipopolysaccharide and anti-cholera toxin immunoglobulin A(IgA) antibodies in protection against *Vibrio cholerae* and cholera toxin by use of monoclonal IgA antibodies in vivo. Infect Immun 61:5279–5285

Arita M, Honda T, Miwatani T, Ohmori K, Takao T, Shimonishi Y (1991) Purification and characterization of a new heat-stable enterotoxin produced by *Vibrio cholerae* non-O1 serogroup Hakata. Infect Immun 59:2186–2188

Arita M, Takeda T, Honda T, Miwatani T (1986) Purification and characterization of *Vibrio cholerae* non-O1 heat-stable enterotoxin. Infect Immun 52:45–49

Armstrong IL, Redmond JW (1974) The fatty acids present in the lipopolysaccharide of *Vibrio cholerae* 569B (Inaba). Biochim Biophys Acta 348:302–305

Armstrong ME, Lavelle EC, Loscher CE, Lynch MA, Mills KH (2005) Proinflammatory responses in the murine brain after intranasal delivery of cholera toxin: Implications for the use of AB toxins as adjuvants in intranasal vaccines. J Infect Dis 192:1628–1633

Attridge SR, Dearlove C, Beyer L, van den Bosch L, Howles A et al. (1991) Characterization and immunogenicity of EX880, a *Salmonella typhi* Ty21a-based clone which produces *Vibrio cholerae* O antigen. Infect Immun 59:2279–2284

Attridge SR, Fazeli A, Manning PA, Stroeher UH (2001) Isolation and characterization of bacteriophage-resistant mutants of *Vibrio cholerae* O139. Microb Pathog 30:237–246

Attridge SR, Qadri F, Albert MJ, Manning PA (2000) Susceptibility of *Vibrio cholerae* O139 to antibody-dependent, complement-mediated bacteriolysis. Clin Diagn Lab Immunol 7:444–450

Attridge SR, Rowley D (1983) The role of the flagellum in the adherence of *Vibrio cholerae*. J Infect Dis 147:864–872

Attridge SR, Voss E, Manning PA (1993) The role of toxin-coregulated pili in the pathogenesis of *Vibrio cholerae* O1 El Tor. Microb Pathog 15:421–431

Attridge SR, Wallerstrom G, Qadri F, Svennerholm AM (2004) Detection of antibodies to toxin-coregulated pili in sera from cholera patients. Infect Immun 72:1824–1827

Badizadegan K, Wheeler HE, Fujinaga Y, Lencer WI (2004) Trafficking of cholera toxin-ganglioside GM1 complex into Golgi and induction of toxicity depend on actin cytoskeleton. Am J Physiol Cell Physiol 287:C1453–1462

Bailey MJ, Hughes C, Koronakis V (1997) RfaH and the ops element, components of a novel system controlling bacterial transcription elongation. Mol Microbiol 26:845–851

Bandyopadhyay R, Das J (1994) The DNA adenine methyltransferase-encoding gene (dam) of *Vibrio cholerae*. Gene 140:67–71

Banerjee R, Das S, Mukhopadhyay K, Nag S, Chakrabortty A, Chaudhuri K (2002) Involvement of in vivo induced cheY-4 gene of *Vibrio cholerae* in motility, early adherence to intestinal epithelial cells and regulation of virulence factors. FEBS Lett 532:221–226

Banerjee SK, Chatterjee SN (1981) Liquid holding recovery and photoreactivation of the ultraviolet-inactivated Vibrios. Indian J Biochem Biophys 18:60–62

Banerjee SK, Chatterjee SN (1984) Effects of furazolidone on the mutation of *Vibrio cholerae* cells to streptomycin resistance. Curr Microbiol 10:19–22

Barua D, Burrows W (1974) Cholera. WB Saunders & Co., Philadelphia, PA

Barua D, Chatterjee SN (1964) Electron microscopy of El Tor Vibrios. Indian J Med Res 52:828–830

Basak J, Chatterjee SN (1994) Induction of adaptive response by nitrofurantoin against oxidative DNA damage in some bacterial cells. Mutat Res 321:127–132

Basak J, Mukherjee U, Chatterjee SN (1992) Adaptive response of *Vibrio cholerae* and *Escherichia coli* to nitrofurantoin. Environ Mol Mutagen 20:53–60

Baselski V, Briggs R, Parker C (1977) Intestinal fluid accumulation induced by oral challenge with *Vibrio cholerae* or cholera toxin in infant mice. Infect Immun 15:704–712

Basu A, Mukhopadhyay AK, Sharma C, Jyot J, Gupta N et al. (1998) Heterogeneity in the organization of the CTX genetic element in strains of *Vibrio cholerae* O139 Bengal isolated from Calcutta, India and Dhaka, Bangladesh and its possible link to the dissimilar incidence of O139 cholera in the two locales. Microb Pathog 24:175–183

Baudry B, Fasano A, Ketley J, Kaper JB (1992) Cloning of a gene (zot) encoding a new toxin produced by *Vibrio cholerae*. Infect Immun 60:428–434

Baum JA (1994) Tn5401, a new class II transposable element from *Bacillus thuringiensis*. J Bacteriol 176:2835–2845

Baumann P, Furniss AL, Lee JV (1984) Genus 1. Vibrio. In: Kricg NR and Holt JG (eds.) Bergey's Manual of Systematic Bacteriology. Williams and Wilkins, Baltimore, MD, pp 518–538

Bear CE, Li CH, Kartner N, Bridges RJ, Jensen TJ, Ramjeesingh M, Riordan JR (1992) Purification and functional reconstitution of the cystic fibrosis transmembrane conductance regulator (CFTR). Cell 68:809–818

Beck NA, Krukonis ES, DiRita VJ (2004) TcpH influences virulence gene expression in *Vibrio cholerae* by inhibiting degradation of the transcription activator TcpP. J Bacteriol 186:8309–8316

Bellamy JE, Knop J, Steele EJ, Chaicumpa W, Rowley D (1975) Antibody cross-linking as a factor in immunity to cholera in infant mice. J Infect Dis 132:181–188

Benitez JA, Garcia L, Silva A, Garcia H, Fando R et al. (1999) Preliminary assessment of the safety and immunogenicity of a new CTXPhi-negative, hemagglutinin/protease-defective El Tor strain as a cholera vaccine candidate. Infect Immun 67:539–545

Benitez JA, Spelbrink RG, Silva A, Phillips TE, Stanley CM, Boesman-Finkelstein M, Finkelstein RA (1997) Adherence of *Vibrio cholerae* to cultured differentiated human intestinal cells: An in vitro colonization model. Infect Immun 65:3474–3477

Berche P, Poyart C, Abachin E, Lelievre H, Vandepitte J, Dodin A, Fournier JM (1994) The novel epidemic strain O139 is closely related to the pandemic strain O1 of *Vibrio cholerae*. J Infect Dis 170:701–704

Bergstrom N, Nair GB, Weintraub A, Jansson PE (2002) Structure of the O-polysaccharide from the lipopolysaccharide from *Vibrio cholerae* O6. Carbohydr Res 337:813–817

Berschneider HM, Knowles MR, Azizkhan RG, Boucher RC, Tobey NA, Orlando RC, Powell DW (1988) Altered intestinal chloride transport in cystic fibrosis. FASEB J 2:2625–2629

Betley MJ, Miller VL, Mekalanos JJ (1986) Genetics of bacterial enterotoxins. Annu Rev Microbiol 40:577–605

Bhadra RK, Roychoudhury S, Banerjee RK, Kar S, Majumdar R et al. (1995) Cholera toxin (CTX) genetic element in *Vibrio cholerae* O139. Microbiology 141 (Pt 8):1977–1983

Bhanumathi R, Sabeena F, Isac SR, Shukla BN, Singh DV (2003) Molecular characterization of *Vibrio cholerae* O139 Bengal isolated from water and the aquatic plant *Eichhornia crassipes* in the River Ganga, Varanasi, India. Appl Environ Microbiol 69:2389–2394

Bhaskaran K (1958) Genetic recombination in *Vibrio cholerae*. J Gen Microbiol 19:71–75

Bhattacharya R, Chatterjee SN (1994) Effect of rehydrating fluid "Electral" on *Vibrio cholerae* cells. Ind J Exp Biol 32:44–48

Bhattacharya T, Chatterjee S, Maiti D, Bhadra RK, Takeda Y, Nair GB, Nandy RK (2006) Molecular analysis of the rstR and orfU genes of the CTX prophages integrated in the small chromosomes of environmental *Vibrio cholerae* non-O1, non-O139 strains. Environ Microbiol 8:526–634

Bhattacharyya S, Ghosh S, Shant J, Ganguly NK, Majumdar S (2004) Role of the W07-toxin on *Vibrio cholerae*-induced diarrhoea. Biochim Biophys Acta 1670:69–80

Bhattacharyya S, Shant J, Ganguly NK, Majumdar S, Ghosh S (2008) A potential epidemic factor from the bacteria, *Vibrio cholerae* WO7. Curr Microbiol 56:98–103

Bhattacharyya SC, Samad SA, Mandal JC, Chatterjee SN (1991) X-ray inactivation, Weigle reactivation, and Weigle mutagenesis of the lysogenic Vibrio kappa phage. Can J Microbiol 37:265–269

Bignold LP, Rogers SD, Siaw TM, Bahnisch J (1991) Inhibition of chemotaxis of neutrophil leukocytes to interleukin-8 by endotoxins of various bacteria. Infect Immun 59:4255–4258

Bik EM, Bunschoten AE, Gouw RD, Mooi FR (1995) Genesis of the novel epidemic *Vibrio cholerae* O139 strain: Evidence for horizontal transfer of genes involved in polysaccharide synthesis. Embo J 14:209–216

Bik EM, Bunschoten AE, Willems RJ, Chang AC, Mooi FR (1996) Genetic organization and functional analysis of the otn DNA essential for cell-wall polysaccharide synthesis in *Vibrio cholerae* O139. Mol Microbiol 20:799–811

Bilge SS, Vary JC Jr, Dowell SF, Tarr PI (1996) Role of the *Escherichia coli* O157:H7 O side chain in adherence and analysis of an rfb locus. Infect Immun 64:4795–4801

Bina J, Zhu J, Dziejman M, Faruque S, Calderwood S, Mekalanos J (2003) ToxR regulon of *Vibrio cholerae* and its expression in vibrios shed by cholera patients. Proc Natl Acad Sci USA 100:2801–2806

Bina JE, Mekalanos JJ (2001) *Vibrio cholerae* tolC is required for bile resistance and colonization. Infect Immun 69:4681–4685

Blake PA (1994) Historical perspectives on pandemic cholera. In: Wachsmuth IK, Blake PA, and Olsvik O (eds.) Vibrio cholerae and Cholera. ASM Press, Washington, DC, pp 293–295

Blake PA, Weaver RE, Hollis DG (1980) Diseases of humans (other than cholera) caused by vibrios. Annu Rev Microbiol 34:341–367

Blakely GW (2004) Smarter than the average phage. Mol Microbiol 54:851–854

Blokesch M, Schoolnik GK (2007) Serogroup conversion of *Vibrio cholerae* in aquatic reservoirs. PLoS Pathog 3:e81

Blum G, Ott M, Lischewski A, Ritter A, Imrich H, Tschape H, Hacker J (1994) Excision of large DNA regions termed pathogenicity islands from tRNA-specific loci in the chromosome of an *Escherichia coli* wild-type pathogen. Infect Immun 62:606–614

Boardman BK, Satchell KJ (2004) *Vibrio cholerae* strains with mutations in an atypical type I secretion system accumulate RTX toxin intracellularly. J Bacteriol 186:8137–8143

Boesman-Finkelstein M, Walton NE, Finkelstein RA (1989) Bovine lactogenic immunity against cholera toxin-related enterotoxins and *Vibrio cholerae* outer membranes. Infect Immun 57:1227–1234

Booth BA, Sciortino CV, Finkelstein RA (1985) Adhesins of *Vibrio cholerae*. In: Mirelman D (ed.) Microbial Lectins and Agglutinins. Wiley, New York, pp 169–182

Bougoudogo F, Vely F, Nato F (1995) Protective activities of serum immunoglobulin G on the mucosal surface to *Vibrio cholerae O1*. Bull Inst Pasteur 93:273–283

Boutonnier A, Dassy B, Dumenil R, Guenole A, Ratsitorahina M, Migliani R, Fournier JM (2003) A simple and convenient microtiter plate assay for the detection of bactericidal antibodies to *Vibrio cholerae* O1 and *Vibrio cholerae* O139. J Microbiol Methods 55:745–753

Boutonnier A, Villeneuve S, Nato F, Dassy B, Fournier JM (2001) Preparation, immunogenicity, and protective efficacy, in a murine model, of a conjugate vaccine composed of the polysaccharide moiety of the lipopolysaccharide of *Vibrio cholerae* O139 bound to tetanus toxoid. Infect Immun 69:3488–3493

Brade H (1985) Occurrence of 2-keto-deoxyoctonic acid 5-phosphate in lipopolysaccharides of *Vibrio cholerae* Ogawa and Inaba. J Bacteriol 161:795–798

Broady KW, Rietschel ET, Luderitz O (1981) The chemical structure of the lipid A component of lipopolysaccharides from *Vibrio cholerae*. Eur J Biochem 115:463–468

Bromander AK, Kjerrulf M, Holmgren J, Lycke N (1993) Cholera toxin enhances alloantigen presentation by cultured intestinal epithelial cells. Scand J Immunol 37:452–458

Brown DA, London E (2000) Structure and function of sphingolipid- and cholesterol-rich membrane rafts. J Biol Chem 275:17221–17224

Brown RC, Taylor RK (1995) Organization of tcp, acf, and toxT genes within a ToxT-dependent operon. Mol Microbiol 16:425–439

Burrows W, Mather AN, McGann VG, Wagner SM (1946) Studies on immunity to Asiatic cholera. J Infect Dis 79:159–167

Cabral-Lilly D, Sosinsky GE, Reed RA, McDermott MR, Shipley GG (1994) Orientation of cholera toxin bound to model membranes. Biophys J 66:935–941

Calia KE, Murtagh M, Ferraro MJ, Calderwood SB (1994) Comparison of *Vibrio cholerae* O139 with *V. cholerae* O1 classical and El Tor biotypes. Infect Immun 62:1504–1506

Callahan LT 3rd, Richardson SH (1973) Biochemistry of *Vibrio cholerae* virulence. 3. Nutritional requirements for toxin production and the effects of pH on toxin elaboration in chemically defined media. Infect Immun 7:567–572

Camberg JL, Sandkvist M (2005) Molecular analysis of the *Vibrio cholerae* type II secretion ATPase EpsE. J Bacteriol 187:249–256

Campos J, Martinez E, Marrero K, Silva Y, Rodriguez BL et al. (2003a) Novel type of specialized transduction for CTX phi or its satellite phage RS1 mediated by filamentous phage VGJ phi in *Vibrio cholerae*. J Bacteriol 185:7231–7240

Campos J, Martinez E, Suzarte E, Rodriguez BL, Marrero K et al. (2003b) VGJ phi, a novel filamentous phage of *Vibrio cholerae*, integrates into the same chromosomal site as CTX phi. J Bacteriol 185:5685–5696

Carillo L, Gilman RH, Mantle RE, Nunez N, Watanabe J et al. (1994) Rapid detection of *Vibrio cholerae* O1 in stools of Peruvian cholera patients by using monoclonal immunodiagnostic kits (Loyaza Cholera Working Group in Peru). J Clin Microbiol 32:856–857

Caroff M, Karibian D, Cavaillon JM, Haeffner-Cavaillon N (2002) Structural and functional analyses of bacterial lipopolysaccharides. Microbes Infect 4:915–926

Caroff M, Lebbar S, Szabo L (1987) Do endotoxins devoid of 3-deoxy-D-manno-2-octulosonic acid exist? Biochem Biophys Res Commun 143:845–847

Carroll PA, Tashima KT, Rogers MB, DiRita VJ, Calderwood SB (1997) Phase variation in tcpH modulates expression of the ToxR regulon in *Vibrio cholerae*. Mol Microbiol 25:1099–1111

Cash RA, Forrest JN, Nalin DR, Abrutyn E (1970) Rapid correction of acidosis and dehydration of cholera with oral electrolyte and glucose solution. Lancet 2:549–550

Cassel D, Selinger Z (1977) Mechanism of adenylate cyclase activation by cholera toxin: Inhibition of GTP hydrolysis at the regulatory site. Proc Natl Acad Sci USA 74:3307–3311

CDC (1994) Laboratory methods for the diagnosis of *Vibrio cholerae*. Centers for Disease Control and Prevention, Atlanta, GA

Ceska M, Effenberger F, Grossmuller F (1978) Highly sensitive solid-phase radioimmunoassay suitable for determination of low amounts of cholera toxin and cholera toxin antibodies. J Clin Microbiol 7:209–213

Chakrabarti SR, Chaudhuri K, Sen K, Das J (1996) Porins of *Vibrio cholerae*: Purification and characterization of OmpU. J Bacteriol 178:524–530

Chakrabortty A, Das S, Majumdar S, Mukhopadhyay K, Roychoudhury S, Chaudhuri K (2000) Use of RNA arbitrarily primed-PCR fingerprinting to identify *Vibrio cholerae* genes differentially expressed in the host following infection. Infect Immun 68:3878–3887

Champion GA, Neely MN, Brennan MA, DiRita VJ (1997) A branch in the ToxR regulatory cascade of *Vibrio cholerae* revealed by characterization of toxT mutant strains. Mol Microbiol 23:323–331

Chanda PK, Chatterjee SN (1976) Photoreactivating property of the *Vibrio cholerae* cell system. Can J Microbiol 22:1186–1187

Chang HS, Sack DA (2001) Detection of anti-lipopolysaccharide antibodies to *Vibrio cholerae* O1 and O139 using a novel microtiter limulus amebocyte lysate (LAL) assay. Clin Chim Acta 312:49–54

Chatterjee SN (1990) Some aspects of cholera research. Curr Sci 59:677–687.

Chatterjee SN, Adhikari PC, Maiti M, Chaudhuri CR, Sur P (1974) Growth of *Vibrio cholerae* cells: Biochemical and electron microscopic study. Indian J Exp Biol 12:35–45

Chatterjee SN, Chaudhuri K (2003) Lipopolysaccharides of *Vibrio cholerae*. I. Physical and chemical characterization. Biochim Biophys Acta 1639:65–79

Chatterjee SN, Chaudhuri K (2004) Lipopolysaccharides of *Vibrio cholerae* II. Genetics of biosynthesis. Biochim Biophys Acta 1690:93–109

Chatterjee SN, Chaudhuri K (2006) Lipopolysaccharides of *Vibrio cholerae*: III. Biological functions. Biochim Biophys Acta 1762:1–16

Chatterjee SN, Das J (1966) Secretory activity of *Vibrio cholerae* as evidenced by electron microscopy. In: Uyeda R (ed.) Electron Microscopy 1966. Maruzen Co. Ltd., Tokyo, pp 139–140

Chatterjee SN, Das J (1967) Electron microscopic observations on the excretion of cell-wall material by *Vibrio cholerae*. J Gen Microbiol 49:1–11

Chatterjee SN, Das J, Barua D (1965) Electron microscopy of cholera phages. Indian J Med Res 53:934–937

Chatterjee SN, Maiti M (1973) Effects of furazolidone on the infection of *Vibrio cholerae* by the phage ϕ149. J Virol 11:872–878

Chatterjee SN, Maiti M (1984) Vibriophages and vibriocins: Physical, chemical, and biological properties. Adv Virus Res 29:263–312

Chatterjee SN, Sur P (1974) Surface blebs on *Vibrio cholerae* cells. In: Saders JV and Goodchild DJ (ed) Proc 8th Int Congr Electron Microsc, Canberra, Australia, 25–31 Aug 1974, pp 652–653

Chaudhary VK, Jinno Y, FitzGerald D, Pastan I (1990) *Pseudomonas* exotoxin contains a specific sequence at the carboxyl terminus that is required for cytotoxicity. Proc Natl Acad Sci USA 87:308–312

Chaudhuri AG, Bhattacharya J, Nair GB, Takeda T, Chakrabarti MK (1998) Rise of cytosolic Ca^{2+} and activation of membrane-bound guanylyl cyclase activity in rat enterocytes by heat-stable enterotoxin of *Vibrio cholerae* non-01. FEMS Microbiol Lett 160:125–129

Chedid L, Audibert F, Bona C, Damais C, Parant F, Parant M (1975) Biological activities of endotoxins detoxified by alkylation. Infect Immun 12:714–721

Chen Y, Bystricky P, Adeyeye J, Panigrahi P, Ali A et al. (2007) The capsule polysaccharide structure and biogenesis for non-O1 *Vibrio cholerae* NRT36S: Genes are embedded in the LPS region. BMC Microbiol 7:20

Chernyak A, Karavanov A, Ogawa Y, Kovac P (2001) Conjugating oligosaccharides to proteins by squaric acid diester chemistry: Rapid monitoring of the progress of conjugation, and recovery of the unused ligand. Carbohydr Res 330:479–486

Chernyak A, Kondo S, Wade TK, Meeks MD, Alzari PM et al. (2002) Induction of protective immunity by synthetic *Vibrio cholerae* hexasaccharide derived from *V. cholerae* O1 Ogawa lipopolysaccharide bound to a protein carrier. J Infect Dis 185:950–962

Chiang SL, Mekalanos JJ (1998) Use of signature-tagged transposon mutagenesis to identify *Vibrio cholerae* genes critical for colonization. Mol Microbiol 27:797–805

Chiang SL, Mekalanos JJ (1999) rfb mutations in *Vibrio cholerae* do not affect surface production of toxin-coregulated pili but still inhibit intestinal colonization. Infect Immun 67:976–980

Chiang SL, Mekalanos JJ, Holden DW (1999) In vivo genetic analysis of bacterial virulence. Annu Rev Microbiol 53:129–154

Chiang SL, Taylor RK, Koomey M, Mekalanos JJ (1995) Single amino acid substitutions in the N-terminus of *Vibrio cholerae* TcpA affect colonization, autoagglutination, and serum resistance. Mol Microbiol 17:1133–1142

Chiavelli DA, Marsh JW, Taylor RK (2001) The mannose-sensitive hemagglutinin of *Vibrio cholerae* promotes adherence to zooplankton. Appl Environ Microbiol 67:3220–3225

Childers BM, Klose KE (2007) Regulation of virulence in *Vibrio cholerae*: The ToxR regulon. Future Microbiol 2:335–344

Childers BM, Weber GG, Prouty MG, Castaneda MM, Peng F, Klose KE (2007) Identification of residues critical for the function of the *Vibrio cholerae* virulence regulator ToxT by scanning alanine mutagenesis. J Mol Biol 367:1413–1430

Chitnis DS, Sharma KD, Kamat RS (1982a) Role of bacterial adhesion in the pathogenesis of cholera. J Med Microbiol 15:43–51

Chitnis DS, Sharma KD, Kamat RS (1982b) Role of somatic antigen of *Vibrio cholerae* in adhesion to intestinal mucosa. J Med Microbiol 15:53–61

Chow KH, Ng TK, Yuen KY, Yam WC (2001) Detection of RTX toxin gene in *Vibrio cholerae* by PCR. J Clin Microbiol 39:2594–2597

Chowdhury TA, Jansson PE, Lindberg B, Lindberg J, Gustafsson B, Holme T (1991) Structural studies of the *Vibrio cholerae* O:3 O-antigen polysaccharide. Carbohydr Res 215:303–314

Clemens JD, van Loon F, Sack DA, Chakraborty J, Rao MR et al. (1991) Field trial of oral cholera vaccines in Bangladesh: Serum vibriocidal and antitoxic antibodies as markers of the risk of cholera. J Infect Dis 163:1235–1242

Colwell RR, Hasan JA, Huq A, Loomis L, Siebeling RJ et al. (1992) Development and evaluation of a rapid, simple, sensitive, monoclonal antibody-based co-agglutination test for direct detection of *Vibrio cholerae* 01. FEMS Microbiol Lett 76:215–219

Comstock LE, Johnson JA, Michalski JM, Morris JG, Jr., Kaper JB (1996) Cloning and sequence of a region encoding a surface polysaccharide of *Vibrio cholerae* O139 and characterization of the insertion site in the chromosome of *Vibrio cholerae* O1. Mol Microbiol 19:815–826

Comstock LE, Maneval D, Jr., Panigrahi P, Joseph A, Levine MM et al. (1995) The capsule and O antigen in *Vibrio cholerae* O139 Bengal are associated with a genetic region not present in *Vibrio cholerae* O1. Infect Immun 63:317–323

Connell TD (2007) Cholera toxin, LT-I, LT-IIa and LT-IIb: the critical role of ganglioside binding in immunomodulation by type I and type II heat-labile enterotoxins. Expert Rev Vaccines 6:821–834

Connell TD, Metzger D, Sfintescu C, Evans RT (1998a) Immunostimulatory activity of LT-IIa, a type II heat-labile enterotoxin of *Escherichia coli*. Immunol Lett 62:117–120

Connell TD, Metzger DJ, Lynch J, Folster JP (1998b) Endochitinase is transported to the extracellular milieu by the eps-encoded general secretory pathway of *Vibrio cholerae*. J Bacteriol 180:5591–5600

Connell TD, Metzger DJ, Wang M, Jobling MG, Holmes RK (1995) Initial studies of the structural signal for extracellular transport of cholera toxin and other proteins recognized by *Vibrio cholerae*. Infect Immun 63:4091–4098

Cordero CL, Sozhamannan S, Satchell KJ (2007) RTX toxin actin cross-linking activity in clinical and environmental isolates of *Vibrio cholerae*. J Clin Microbiol 45:2289–2292

Coster TS, Killeen KP, Waldor MK, Beattie DT, Spriggs DR et al. (1995) Safety, immunogenicity, and efficacy of live attenuated *Vibrio cholerae* O139 vaccine prototype. Lancet 345: 949–952

Cox AD, Brisson JR, Thibault P, Perry MB (1997) Structural analysis of the lipopolysaccharide from *Vibrio cholerae* serotype O22. Carbohydr Res 304:191–208

Cox AD, Brisson JR, Varma V, Perry MB (1996) Structural analysis of the lipopolysaccharide from *Vibrio cholerae* O139. Carbohydr Res 290:43–58

Cox AD, Perry MB (1996) Structural analysis of the O-antigen-core region of the lipopolysaccharide from *Vibrio cholerae* O139. Carbohydr Res 290:59–65

Cox DS, Gao H, Raje S, Scott KR, Eddington ND (2001) Enhancing the permeation of marker compounds and enaminone anticonvulsants across Caco-2 monolayers by modulating tight junctions using zonula occludens toxin. Eur J Pharm Biopharm 52:145–150

Cox DS, Raje S, Gao H, Salama NN, Eddington ND (2002) Enhanced permeability of molecular weight markers and poorly bioavailable compounds across Caco-2 cell monolayers using the absorption enhancer, zonula occludens toxin. Pharm Res 19:1680–1688

Craig JP (1965) A permeability factor (toxin) found in cholera stools and culture filtrates and its neutralization by convalescent cholera sera. Nature 207:614–616

Craig JP (1966) Preparation of the vascular permeability factor of *Vibrio cholerae*. J Bacteriol 92:793–795

Craig L, Pique ME, Tainer JA (2004) Type IV pilus structure and bacterial pathogenicity. Nat Rev Microbiol 2:363–378

Craig L, Taylor RK, Pique ME, Adair BD, Arvai AS et al. (2003) Type IV pilin structure and assembly: X-ray and EM analyses of *Vibrio cholerae* toxin-coregulated pilus and *Pseudomonas aeruginosa* PAK pilin. Mol Cell 11:1139–1150

Crawford JA, Kaper JB, DiRita VJ (1998) Analysis of ToxR-dependent transcription activation of ompU, the gene encoding a major envelope protein in *Vibrio cholerae*. Mol Microbiol 29:235–246

Cryz SJ, Jr., Furer E, Germanier R (1984) Protection against fatal *Pseudomonas aeruginosa* burn wound sepsis by immunization with lipopolysaccharide and high-molecular-weight polysaccharide. Infect Immun 43:795–799

Cryz SJ, Jr., Kaper J, Tacket C, Nataro J, Levine MM (1995) *Vibrio cholerae* CVD103-HgR live oral attenuated vaccine: Construction, safety, immunogenicity, excretion and non-target effects. Dev Biol Stand 84:237–244

Cuatrecasas P (1973a) Gangliosides and membrane receptors for cholera toxin. Biochemistry 12:3558–3566

Cuatrecasas P (1973b) Interaction of *Vibrio cholerae* enterotoxin with cell membranes. Biochemistry 12:3547–3558

Curlin GT, Craig JP, Subong A, Carpenter CC (1970) Antitoxic immunity in experimental canine cholera. J Infect Dis 121:463–470

Dalbey RE, Chen M (2004) Sec-translocase mediated membrane protein biogenesis. Biochim Biophys Acta 1694:37–53

Dallas WS, Falkow S (1979) The molecular nature of heat-labile enterotoxin (LT) of *Escherichia coli*. Nature 277:406–407

Dallas WS, Falkow S (1980) Amino acid sequence homology between cholera toxin and *Escherichia coli* heat-labile toxin. Nature 288:499–501

Dalsgaard A, Serichantalergs O, Forslund A, Lin W, Mekalanos J et al. (2001) Clinical and environmental isolates of *Vibrio cholerae* serogroup O141 carry the CTX phage and the genes encoding the toxin-coregulated pili. J Clin Microbiol 39:4086–4092

Das B, Ghosh RK, Sharma C, Vasin N, Ghosh A (1993) Tandem repeats of cholera toxin gene in *Vibrio cholerae* O139. Lancet 342:1173–1174

Das B, Halder K, Pal P, Bhadra RK (2007) Small chromosomal integration site of classical CTX prophage in Mozambique *Vibrio cholerae* O1 biotype El Tor strain. Arch Microbiol 188:677–683

Das J, Chatterjee SN (1966a) Electron microscopic studies on some ultrastructural aspects of *Vibrio cholerae*. Ind J Med Res 54:330–338

Das J, Chatterjee SN (1966b) Electron microscopy of cholerae phages. In: Uyeda R (ed.) Electron Microscopy 1966. Maruzen Co. Ltd., Tokyo, pp 259–260

Das S, Chakrabortty A, Banerjee R, Chaudhuri K (2002) Involvement of in vivo induced icmF gene of *Vibrio cholerae* in motility, adherence to epithelial cells, and conjugation frequency. Biochem Biophys Res Commun 295:922–928

Das S, Chakrabortty A, Banerjee R, Roychoudhury S, Chaudhuri K (2000) Comparison of global transcription responses allows identification of *Vibrio cholerae* genes differentially expressed following infection. FEMS Microbiol Lett 190:87–91

Datta A, Parker CD, Wohlhieter JA, Baron LS (1973) Isolation and characterization of the fertility factor P of *Vibrio cholerae*. J Bacteriol 113:763–771

Davis BM, Kimsey HH, Chang W, Waldor MK (1999) The *Vibrio cholerae* O139 Calcutta bacteriophage CTXphi is infectious and encodes a novel repressor. J Bacteriol 181:6779–6787

Davis BM, Kimsey HH, Kane AV, Waldor MK (2002) A satellite phage-encoded antirepressor induces repressor aggregation and cholera toxin gene transfer. Embo J 21:4240–4249

Davis BM, Moyer KE, Boyd EF, Waldor MK (2000a) CTX prophages in classical biotype *Vibrio cholerae*: Functional phage genes but dysfunctional phage genomes. J Bacteriol 182:6992–6998

Davis BM, Lawson EH, Sandkvist M, Ali A, Sozhamannan S, Waldor MK (2000b) Convergence of the secretory pathways for cholera toxin and the filamentous phage, CTXphi. Science 288:333–335

Davis BM, Waldor MK (2000) CTXphi contains a hybrid genome derived from tandemly integrated elements. Proc Natl Acad Sci USA 97:8572–8577

Davis BM, Waldor MK (2003) Filamentous phages linked to virulence of *Vibrio cholerae*. Curr Opin Microbiol 6:35–42

Deneke CF, Colwell RR (1973) Lipopolysaccharide and proteins of the cell envelope of *Vibrio marinus*, a marine bacterium. Can J Microbiol 19:1211–1217

De Silva RS, Kovacikova G, Lin W, Taylor RK, Skorupski K, Kull FJ (2007) Crystal structure of the *Vibrio cholerae* quorum-sensing regulatory protein HapR. J Bacteriol 189:5683–5691

De SN (1959) Enterotoxicity of bacteria-free culture-filtrate of *Vibrio cholerae*. Nature 183:1533–1534

De SN, Chatterje DN (1953) An experimental study of the mechanism of action of *Vibrio cholerae* on the intestinal mucous membrane. J Pathol Bacteriol 66:559–562

Devoe IW, Gilchrist JE (1973) Release of endotoxin in the form of cell wall blebs during in vitro growth of *Neisseria meningitides*. J Exp Med 138:1156–1167

de Wolf MJ, Fridkin M, Epstein M, Kohn LD (1981a) Structure–function studies of cholera toxin and its A and B protomers. Modification of tryptophan residues. J Biol Chem 256:5481–5488

de Wolf MJ, Fridkin M, Kohn LD (1981b) Tryptophan residues of cholera toxin and its A and B protomers. Intrinsic fluorescence and solute quenching upon interacting with the ganglioside GM1, oligo-GM1, or dansylated oligo-GM1. J Biol Chem 256:5489–5496

Desmarchelier PM, Senn CR (1989) A molecular epidemiological study of *Vibrio cholerae* in Australia. Med J Aust 150:631–634

Dharmasena MN, Jewell DA, Taylor RK (2007) Development of peptide mimics of a protective epitope of *Vibrio cholerae* Ogawa O-antigen and investigation of the structural basis of peptide mimicry. J Biol Chem 282:33805–33816

Di Pierro M, Lu R, Uzzau S, Wang W, Margaretten K et al. (2001) Zonula occludens toxin structure-function analysis. Identification of the fragment biologically active on tight junctions and of the zonulin receptor binding domain. J Biol Chem 276:19160–19165

Diks SH, Richel DJ, Peppelenbosch MP (2004) LPS signal transduction: The picture is becoming more complex. Curr Top Med Chem 4:1115–1126

DiRita VJ (1992) Co-ordinate expression of virulence genes by ToxR in *Vibrio cholerae*. Mol Microbiol 6:451–458

DiRita VJ, Mekalanos JJ (1991) Periplasmic interaction between two membrane regulatory proteins, ToxR and ToxS, results in signal transduction and transcriptional activation. Cell 64:29–37

DiRita VJ, Neely M, Taylor RK, Bruss PM (1996) Differential expression of the ToxR regulon in classical and E1 Tor biotypes of *Vibrio cholerae* is due to biotype-specific control over toxT expression. Proc Natl Acad Sci USA 93:7991–7995

DiRita VJ, Parsot C, Jander G, Mekalanos JJ (1991) Regulatory cascade controls virulence in *Vibrio cholerae*. Proc Natl Acad Sci USA 88:5403–5407

Donta ST (1974) Comparison of the effects of cholera enterotoxin and ACTH on adrenal cells in tissue culture. Am J Physiol 227:109–113

Donta ST, Beristain S, Tomicic TK (1993) Inhibition of heat-labile cholera and *Escherichia coli* enterotoxins by brefeldin A. Infect Immun 61:3282–3286

Donta ST, Moon HW, Whipp SC (1974) Detection of heat-labile *Escherichia coli* enterotoxin with the use of adrenal cells in tissue culture. Science 183:334–336

Dorward DW, Garon CF (1989) Export and intracellular transfer of DNA via membrane blebs of *Neisseria gonorrhoeae*. J Bacteriol 171:2499–2505

Dreisbach VC, Cowley S, Elkins KL (2000) Purified lipopolysaccharide from *Francisella tularensis* live vaccine strain (LVS) induces protective immunity against LVS infection that requires B cells and gamma interferon. Infect Immun 68:1988–1996

Dumontier S, Berche P (1998) *Vibrio cholerae* O22 might be a putative source of exogenous DNA resulting in the emergence of the new strain of *Vibrio cholerae* O139. FEMS Microbiol Lett 164:91–98

Duong F, Eichler J, Price A, Leonard MR, Wickner W (1997) Biogenesis of the Gram-negative bacterial envelope. Cell 91:567–573

Duret G, Delcour AH (2006) Deoxycholic acid blocks *Vibrio cholerae* OmpT but not OmpU porin. J Biol Chem 281:19899–19905

Dutta NK, Panse MV, Kulkarni DR (1959) Role of cholera a toxin in experimental cholera. J Bacteriol 78:594–595

Dziejman M, Balon E, Boyd D, Fraser CM, Heidelberg JF, Mekalanos JJ (2002) Comparative genomic analysis of *Vibrio cholerae*: Genes that correlate with cholera endemic and pandemic disease. Proc Natl Acad Sci USA 99:1556–1561

Edwards KA, March JC (2007) GM(1)-functionalized liposomes in a microtiter plate assay for cholera toxin in *Vibrio cholerae* culture samples. Anal Biochem 368:39–48

Egan ES, Waldor MK (2003) Distinct replication requirements for the two *Vibrio cholerae* chromosomes. Cell 114:521–530

Eidels L, Proia RL, Hart DA (1983) Membrane receptors for bacterial toxins. Microbiol Rev 47:596–620

Elson CO, Ealding W (1984) Cholera toxin feeding did not induce oral tolerance in mice and abrogated oral tolerance to an unrelated protein antigen. J Immunol 133:2892–2897

Ewalt KL, Haigis RW, Rooney R, Ackley D, Krihak M (2001) Detection of biological toxins on an active electronic microchip. Anal Biochem 289:162–172

Evans DJ Jr, Richardson SH (1968) In vitro production of choleragen and vascular permeability factor by *Vibrio cholerae*. J Bacteriol 96:126–130

Everiss KD, Hughes KJ, Kovach ME, Peterson KM (1994) The *Vibrio cholerae* acfB colonization determinant encodes an inner membrane protein that is related to a family of signal-transducing proteins. Infect Immun 62:3289–3298

Fallarino A, Mavrangelos C, Stroeher UH, Manning PA (1997) Identification of additional genes required for O-antigen biosynthesis in *Vibrio cholerae* O1. J Bacteriol 179: 2147–2153

Fan E, O'Neal CJ, Mitchell DD, Robien MA, Zhang Z et al. (2004) Structural biology and structure-based inhibitor design of cholera toxin and heat-labile enterotoxin. Int J Med Microbiol 294:217–223

Faruque SM, Ahmed KM, Abdul Alim AR, Qadri F, Siddique AK, Albert MJ (1997) Emergence of a new clone of toxigenic *Vibrio cholerae* O1 biotype El Tor displacing *V. cholerae* O139 Bengal in Bangladesh. J Clin Microbiol 35:624–630

Faruque SM, Albert MJ, Mekalanos JJ (1998) Epidemiology, genetics, and ecology of toxigenic *Vibrio cholerae*. Microbiol Mol Biol Rev 62:1301–1314

Faruque SM, Asadulghani, Kamruzzaman M, Nandi RK, Ghosh AN et al. (2002) RS1 element of *Vibrio cholerae* can propagate horizontally as a filamentous phage exploiting the morphogenesis genes of CTXphi. Infect Immun 70:163–170

Faruque SM, Asadulghani, Rahman MM, Waldor MK, Sack DA (2000) Sunlight-induced propagation of the lysogenic phage encoding cholera toxin. Infect Immun 68:4795–4801

Faruque SM, Chowdhury N, Kamruzzaman M, Ahmad QS, Faruque AS et al. (2003) Reemergence of epidemic *Vibrio cholerae* O139, Bangladesh. Emerg Infect Dis 9:1116–1122

Faruque SM, Mekalanos JJ (2003) Pathogenicity islands and phages in *Vibrio cholerae* evolution. Trends Microbiol 11:505–510

Faruque SM, Naser IB, Islam MJ, Faruque AS, Ghosh AN et al. (2005a) Seasonal epidemics of cholera inversely correlate with the prevalence of environmental cholera phages. Proc Natl Acad Sci USA 102:1702–1707

Faruque SM, Islam MJ, Ahmad QS, Faruque AS, Sack DA, Nair GB, Mekalanos JJ (2005b) Self-limiting nature of seasonal cholera epidemics: Role of host-mediated amplification of phage. Proc Natl Acad Sci USA 102:6119–6124

Faruque SM, Siddique AK, Saha MN, Asadulghani, Rahman MM et al. (1999) Molecular characterization of a new ribotype of *Vibrio cholerae* O139 Bengal associated with an outbreak of cholera in Bangladesh. J Clin Microbiol 37:1313–1318

Fasano A (1999a) Cellular microbiology: can we learn cell physiology from microorganisms? Am J Physiol 276:C765–776

Fasano A (1999b) Intestinal toxins. Curr Opin Gastroenterol 15:523–528

Fasano A (2001) The genome of the smartest pathogen decoded: Is the cholera war over? J Pediatr Gastroenterol Nutr 32:1–3

Fasano A, Baudry B, Pumplin DW, Wasserman SS, Tall BD, Ketley JM, Kaper JB (1991) *Vibrio cholerae* produces a second enterotoxin, which affects intestinal tight junctions. Proc Natl Acad Sci USA 88:5242–5246

Fasano A, Fiorentini C, Donelli G, Uzzau S, Kaper JB et al. (1995) Zonula occludens toxin modulates tight junctions through protein kinase C-dependent actin reorganization, in vitro. J Clin Invest 96:710–720

Fasano A, Uzzau S (1997) Modulation of intestinal tight junctions by zonula occludens toxin permits enteral administration of insulin and other macromolecules in an animal model. J Clin Invest 99:1158–1164

Fasano A, Uzzau S, Fiore C, Margaretten K (1997) The enterotoxic effect of zonula occludens toxin on rabbit small intestine involves the paracellular pathway. Gastroenterology 112:839–846

Favre D, Cryz SJ Jr, Viret JF (1996) Construction and characterization of a potential live oral carrier-based vaccine against *Vibrio cholerae* O139. Infect Immun 64:3565–3570

Fernandes PB, Smith HL Jr (1977) The effect of anaerobiosis and bile salts on the growth and toxin production by *Vibrio cholerae*. J Gen Microbiol 98:77–86

Field M, Fromm D, McColl I (1971) Ion transport in rabbit ileal mucosa. I. Na and Cl fluxes and short-circuit current. Am J Physiol 220:1388–1396

Fields PI, Popovic T, Wachsmuth K, Olsvik O (1992) Use of polymerase chain reaction for detection of toxigenic *Vibrio cholerae* O1 strains from the Latin American cholera epidemic. J Clin Microbiol 30:2118–2121

Figueroa-Arredondo P, Heuser JE, Akopyants NS, Morisaki JH, Giono-Cerezo S, Enriquez-Rincon F, Berg DE (2001) Cell vacuolation caused by *Vibrio cholerae* hemolysin. Infect Immun 69:1613–1624

Filloux A, Bally M, Ball G, Akrim M, Tommassen J, Lazdunski A (1990) Protein secretion in gram-negative bacteria: Transport across the outer membrane involves common mechanisms in different bacteria. Embo J 9:4323–4329

Finkelstein RA (1970) Antitoxic immunity in experimental cholera: Observations with purified antigens and the ligated ileal loop model. Infect Immun 1:464–467

Finkelstein RA, LoSpalluto JJ (1969) Pathogenesis of experimental cholera. Preparation and isolation of choleragen and choleragenoid. J Exp Med 130:185–202

Finkelstein RA, LoSpalluto JJ (1970) Production of highly purified choleragen and choleragenoid. J Infect Dis 121(Suppl 121):163+
Finkelstein RA, Norris HT, Dutta NK (1964) Pathogenesis experimental cholera in infant rabbits. I. Observations on the intraintestinal infection and experimental cholera produced with cell-free products. J Infect Dis 114:203–216
Finkelstein RA, Peterson JW (1970) In vitro detection of antibody to cholera enterotoxin in cholera patients and laboratory animals. Infect Immun 1:21–29
Finkelstein RA, Vasil ML, Holmes RK (1974) Studies on toxinogenesis in *Vibrio cholerae*. I. Isolation of mutants with altered toxinogenicity. J Infect Dis 129:117–123
Fiorentini C, Donelli G, Matarrese P, Fabbri A, Paradisi S, Boquet P (1995) *Escherichia coli* cytotoxic necrotizing factor 1: Evidence for induction of actin assembly by constitutive activation of the p21 Rho GTPase. Infect Immun 63:3936–3944
Fischer R, McGhee JR, Vu HL, Atkinson TP, Jackson RJ, Tome D, Boyaka PN (2005) Oral and nasal sensitization promote distinct immune responses and lung reactivity in a mouse model of peanut allergy. Am J Pathol 167:1621–1630
Fishman PH, Moss J, Osborne JC Jr (1978) Interaction of choleragen with the oligosaccharide of ganglioside GM1: Evidence for multiple oligosaccharide binding sites. Biochemistry 17:711–716
Folster JP, Connell TD (2002) The extracellular transport signal of the *Vibrio cholerae* endochitinase (ChiA) is a structural motif located between amino acids 75 and 555. J Bacteriol 184:2225–2234
Forrest BD, LaBrooy JT, Attridge SR, Boehm G, Beyer L et al. (1989) Immunogenicity of a candidate live oral typhoid/cholera hybrid vaccine in humans. J Infect Dis 159:145–146
Freitas FS, Momen H, Salles CA (2002) The zymorars of *Vibrio Cholerae*: Multilocus enzyme electrophoresis of *Vibrio Cholerae*. Mem Inst Oswaldo Cruz 97:511–516
Freter R, Jones GW (1976) Adhesive properties of *Vibrio cholerae*: Nature of the interaction with intact mucosal surfaces. Infect Immun 14:246–256
Fuerst JA, Perry JW (1988) Demonstration of lipopolysaccharide on sheathed flagella of *Vibrio cholerae* O:1 by protein A-gold immunoelectron microscopy. J Bacteriol 170:1488–1494
Fujita K, Finkelstein RA (1972) Antitoxic immunity in experimental cholera: Comparison of immunity induced perorally and parenterally in mice. J Infect Dis 125:647–655
Fukuta S, Magnani JL, Twiddy EM, Holmes RK, Ginsburg V (1988) Comparison of the carbohydrate-binding specificities of cholera toxin and *Escherichia coli* heat-labile enterotoxins LTh-I, LT-IIa, and LT-IIb. Infect Immun 56:1748–1753
Fullner KJ, Boucher JC, Hanes MA, Haines GK 3rd, Meehan BM et al. (2002) The contribution of accessory toxins of *Vibrio cholerae* O1 El Tor to the proinflammatory response in a murine pulmonary cholera model. J Exp Med 195:1455–1462
Fullner KJ, Lencer WI, Mekalanos JJ (2001) *Vibrio cholerae*-induced cellular responses of polarized T84 intestinal epithelial cells are dependent on production of cholera toxin and the RTX toxin. Infect Immun 69:6310–6317
Fullner KJ, Mekalanos JJ (2000) In vivo covalent cross-linking of cellular actin by the *Vibrio cholerae* RTX toxin. Embo J 19:5315–5323
Gagliardi MC, Sallusto F, Marinaro M, Langenkamp A, Lanzavecchia A, De Magistris MT (2000) Cholera toxin induces maturation of human dendritic cells and licences them for Th2 priming. Eur J Immunol 30:2394–2403
Galanos C, Luderitz O, Rietschel ET, Westphal O, Brade H et al. (1985) Synthetic and natural *Escherichia coli* free lipid A express identical endotoxic activities. Eur J Biochem 148:1–5
Galanos C, Rietschel ET, Luderitz O, Westphal O, Kim YB, Watson DW (1972) Biological activities of lipid A complexed with bovine-serum albumin. Eur J Biochem 31:230–233
Galen JE, Ketley JM, Fasano A, Richardson SH, Wasserman SS, Kaper JB (1992) Role of *Vibrio cholerae* neuraminidase in the function of cholera toxin. Infect Immun 60:406–415
Gallut J (1971) The 7th cholera pandemia (1961–1966, 1970). Bull Soc Pathol Exot Filiales 64:551–560

Garcia L, Jidy MD, Garcia H, et al. (2005) The vaccine candidate *Vibrio cholerae* 638 is protective against cholera in healthy volunteers. Infect Immun 73:3018–3024

Gershman LC, Selden LA, Estes JE (1986) High affinity binding of divalent cation to actin monomer is much stronger than previously reported. Biochem Biophys Res Commun 135:607–614

Ghosh C, Nandy RK, Dasgupta SK, Nair GB, Hall RH, Ghose AC (1997) A search for cholera toxin (CT), toxin coregulated pilus (TCP), the regulatory element ToxR and other virulence factors in non-01/non-0139 *Vibrio cholerae*. Microb Pathog 22:199–208

Ghosh S, Maiti M, Chatterjee SN (1976) Spectrophotometric studies of base composition & helix-coil transition of *Vibrio cholerae* DNA. Indian J Biochem Biophys 13:297–302

Giannella RA (1976) Suckling mouse model for detection of heat-stable *Escherichia coli* enterotoxin: Characteristics of the model. Infect Immun 14:95–99

Gill DM (1976) The arrangement of subunits in cholera toxin. Biochemistry 15:1242–1248

Gill DM (1978) Seven peptides that cross cell membranes. In: Jeljaszewicz I and WadstrÖm T (eds) Bacterial Toxins and Cell Membranes. Academic, San Diego, CA, pp 291–332

Gill DM, Clements JD, Robertson DC, Finkelstein RA (1981) Subunit number and arrangement in *Escherichia coli* heat-labile enterotoxin. Infect Immun 33:677–682

Gill DM, Coburn J (1987) ADP-ribosylation by cholera toxin: functional analysis of a cellular system that stimulates the enzymic activity of cholera toxin fragment A1. Biochemistry 26:6364–6371

Gill DM, King CA (1975) The mechanism of action of cholera toxin in pigeon erythrocyte lysates. J Biol Chem 250:6424–6432

Gizurarson S, Tamura S, Kurata T, Hasiguchi K, Ogawa H (1991) The effect of cholera toxin and cholera toxin B subunit on the nasal mucosal membrane. Vaccine 9:825–832

Glass RI, Svennerholm AM, Khan MR, Huda S, Huq MI, Holmgren J (1985) Seroepidemiological studies of El Tor cholera in Bangladesh: Association of serum antibody levels with protection. J Infect Dis 151:236–242

Goldberg I, Mekalanos JJ (1986) Cloning of the *Vibrio cholerae* recA gene and construction of a *Vibrio cholerae* recA mutant. J Bacteriol 165:715–722

Goldberg SL, Murphy JR (1984) Molecular cloning of the hemolysin determinant from *Vibrio cholerae* El Tor. J Bacteriol 160:239–244

Gotoh M, Barnes CN, Kovac P (1994) Improved synthesis and the crystal structure of methyl 4,6-dideoxy-4-(3-deoxy-L-glycero-tetronamido)-alpha-D-mannopyrano side, the methyl alpha-glycoside of the intracatenary repeating unit of the O-polysaccharide of *Vibrio cholerae* O:1. Carbohydr Res 260:203–218

Gronow S, Brabetz W, Brade H (2000) Comparative functional characterization in vitro of heptosyltransferase I (WaaC) and II (WaaF) from *Escherichia coli*. Eur J Biochem 267:6602–6611

Guerrant RL, Brunton LL, Schnaitman TC, Rebhun LI, Gilman AG (1974) Cyclic adenosine monophosphate and alteration of Chinese hamster ovary cell morphology: A rapid, sensitive in vitro assay for the enterotoxins of *Vibrio cholerae* and *Escherichia coli*. Infect Immun 10:320–327

Guidolin A, Manning PA (1985) Bacteriophage CP-T1 of *Vibrio cholerae*. Identification of the cell surface receptor. Eur J Biochem 153:89–94

Guidolin A, Manning PA (1987) Genetics of *Vibrio cholerae* and its bacteriophages. Microbiol Rev 51:285–298

Gumbiner B (1987) Structure, biochemistry, and assembly of epithelial tight junctions. Am J Physiol 253:C749–758

Gupta RK, Szu SC, Finkelstein RA, Robbins JB (1992) Synthesis, characterization, and some immunological properties of conjugates composed of the detoxified lipopolysaccharide of *Vibrio cholerae* O1 serotype Inaba bound to cholera toxin. Infect Immun 60:3201–3208

Gupta RK, Taylor DN, Bryla DA, Robbins JB, Szu SC (1998) Phase 1 evaluation of *Vibrio cholerae* O1, serotype Inaba, polysaccharide-cholera toxin conjugates in adult volunteers. Infect Immun 66:3095–3099

Gupta S, Chowdhury R (1997) Bile affects production of virulence factors and motility of *Vibrio cholerae*. Infect Immun 65:1131–1134

Gustafsson B, Holme T (1985) Immunological characterization of *Vibrio cholerae* O:1 lipopolysaccharide, O-side chain, and core with monoclonal antibodies. Infect Immun 49: 275–280

Hacker J, Blum-Oehler G, Muhldorfer I, Tschape H (1997) Pathogenicity islands of virulent bacteria: Structure, function and impact on microbial evolution. Mol Microbiol 23:1089–1097

Hacker J, Kaper JB (1999) The concept of pathogenicity islands. In: Kaper JB and Hacker J (eds) Pathogenicity Islands and Other Mobile Virulence Elements. American Society for Microbiology, Washington, DC, pp 1–11

Hagiwara Y, Kawamura YI, Kataoka K, Rahima B, Jackson RJ et al. (2006) A second generation of double mutant cholera toxin adjuvants: Enhanced immunity without intracellular trafficking. J Immunol 177:3045–3054

Haishima Y, Kondo S, Hisatsune K (1988) O-antigenic lipopolysaccharides isolated from a marine Vibrio, bio-serogroup 1875, possessing an antigenic factor in common with O1 *Vibrio cholerae*. J Gen Microbiol 134:1827–1833

Hall A, Nobes CD (2000) Rho GTPases: Molecular switches that control the organization and dynamics of the actin cytoskeleton. Philos Trans R Soc Lond B Biol Sci 355:965–970

Hall RH, Khambaty FM, Kothary M, Keasler SP (1993) Non-O1 *Vibrio cholerae*. Lancet 342:430

Hango AJ, Watnick PI (2002) *Vibrio cholerae* CytR is a repressor of biofilm development. Mol Microbiol. 45:471–483

Hansen JK, Forest KT (2006) Type IV pilin structures: Insights on shared architecture, fiber assembly, receptor binding and type II secretion. J Mol Microbiol Biotechnol 11:192–207

Hardy SJ, Holmgren J, Johansson S, Sanchez J, Hirst TR (1988) Coordinated assembly of multisubunit proteins: Oligomerization of bacterial enterotoxins in vivo and in vitro. Proc Natl Acad Sci USA 85:7109–7113

Harlocker SL, Bergstrom L, Inouye M (1995) Tandem binding of six OmpR proteins to the ompF upstream regulatory sequence of *Escherichia coli*. J Biol Chem 270:26849–26856

Harris JR, Bhakdi S, Meissner U, Scheffler D, Bittman R et al. (2002) Interaction of the *Vibrio cholerae* cytolysin (VCC) with cholesterol, some cholesterol esters, and cholesterol derivatives: A TEM study. J Struct Biol 139:122–135

Hasan JA, Huq A, Nair GB, Garg S, Mukhopadhyay AK et al. (1995) Development and testing of monoclonal antibody-based rapid immunodiagnostic test kits for direct detection of *Vibrio cholerae* O139 synonym Bengal. J Clin Microbiol 33:2935–2939

Hasan JA, Huq A, Tamplin ML, Siebeling RJ, Colwell RR (1994) A novel kit for rapid detection of *Vibrio cholerae* O1. J Clin Microbiol 32:249–252

Hase S, Rietschel ET (1976) Isolation and analysis of the lipid A backbone: lipid A structures from lipopolysaccharides from various bacterial groups. Eur J Biochem 63:101–107

Hase CC, Mekalanos JJ (1998) TcpP protein is a positive regulator of virulence gene expression in *Vibrio cholerae*. Proc Natl Acad Sci USA 95:730–734

Hazes B, Read RJ (1997) Accumulating evidence suggests that several AB-toxins subvert the endoplasmic reticulum-associated protein degradation pathway to enter target cells. Biochemistry 36:11051–11054

Heidelberg JF, Eisen JA, Nelson WC, Clayton RA, Gwinn ML et al. (2000) DNA sequence of both chromosomes of the cholera pathogen *Vibrio cholerae*. Nature 406:477–483

Heilpern AJ, Waldor MK (2000) CTXphi infection of *Vibrio cholerae* requires the tolQRA gene products. J Bacteriol 182:1739–1747

Heilpern AJ, Waldor MK (2003) pIIICTX, a predicted CTXphi minor coat protein, can expand the host range of coliphage fd to include *Vibrio cholerae*. J Bacteriol 185:1037–1044

Heinrichs DE, Yethon JA, Whitfield C (1998) Molecular basis for structural diversity in the core regions of the lipopolysaccharides of *Escherichia coli* and *Salmonella enterica*. Mol Microbiol 30:221–232

Henning S, Rubin D, Shulman R (1994) Ontogeny of the intestinal mucosa. In: Johnson L (ed) Physiology of the Gastrointestinal Tract. Raven, New York, pp 571–610

Hepler JR, Gilman AG (1992) G proteins. Trends Biochem Sci 17:383–387

Herrington DA, Hall RH, Losonsky G, Mekalanos JJ, Taylor RK, Levine MM (1988) Toxin, toxin-coregulated pili and ToxR regulon are essential for *Vibrio cholerae* pathogenesis in humans. J Exp Med 168:1487–1492

Higgins DE, DiRita VJ (1994) Transcriptional control of toxT, a regulatory gene in the ToxR regulon of *Vibrio cholerae*. Mol Microbiol 14:17–29

Higgins DE, Nazareno E, DiRita VJ (1992) The virulence gene activator ToxT from *Vibrio cholerae* is a member of the AraC family of transcriptional activators. J Bacteriol 174: 6974–6980

Hill CW, Sandt CH, Vlazny DA (1994) Rhs elements of *Escherichia coli*: A family of genetic composites each encoding a large mosaic protein. Mol Microbiol 12:865–871

Hirst TR, Hardy SJ, Randall LL (1983) Assembly in vivo of enterotoxin from *Escherichia coli*: Formation of the B subunit oligomer. J Bacteriol 153:21–26

Hirst TR, Holmgren J (1987a) Conformation of protein secreted across bacterial outer membranes: A study of enterotoxin translocation from *Vibrio cholerae*. Proc Natl Acad Sci USA 84:7418–7422

Hirst TR, Holmgren J (1987b) Transient entry of enterotoxin subunits into the periplasm occurs during their secretion from *Vibrio cholerae*. J Bacteriol 169:1037–1045

Hirst TR, Sanchez J, Kaper JB, Hardy SJ, Holmgren J (1984a) Mechanism of toxin secretion by *Vibrio cholerae* investigated in strains harboring plasmids that encode heat-labile enterotoxins of *Escherichia coli*. Proc Natl Acad Sci USA 81:7752–7756

Hirst TR, Randall LL, Hardy SJ (1984b) Cellular location of heat-labile enterotoxin in *Escherichia coli*. J Bacteriol 157:637–642

Hisatsune K, Hayashi M, Haishima Y, Kondo S (1989) Relationship between structure and antigenicity of O1 *Vibrio cholerae* lipopolysaccharides. J Gen Microbiol 135:1901–1907

Hisatsune K, Kondo S (1980) Lipopolysaccharides of R mutants isolated from *Vibrio cholerae*. Biochem J 185:77–81

Hisatsune K, Kondo S, Iguchi T, Machida M, Asou S, Inaguma M, Yamamoto F (1982) Sugar composition of lipopolysaccharides of family Vibrionaceae. Absence of 2-keto-3-deoxyoctonate (KDO) except in *Vibrio parahaemolyticus* O6. Microbiol Immunol 26:649–664

Hisatsune K, Kondo S, Iguchi T, Machida M, Asou S, Inaguma M, Yamamoto F (1983) Sugar compositions of lipopolysaccharides of family Vibrionaceae: absence of 2-keto-3-deoxyoctonate (KDO) except in *Vibrio parahaemolyticus* O6 and *Plesiomonas spigelloides*. In: Kuwahara S and Pierce NF (eds) Advances in Research on Cholera and Related Diarrhea. Martinus Nijhoff, The Hague, Netherlands, pp 59–74

Hisatsune K, Kondo S, Isshiki Y, Iguchi T, Haishima Y (1993a) Occurrence of 2-O-methyl-N-(3-deoxy-L-glycero-tetronyl)-D-perosamine (4-amino-4,6-dideoxy-D-manno-pyranose) in lipopolysaccharide from Ogawa but not from Inaba O forms of O1 *Vibrio cholerae*. Biochem Biophys Res Commun 190:302–307

Hisatsune K, Kondo S, Isshiki Y, Iguchi T, Kawamata Y, Shimada T (1993b) O-antigenic lipopolysaccharide of *Vibrio cholerae* O139 Bengal, a new epidemic strain for recent cholera in the Indian subcontinent. Biochem Biophys Res Commun 196:1309–1315

Hisatsune K, Kondo S, Kawata T, Kishimoto Y (1979) Fatty acid composition of lipopolysaccharides of *Vibrio cholerae* 35A3 (Inaba), NIB 90 (Ogawa), and 4715 (Nag). J Bacteriol 138:288–290

Hisatsune K, Kondo S, Kobayashi K (1978) Lipopolysaccharides of *Vibrio cholerae* (II): an immunochemical study on O-antigenic structure (proceedings). Jpn J Med Sci Biol 31:181–184

Ho AS, Mietzner TA, Smith AJ, Schoolnik GK (1990) The pili of *Aeromonas hydrophila*: Identification of an environmentally regulated "mini pilin." J Exp Med 172:795–806

Hobbs M, Mattick JS (1993) Common components in the assembly of type 4 fimbriae, DNA transfer systems, filamentous phage and protein-secretion apparatus: A general system for the formation of surface-associated protein complexes. Mol Microbiol 10:233–243

Hobbs M, Reeves PR (1994) The JUMPstart sequence: A 39 bp element common to several polysaccharide gene clusters. Mol Microbiol 12:855–856

Hoge CW, Sethabutr O, Bodhidatta L, Echeverria P, Robertson DC, Morris JG Jr (1990) Use of a synthetic oligonucleotide probe to detect strains of non-serovar O1 *Vibrio cholerae* carrying the gene for heat-stable enterotoxin (NAG-ST). J Clin Microbiol 28:1473–1476

Holland IB, Schmitt L, Young J (2005) Type 1 protein secretion in bacteria, the ABC-transporter dependent pathway (review). Mol Membr Biol 22:29–39

Holmgren J, Andersson A, Wallerstrom G, Ouchterlony O (1972) Experimental studies on cholera immunization. II. Evidence for protective antitoxic immunity mediated by serum antibodies as well as local antibodies. Infect Immun 5:662–667

Holmgren J, Lonnroth I, Svennerholm L (1973) Tissue receptor for cholera exotoxin: Postulated structure from studies with GM1 ganglioside and related glycolipids. Infect Immun 8:208–214

Holmgren J, Svennerholm AM (1979) Local IgA antibody formation after vaccination against cholera and poliomyelitis. Lakartidningen 76:3401–3403

Holmgren J, Svennerholm AM, Ouchterlony O, Anderson A, Walletstrom G, Westerberg-Berndtsson U (1975) Antitoxic immunity in experimental cholera: Protection, and serum and local antibody responses in rabbits after enteral and parenteral immunization. Infect Immun 12:1331–1340

Holst O, Brade H (1992) Molecular biochemistry and cellular biology. In: Morrison DC and Ryan JL (eds) Bacterial Endotoxic Lipopolysaccharides. CRC Press, Boca Raton, FL, pp 135–169

Honda T, Arita M, Takeda T, Yoh M, Miwatani T (1985) Non-O1 *Vibrio cholerae* produces two newly identified toxins related to *Vibrio parahaemolyticus* haemolysin and *Escherichia coli* heat-stable enterotoxin. Lancet 2:163–164

Honda T, Finkelstein RA (1979) Selection and characteristics of a *Vibrio cholerae* mutant lacking the A (ADP-ribosylating) portion of the cholera enterotoxin. Proc Natl Acad Sci USA 76:2052–2056

Hoque KM, Saha S, Gupta DD, Chakrabarti MK (2004) Role of nitric oxide in NAG-ST induced store-operated calcium entry in rat intestinal epithelial cells. Toxicology 201:95–103

Horstman AL, Bauman SJ, Kuehn MJ (2004) Lipopolysaccharide 3-deoxy-D-manno-octulosonic acid (Kdo) core determines bacterial association of secreted toxins. J Biol Chem 279:8070–8075

Hoshino K, Yamasaki S, Mukhopadhyay AK, Chakraborty S, Basu A et al. (1998) Development and evaluation of a multiplex PCR assay for rapid detection of toxigenic *Vibrio cholerae* O1 and O139. FEMS Immunol Med Microbiol 20:201–207

Hu SH, Peek JA, Rattigan E, Taylor RK, Martin JL (1997) Structure of TcpG, the DsbA protein folding catalyst from *Vibrio cholerae*. J Mol Biol 268:137–146

Hugh R (1965b) A comparison of *Vibrio cholerae* Pacini and *Vibrio eltor* Pribram. Int Bull Bacteriol Nomencl Taxon 15:61–68

Hugh R (1965a) Nomenclature and taxonomy of *Vibrio cholerae* Pacini 1854 and *Vibrio eltor* Pribram 1933. In: Bushell OA and Brookhyser CS (eds) Proc Cholera Res Symp, Honululu, Hawaii, US Govt Printing Office, Washington, 24–29 Jan 1965, pp 1–4

Hung AY, Sheng M (2002) PDZ domains: Structural modules for protein complex assembly. J Biol Chem 277:5699–5702

Ichinose Y, Yamamoto K, Nakasone N, Tanabe MJ, Takeda T, Miwatani T, Iwanaga M (1987) Enterotoxicity of El Tor-like hemolysin of non-O1 *Vibrio cholerae*. Infect Immun 55:1090–1093

Ikigai H, Akatsuka A, Tsujiyama H, Nakae T, Shimamura T (1996) Mechanism of membrane damage by El Tor hemolysin of *Vibrio cholerae* O1. Infect Immun 64:2968–2973

Iredell JR, Manning PA (1994) The toxin-co-regulated pilus of *Vibrio cholerae* O1: A model for type 4 pilus biogenesis? Trends Microbiol 2:187–192

Iredell JR, Stroeher UH, Ward HM, Manning PA (1998) Lipopolysaccharide O-antigen expression and the effect of its absence on virulence in rfb mutants of *Vibrio cholerae* O1. FEMS Immunol Med Microbiol 20:45–54

Islam LN, Nabi AH, Ahmed KM, Sultana N (2002) Endotoxins of enteric pathogens are chemotactic factors for human neutrophils. J Biochem Mol Biol 35:482–487

Islam LN, Wilkinson PC (1988) Chemotactic factor-induced polarization, receptor redistribution, and locomotion of human blood monocytes. Immunology 64:501–507

Isshiki Y, Kondo S, Iguchi T, Sano Y, Shimada T, Hisatsune K (1996) An immunochemical study of serological cross-reaction between lipopolysaccharides from *Vibrio cholerae* O22 and O139. Microbiology 142 (Pt 6):1499–1504

Ito T, Higuchi T, Hirobe M, Hiramatsu K, Yokota T (1994) Identification of a novel sugar, 4-amino-4,6-dideoxy-2-O-methylmannose in the lipopolysaccharide of *Vibrio cholerae* O1 serotype Ogawa. Carbohydr Res 256:113–128

Ito T, Kuwahara S, Yokota T (1983) Automatic and manual latex agglutination tests for measurement of cholera toxin and heat-labile enterotoxin of *Escherichia coli*. J Clin Microbiol 17:7–12

Iwanaga M, Yamamoto K, Higa N, Ichinose Y, Nakasone N, Tanabe M (1986) Culture conditions for stimulating cholera toxin production by *Vibrio cholerae* O1 El Tor. Microbiol Immunol 30:1075–1083

Izquierdo L, Merino S, Coderch N, Regue M, Tomas JM (2002) The wavB gene of *Vibrio cholerae* and the waaE of *Klebsiella pneumoniae* codify for a beta-1,4-glucosyltransferase involved in the transfer of a glucose residue to the L-glycero-D-manno-heptose I in the lipopolysaccharide inner core. FEMS Microbiol Lett 216:211–216

Jackson GD, Redmond JW (1971) Immunochemical studies of the O-antigens of *Vibrio cholerae*. The constitution of a lipopolysaccharide from *V. cholerae* 569B (Inaba). FEBS Lett 13:117–120

Jann B, Jann K, Beyaert GO (1973) 2-Amino-2,6-dideoxy-D-glucose (D-quinovosamine): A constituent of the lipopolysaccharides of *Vibrio cholerae*. Eur J Biochem 37:531–534

Jedani KE, Stroeher UH, Manning PA (1998) Distribution of IS1358 and linkage to rfb-related genes in *V. cholerae* and *V. anguillarum*, cited in U.H. Stroeher et al. Gene 223:269–282

Jermyn WS, Boyd EF (2002) Characterization of a novel Vibrio pathogenicity island (VPI-2) encoding neuraminidase (nanH) among toxigenic *Vibrio cholerae* isolates. Microbiology 148:3681–3693

Jermyn WS, Boyd EF (2005) Molecular evolution of Vibrio pathogenicity island-2 (VPI-2): mosaic structure among *Vibrio cholerae* and *Vibrio mimicus* natural isolates. Microbiology 151:311–322

Jertborn M, Svennerholm AM, Holmgren J (1988) IgG and IgA subclass distribution of antitoxin antibody responses after oral cholera vaccination or cholera disease. Int Arch Allergy Appl Immunol 85:358–363

Jertborn M, Svennerholm AM, Holmgren J (1994) Immunological memory after immunization with oral cholera B subunit–whole-cell vaccine in Swedish volunteers. Vaccine 12:1078–1082

Jesudason MV, Cherian AM, John TJ (1993) Blood stream invasion by *Vibrio cholerae* O139. Lancet 342:431

Jetborn M, Svennerholm AM, Holmgren J (1986) Saliva, breast milk and serum antibody responses as direct measures of intestinal immunity after oral cholera vaccination or natural disease. J Clin Microbiol 24:203–209

Jobling MG, Holmes RK (1991) Analysis of structure and function of the B subunit of cholera toxin by the use of site-directed mutagenesis. Mol Microbiol 5:1755–1767

Jobling MG, Holmes RK (1997) Characterization of hapR, a positive regulator of the *Vibrio cholerae* HA/protease gene hap, and its identification as a functional homologue of the *Vibrio harveyi* luxR gene. Mol Microbiol 26:1023–1034

Johnson JA, Joseph A, Morris JG (1995) Capsular polysaccharide–protein conjugate vaccines against *Vibrio cholerae* O139 Bengal. Bull Inst Pasteur 93 285–290

Johnson JA, Panigrahi P, Morris JG Jr (1992) Non-O1 *Vibrio cholerae* NRT36S produces a polysaccharide capsule that determines colony morphology, serum resistance, and virulence in mice. Infect Immun 60:864–869

Johnson JA, Salles CA, Panigrahi P, Albert MJ, Wright AC, Johnson RJ, Morris JG Jr (1994) *Vibrio cholerae* O139 synonym bengal is closely related to *Vibrio cholerae* El Tor but has important differences. Infect Immun 62:2108–2110

Johnson TL, Abendroth J, Hol WG, Sandkvist M (2006) Type II secretion: From structure to function. FEMS Microbiol Lett 255:175–186

Jonson G, Holmgren J, Svennerholm AM (1991) Identification of a mannose-binding pilus on *Vibrio cholerae* El Tor. Microb Pathog 11:433–441

Jonson G, Osek J, Svennerholm AM, Holmgren J (1996) Immune mechanisms and protective antigens of *Vibrio cholerae* serogroup O139 as a basis for vaccine development. Infect Immun 64:3778–3785

Jonson G, Svennehohn AM, Holmgren J (1990) Expression of virulence factors by classical and El Tor *Vibrio cholerae* in-vivo and in-vitro. FEMS Microbiol Ecol 74:221–228

Jouravleva EA, McDonald GA, Marsh JW, Taylor RK, Boesman-Finkelstein M, Finkelstein RA (1998) The *Vibrio cholerae* mannose-sensitive hemagglutinin is the receptor for a filamentous bacteriophage from *V. cholerae* O139. Infect Immun 66:2535–2539

Kabir S (1982) Characterization of the lipopolysaccharide from *Vibrio cholerae* 395 (Ogawa). Infect Immun 38:1263–1272

Kabir S (2005) Cholera vaccines: the current status and problems. Rev Med Microbiol 16:110–116

Kabir S, Mann P (1980) Immunological properties of the cell envelope components of *Vibrio cholerae*. J Gen Microbiol 119:517–525

Kaca W, Brade L, Rietschel ET, Brade H (1986) The effect of removal of D-fructose on the antigenicity of the lipopolysaccharide from a rough mutant of *Vibrio cholerae* Ogawa. Carbohydr Res 149:293–298

Kadokura H, Katzen F, Beckwith J (2003) Protein disulfide bond formation in prokaryotes. Annu Rev Biochem 72:111–135

Kadurugamuwa JL, Beveridge TJ (1996) Bacteriolytic effect of membrane vesicles from *Pseudomonus aeruginosa* on other bacteria including pathogens: conceptually new antibiotics. J Bacteriol 178:2267–2274

Kahn RA, Gilman AG (1984) ADP-ribosylation of Gs promotes the dissociation of its alpha and beta subunits. J Biol Chem 259:6235–6240

Kamal AM (1971) Outbreak of gastro-enteritis by non-agglutinable (NAG) vibrios in the Republic of Sudan. J Egypt Public Health Assoc XLVI: 125–159

Kaper JB (1989) *Vibrio cholerae* vaccines. Rev Infect Dis 11(Suppl 3):S568–573

Kaper JB, Bradford HB, Roberts NC, Falkow S (1982) Molecular epidemiology of *Vibrio cholerae* in the US Gulf Coast. J Clin Microbiol 16:129–134

Kaper JB, Levine MM (1981) Cloned cholera enterotoxin genes in study and prevention of cholera. Lancet 2:1162–1163

Kaper JB, Morris JG Jr, Levine MM (1995) Cholera. Clin Microbiol Rev 8:48–86

Kaper JB, Nataro JP, Roberts NC, Siebeling RJ, Bradford HB (1986) Molecular epidemiology of non-O1 *Vibrio cholerae* and *Vibrio mimicus* in the U.S. Gulf Coast region. J Clin Microbiol 23:652–654

Kapfhammer D, Blass J, Evers S, Reidl J (2002) *Vibrio cholerae* phage K139: Complete genome sequence and comparative genomics of related phages. J Bacteriol 184:6592–6601

Karaolis DK, Johnson JA, Bailey CC, Boedeker EC, Kaper JB, Reeves PR (1998) A *Vibrio cholerae* pathogenicity island associated with epidemic and pandemic strains. Proc Natl Acad Sci USA 95:3134–3139

Karaolis DK, Lan R, Kaper JB, Reeves PR (2001) Comparison of *Vibrio cholerae* pathogenicity islands in sixth and seventh pandemic strains. Infect Immun 69:1947–1952

Karaolis DK, Somara S, Maneval DR Jr, Johnson JA, Kaper JB (1999) A bacteriophage encoding a pathogenicity island, a type-IV pilus and a phage receptor in cholera bacteria. Nature 399:375–379

Kasper DL, Weintraub A, Lindberg AA, Lonngren J (1983) Capsular polysaccharides and lipopolysaccharides from two *Bacteroides fragilis* reference strains: chemical and immunochemical characterization. J Bacteriol 153:991–997

Katagiri YU, Mori T, Nakajima H, Katagiri C, Taguchi T et al. (1999) Activation of Src family kinase yes induced by Shiga toxin binding to globotriaosyl ceramide (Gb3/CD77) in low density, detergent-insoluble microdomains. J Biol Chem 274:35278–35282

Kenne L, Lindberg B, Unger P, Holme T, Holmgren J (1979) Structural studies of the *Vibrio cholerae* O-antigen. Carbohydr Res 68:C14–16

Kenne L, Lindberg B, Unger P, Gustafsson B, Holme T (1982) Structural studies of the *Vibrio cholerae* O-antigen. Carbohydr Res 100:341–349

Kenne L, Lindberg B (1983) Bacterial lipopolysaccharides. In: Aspinell GO (ed.) The Polysaccharides. Academic, New York, pp 287–363

Kenne L, Unger P, Wehler T (1988a) Synthesis and nuclear magnetic resonance studies of some N-acylated methyl 4-amino-4,6-dideoxy-α-D-mannopyranosides. J Chem Soc Perkin Trans 1:1183–1186

Kenne L, Lindberg B, Schweda E, Gustafsson B, Holme T (1988b) Structural studies of the O-antigen from *Vibrio cholerae* O:2. Carbohydr Res 180:285–294

Kenner JR, Coster TS, Taylor DN, Trofa AF, Barrera-Oro M et al. (1995) Peru-15, an improved live attenuated oral vaccine candidate for *Vibrio cholerae* O1. J Infect Dis 172:1126–1129

Ketley JM, Michalski J, Galen J, Levine MM, Kaper JB (1993) Construction of genetically marked *Vibrio cholerae* O1 vaccine strains. FEMS Microbiol Lett 111:15–21

Khetawat G, Bhadra RK, Nandi S, Das J (1999) Resurgent *Vibrio cholerae* O139: Rearrangement of cholera toxin genetic elements and amplification of rrn operon. Infect Immun 67:148–154

Kierek K, Watnick PI (2003a) Environmental determinants of *Vibrio cholerae* biofilm development. Appl Environ Microbiol 69:5079–5088

Kierek K, Watnick PI (2003b) The *Vibrio cholerae* O139 O-antigen polysaccharide is essential for Ca^{2+}-dependent biofilm development in sea water. Proc Natl Acad Sci USA 100:14357–14362

Kimsey HH, Waldor MK (1998) CTXphi immunity: Application in the development of cholera vaccines. Proc Natl Acad Sci USA 95:7035–7039

Kimsey HH, Waldor MK (2004) The CTXphi repressor RstR binds DNA cooperatively to form tetrameric repressor-operator complexes. J Biol Chem 279:2640–2647

King CA, Van Heyningen WE (1973) Deactivation of cholera toxin by a sialidase-resistant monosialosylganglioside. J Infect Dis 127:639–647

Kinosian HJ, Selden LA, Estes JE, Gershman LC (1993) Nucleotide binding to actin. Cation dependence of nucleotide dissociation and exchange rates. J Biol Chem 268:8683–8691

Kjellberg A, Weintraub A, Albert MJ, Widmalm G (1997) Structural analysis of the O-antigenic polysaccharide from *Vibrio cholerae* O10. Eur J Biochem 249:758–761

Klose KE (2001) Regulation of virulence in *Vibrio cholerae*. Int J Med Microbiol 291:81–88

Knirel YA, Paredes L, Jansson PE, Weintraub A, Widmalm G, Albert MJ (1995) Structure of the capsular polysaccharide of *Vibrio cholerae* O139 synonym Bengal containing D-galactose 4,6-cyclophosphate. Eur J Biochem 232:391–396

Knirel YA, Senchenkova SN, Jansson PE, Weintraub A (1998) More on the structure of *Vibrio cholerae* O22 lipopolysaccharide. Carbohydr Res 310:117–119

Knirel YA, Widmalm G, Senchenkova SN, Jansson PE, Weintraub A (1997) Structural studies on the short-chain lipopolysaccharide of *Vibrio cholerae* O139 Bengal. Eur J Biochem 247:402–410

Koblavi S, Grimont F, Grimont PA (1990) Clonal diversity of *Vibrio cholerae* O1 evidenced by rRNA gene restriction patterns. Res Microbiol 141:645–657

Koch R (1884) An address on cholera and its bacillus. Br Med J 2:403–407

Kocharova NA, Perepelov AV, Zatonsky GV, Shashkov AS, Knirel YA, Jansson PE, Weintraub A (2001a) Structural studies of the O-specific polysaccharide of *Vibrio cholerae* O8 using solvolysis with triflic acid. Carbohydr Res 330:83–92

Kocharova NA, Knirel YA, Jansson P, Weintraub A (2001b) Structure of the O-specific polysaccharide of *Vibrio cholerae* O9 containing 2-acetamido-2-deoxy-D-galacturonic acid. Carbohydr Res 332:279–284

Kohler R, Schafer K, Muller S, Vignon G, Diederichs K et al. (2004) Structure and assembly of the pseudopilin PulG. Mol Microbiol 54:647–664

Kondo K, Takade A, Amako K (1993a) Release of the outer membrane vesicles from *Vibrio cholerae* and *Vibrio parahaemolyticus*. Microbiol Immunol 37:149–152

Kondo S, Haishima Y, Hisatsune K (1992) Taxonomic implication of the apparent undetectability of 3-deoxy-D-manno-2-octulosonate (Kdo) in lipopolysaccharides of the representatives of the family Vibrionaceae and the occurrence of Kdo 4-phosphate in their inner-core regions. Carbohydr Res 231:55–64

Kondo S, Hisatsune K (1989) Sugar composition of the polysaccharide portion of lipopolysaccharides isolated from non-O1 *Vibrio cholerae* O2 to O41, O44, and O68. Microbiol Immunol 33:641–648

Kondo S, Iguchi T, Hisatsune K (1988) A comparative study of the sugar composition of lipopolysaccharides isolated from *Vibrio cholerae*, '*Vibrio albensis*' and *Vibrio metschnikovii*. J Gen Microbiol 134:1699–1705

Kondo S, Sano Y, Isshiki Y, Hisatsune K (1996) The O polysaccharide chain of the lipopolysaccharide from *Vibrio cholerae* O76 is a homopolymer of N-[(*S*)-(+)-2-hydroxypropionyl]-alpha-L-perosamine. Microbiology 142(Pt 10):2879–2885

Kondo S, Watabe T, Haishima Y, Hisatsune K (1993b) Identification of oligosaccharides consisting of D-glucuronic acid and L-glycero-D-manno- and D-glycero-D-manno-heptose isolated from *Vibrio parahaemolyticus* O2 lipopolysaccharide. Carbohydr Res 245:353–359

Konry T, Novoa A, Cosnier S, Marks RS (2003) Development of an "electroptode" immunosensor: Indium tin oxide-coated optical fiber tips conjugated with an electropolymerized thin film with conjugated cholera toxin B subunit. Anal Chem 75:2633–2639

Koonin EV (1992) The second cholera toxin, Zot, and its plasmid-encoded and phage-encoded homologues constitute a group of putative ATPases with an altered purine NTP-binding motif. FEBS Lett 312:3–6

Koronakis V, Sharff A, Koronakis E, Luisi B, Hughes C (2000) Crystal structure of the bacterial membrane protein TolC central to multidrug efflux and protein export. Nature 405:914–919

Korotkov KV, Hol WG (2008) Structure of the GspK-GspI-GspJ complex from the enterotoxigenic *Escherichia coli* type 2 secretion system. Nat Struct Mol Biol 15:462–468

Korotkov KV, Krumm B, Bagdasarian M, Hol WG (2006) Structural and functional studies of EpsC, a crucial component of the type 2 secretion system from *Vibrio cholerae*. J Mol Biol 363:311–321

Kossaczka Z, Shiloach J, Johnson V, Taylor DN, Finkelstein RA, Robbins JB, Szu SC (2000) *Vibrio cholerae* O139 conjugate vaccines: Synthesis and immunogenicity of *V. cholerae* O139 capsular polysaccharide conjugates with recombinant diphtheria toxin mutant in mice. Infect Immun 68:5037–5043

Kostova Z, Wolf DH (2003) For whom the bell tolls: Protein quality control of the endoplasmic reticulum and the ubiquitin–proteasome connection. Embo J 22:2309–2317

Kothary MH, Richardson SH (1987) Fluid accumulation in infant mice caused by *Vibrio hollisae* and its extracellular enterotoxin. Infect Immun 55:626–630

Kovach ME, Shaffer MD, Peterson KM (1996) A putative integrase gene defines the distal end of a large cluster of ToxR-regulated colonization genes in *Vibrio cholerae*. Microbiology 142(Pt 8):2165–2174

Kovacikova G, Lin W, Skorupski K (2004) *Vibrio cholerae* AphA uses a novel mechanism for virulence gene activation that involves interaction with the LysR-type regulator AphB at the tcpPH promoter. Mol Microbiol 53:129–142

Kovacikova G, Skorupski K (1999) A *Vibrio cholerae* LysR homolog, AphB, cooperates with AphA at the tcpPH promoter to activate expression of the ToxR virulence cascade. J Bacteriol 181:4250–4256

Kovacikova G, Skorupski K (2002) Regulation of virulence gene expression in *Vibrio cholerae* by quorum sensing: HapR functions at the aphA promoter. Mol Microbiol 46:1135–1147

Kovbasnjuk O, Edidin M, Donowitz M (2001) Role of lipid rafts in Shiga toxin 1 interaction with the apical surface of Caco-2 cells. J Cell Sci 114:4025–4031

Krause M, Roudier C, Fierer J, Harwood J, Guiney D (1991) Molecular analysis of the virulence locus of the *Salmonella dublin* plasmid pSDL2. Mol Microbiol 5:307–316

Krukonis ES, DiRita VJ (2003a) DNA binding and ToxR responsiveness by the wing domain of TcpP, an activator of virulence gene expression in *Vibrio cholerae*. Mol Cell 12:157–165

Krukonis ES, DiRita VJ (2003b) From motility to virulence: Sensing and responding to environmental signals in *Vibrio cholerae*. Curr Opin Microbiol 6:186–190

Krukonis ES, Yu RR, Dirita VJ (2000) The *Vibrio cholerae* ToxR/TcpP/ToxT virulence cascade: Distinct roles for two membrane-localized transcriptional activators on a single promoter. Mol Microbiol 38:67–84

Kuehn MJ, Kesty NC (2005) Bacterial outer membrane vesicles and the host–pathogen interaction. Genes Dev 19:2645–2655

Kunkel SL, Robertson DC (1979a) Factors affecting release of heat-labile enterotoxin by enterotoxigenic *Escherichia coli*. Infect Immun 23:652–659

Kunkel SL, Robertson DC (1979b) Purification and chemical characterization of the heat-labile enterotoxin produced by enterotoxigenic *Escherichia coli*. Infect Immun 25:586–596

Lai CY, Xia QC, Salotra PT (1983) Location and amino acid sequence around the ADP-ribosylation site in the cholera toxin active subunit A1. Biochem Biophys Res Commun 116:341–348

Lally ET, Hill RB, Kieba IR, Korostoff J (1999) The interaction between RTX toxins and target cells. Trends Microbiol 7:356–361

Lamaze C, Dujeancourt A, Baba T, Lo CG, Benmerah A, Dautry-Varsat A (2001) Interleukin 2 receptors and detergent-resistant membrane domains define a clathrin-independent endocytic pathway. Mol Cell 7:661–671

Lamaze C, Schmid SL (1995) The emergence of clathrin-independent pinocytic pathways. Curr Opin Cell Biol 7:573–580

Landersjo C, Weintraub A, Ansaruzzaman M, Albert MJ, Widmalm G (1998) Structural analysis of the O-antigenic polysaccharide from *Vibrio mimicus* N-1990. Eur J Biochem 251:986–990

Lange S, Holmgren J (1978) Protective antitoxic cholera immunity in mice: Influence of route and number of immunizations and mode of action of protective antibodies. Acta Pathol Microbiol Scand C 86C:145–152

LaRocque RC, Harris JB, Ryan ET, Qadri F, Calderwood SB (2006) Postgenomic approaches to cholera vaccine development. Expert Rev Vaccines 5:337–346

Lavelle EC, McNeela E, Armstrong ME, Leavy O, Higgins SC, Mills KH (2003) Cholera toxin promotes the induction of regulatory T cells specific for bystander antigens by modulating dendritic cell activation. J Immunol 171:2384–2392

Le PU, Nabi IR (2003) Distinct caveolae-mediated endocytic pathways target the Golgi apparatus and the endoplasmic reticulum. J Cell Sci 116:1059–1071

Lee PA, Tullman-Ercek D, Georgiou G (2006) The bacterial twin-arginine translocation pathway. Annu Rev Microbiol 60:373–395

Lee SH, Butler SM, Camilli A (2001) Selection for in vivo regulators of bacterial virulence. Proc Natl Acad Sci USA 98:6889–6894

Lee SH, Hava DL, Waldor MK, Camilli A (1999) Regulation and temporal expression patterns of *Vibrio cholerae* virulence genes during infection. Cell 99:625–634

Lei PS, Ogawa Y, Flippen-Anderson JL, Kovac P (1995a) Synthesis and crystal structure of methyl 4-6-dideoxy-4-(3-deoxy-L-glycero-tetronamido)-2-O-methyl-alpha-D-mannopyranoside, the methyl alpha-glycoside of the terminal unit, and presumed antigenic determinant, of the O-specific polysaccharide of *Vibrio cholerae* O:1, serotype Ogawa. Carbohydr Res 275:117–129

Lei PS, Ogawa Y, Kovac P (1995b) Synthesis of the methyl alpha-glycoside of a trisaccharide mimicking the terminus of the O antigen of *Vibrio cholerae* O:1, serotype Inaba. Carbohydr Res 279:117–131

Lei P, Ogawa Y, Kovac P (1996) Synthesis of the methyl alpha-glycosides of a di-, tri-, and a tetrasaccharide fragment mimicking the terminus of the O-polysaccharide of *Vibrio cholerae* O:1, serotype Ogawa. Carbohydr Res 281:47–60

Lencer WI, Constable C, Moe S, Jobling MG, Webb HM et al. (1995) Targeting of cholera toxin and *Escherichia coli* heat labile toxin in polarized epithelia: Role of COOH-terminal KDEL. J Cell Biol 131:951–962

Lencer WI, Tsai B (2003) The intracellular voyage of cholera toxin: going retro. Trends Biochem Sci 28:639–645

Lenz DH, Mok KC, Lilley BN, Kulkarni RV, Wingreen NS, Bassler BL (2004) The small RNA chaperone Hfq and multiple small RNAs control quorum sensing in *Vibrio harveyi* and *Vibrio cholerae*. Cell 118:69–82

Lerm M, Selzer J, Hoffmeyer A, Rapp UR, Aktories K, Schmidt G (1999) Deamidation of Cdc42 and Rac by *Escherichia coli* cytotoxic necrotizing factor 1: Activation of c-Jun N-terminal kinase in HeLa cells. Infect Immun 67:496–503

Lesieur C, Cliff MJ, Carter R, James RF, Clarke AR, Hirst TR (2002) A kinetic model of intermediate formation during assembly of cholera toxin B-subunit pentamers. J Biol Chem 277:16697–16704

Levine ML, Pierce NF (1992) Immunity and vaccine development in cholera. In: Barua D and Greenough III WB (eds.) Cholera. Plenum, New York, pp 285–327

Levine MM, Kaper JB (1995) Live oral cholera vaccine: from principle to product. Bull Inst Pasteur 93:243–253

Levine MM, Kaper JB, Black RE, Clements ML (1983) New knowledge on pathogenesis of bacterial enteric infections as applied to vaccine development. Microbiol Rev 47:510–550

Levine MM, Kaper JB, Herrington D, Losonsky G, Morris JG et al. (1988a) Volunteer studies of deletion mutants of *Vibrio cholerae* O1 prepared by recombinant techniques. Infect Immun 56:161–167

Levine MM, Kaper JB, Herrington D, Ketley J, Losonsky G et al. (1988b) Safety, immunogenicity, and efficacy of recombinant live oral cholera vaccines, CVD 103 and CVD 103-HgR. Lancet 2:467–470

Levine MM, Nalin DR, Craig JP, Hoover D, Bergquist EJ et al. (1979) Immunity of cholera in man: Relative role of antibacterial versus antitoxic immunity. Trans R Soc Trop Med Hyg 73:3–9

Levine MM, Tacket CO (1994) Recombinant live oral cholera vaccines. In: Wachsmuth IK, Blake PA, and Olsvik O (eds) Vibrio Cholerae and Cholera: Molecular to Global Perspective. American Society for Microbiology, Washington, DC, pp 395–413

Levine MM, Young CR, Hughes TP, O'Donnell S, Black RE et al. (1981) Duration of serum antitoxin response following *Vibrio cholerae* infection in North Americans: Relevance for seroepidemiology. Am J Epidemiol 114:348–354

Levner MH, Urbano C, Rubin BA (1980) Lincomycin increases synthetic rate and periplasmic pool size for cholera toxin. J Bacteriol 143:441–447

Lewis MJ, Sweet DJ, Pelham HR (1990) The ERD2 gene determines the specificity of the luminal ER protein retention system. Cell 61:1359–1363

Li Z, Clarke AJ, Beveridge TJ (1998) Gram-negative bacteria produce membrane vesicles which are capable of killing other bacteria. J Bacteriol 180:5478–5483

Li CC, Crawford JA, DiRita VJ, Kaper JB (2000) Molecular cloning and transcriptional regulation of ompT, a ToxR-repressed gene in *Vibrio cholerae*. Mol Microbiol 35:189–203

Li CC, Merrell DS, Camilli A, Kaper JB (2002a) ToxR interferes with CRP-dependent transcriptional activation of ompT in *Vibrio cholerae*. Mol Microbiol 43:1577–1589

Li M, Shimada T, Morris JG, Jr., Sulakvelidze A, Sozhamannan S (2002b) Evidence for the emergence of non-O1 and non-O139 *Vibrio cholerae* strains with pathogenic potential by exchange of O-antigen biosynthesis regions. Infect Immun 70:2441–2453

Li TK, Fox BS (1996) Cholera toxin B subunit binding to an antigen-presenting cell directly costimulates cytokine production from a T cell clone. Int Immunol 8:1849–1856

Liang XP, Lamm ME, Nedrud JG (1988) Oral administration of cholera toxin-Sendai virus conjugate potentiates gut and respiratory immunity against Sendai virus. J Immunol 141:1495–1501

Liao X, Poirot E, Chang AH, Zhang X, Zhang J et al. (2002) The binding of synthetic analogs of the upstream, terminal residue of the O-polysaccharides (O-PS) of *Vibrio cholerae* O:1 serotypes Ogawa and Inaba to two murine monoclonal antibodies (MAbs) specific for the Ogawa lipopolysaccharide (LPS). Carbohydr Res 337:2437–2442

Licht TR, Krogfelt KA, Cohen PS, Poulsen LK, Urbance J, Molin S (1996) Role of lipopolysaccharide in colonization of the mouse intestine by *Salmonella typhimurium* studied by in situ hybridization. Infect Immun 64:3811–3817

Lin ECC (1996) Dissimilatory pathways for sugars, polyols and carboxylates. In: Neidhardt FC, Curtiss III R, Ingraham JL, Lin ECC, Low KB, Megasanik B, Reznikoff WC, Riley M, Schaechter M, and Umberger HE (eds) *Escherichia coli* and Salmonella: Cellular and Molecular Biology, 2nd edn. American Society for Microbiology, Washington, DC, pp 307–342

Lin W, Fullner KJ, Clayton R, Sexton JA, Rogers MB et al. (1999) Identification of a vibrio cholerae RTX toxin gene cluster that is tightly linked to the cholera toxin prophage. Proc Natl Acad Sci USA 96:1071–1076

Lin Z, Kumagai K, Baba K, Mekalanos JJ, Nishibuchi M (1993) Vibrio parahaemolyticus has a homolog of the *Vibrio cholerae* toxRS operon that mediates environmentally induced regulation of the thermostable direct hemolysin gene. J Bacteriol 175:3844–3855

Lindberg AA (1967) Studies of a receptor for felix O-1 phage in *Salmonella minnesota.* J Gen Microbiol 48:225–233

Little JW (1984) Autodigestion of lexA and phage lambda repressors. Proc Natl Acad Sci USA 81:1375–1379

Lockman H, Kaper JB (1983) Nucleotide sequence analysis of the A2 and B subunits of *Vibrio cholerae* enterotoxin. J Biol Chem 258:13722–13726

London E, Luongo CL (1989) Domain-specific bias in arginine/lysine usage by protein toxins. Biochem Biophys Res Commun 160:333–339

Lord JM, Roberts LM (1998a) Toxin entry: retrograde transport through the secretory pathway. J Cell Biol 140:733–736

Lord JM, Roberts LM (1998b) Retrograde transport: Going against the flow. Curr Biol 8:R56–58

Lord JM, Roberts LM, Lencer WI (2005) Entry of protein toxins into mammalian cells by crossing the endoplasmic reticulum membrane: Co-opting basic mechanisms of endoplasmic reticulum-associated degradation. Curr Top Microbiol Immunol 300:149–168

Losonsky GA, Lim Y, Motamedi P, Comstock LE, Johnson JA et al. (1997) Vibriocidal antibody responses in North American volunteers exposed to wild-type or vaccine *Vibrio cholerae* O139: Specificity and relevance to immunity. Clin Diagn Lab Immunol 4:264–269

Losonsky GA, Yunyongying J, Lim V, Reymann M, Lim YL, Wasserman SS, Levine MM (1996) Factors influencing secondary vibriocidal immune responses: Relevance for understanding immunity to cholera. Infect Immun 64:10–15

Lu HM, Lory S (1996) A specific targeting domain in mature exotoxin A is required for its extracellular secretion from *Pseudomonas aeruginosa.* Embo J 15:429–436

Lu L, Khan S, Lencer W, Walker WA (2005) Endocytosis of cholera toxin by human enterocytes is developmentally regulated. Am J Physiol Gastrointest Liver Physiol 289:G332–G341

Luderitz O, Westphal O, Staub AM, Nikado H (1971) Isolation and chemical and immunological characterization of bacterial lipopolysaccharides. In: Weinbaum G, Kadis S, and Ajl SJ (eds) Microbial Toxins. Academic, New York, pp 145–233

Ludwig DS, Holmes RK, Schoolnik GK (1985) Chemical and immunochemical studies on the receptor binding domain of cholera toxin B subunit. J Biol Chem 260:12528–12534

Ludwig DS, Ribi HO, Schoolnik GK, Kornberg RD (1986) Two-dimensional crystals of cholera toxin B-subunit-receptor complexes: projected structure at 17-A resolution. Proc Natl Acad Sci USA 83:8585–8588

Luirink J, von Heijne G, Houben E, de Gier JW (2005) Biogenesis of inner membrane proteins in *Escherichia coli.* Annu Rev Microbiol 59:329–355

Lycke N, Bromander AK, Holmgren J (1989) Role of local IgA antitoxin-producing cells for intestinal protection against cholera toxin challenge. Int Arch Allergy Appl Immunol 88:273–279

Lycke N, Holmgren J (1986a) Strong adjuvant properties of cholera toxin on gut mucosal immune responses to orally presented antigens. Immunology 59:301–308

Lycke N, Holmgren J (1986b) Intestinal mucosal memory and presence of memory cells in lamina propria and Peyer's patches in mice 2 years after oral immunization with cholera toxin. Scand J Immunol 23:611–616

Lycke N, Karlsson U, Sjolander A, Magnusson KE (1991) The adjuvant action of cholera toxin is associated with an increased intestinal permeability for luminal antigens. Scand J Immunol 33:691–698

Lycke N, Schon K (2001) The B cell targeted adjuvant, CTA1-DD, exhibits potent mucosal immunoenhancing activity despite pre-existing anti-toxin immunity. Vaccine 19:2542–2548

Lycke N, Tsuji T, Holmgren J (1992) The adjuvant effect of *Vibrio cholerae* and *Escherichia coli* heat-labile enterotoxins is linked to their ADP-ribosyltransferase activity. Eur J Immunol 22:2277–2281

Ma X, Saksena R, Chernyak A, Kovac P (2003) Neoglycoconjugates from synthetic tetra- and hexasaccharides that mimic the terminus of the O-PS of *Vibrio cholerae* O:1, serotype Inaba. Org Biomol Chem 1:775–784

Maiti D, Das B, Saha A, Nandy RK, Nair GB, Bhadra RK (2006) Genetic organization of pre-CTX and CTX prophages in the genome of an environmental *Vibrio cholerae* non-O1, non-O139 strain. Microbiology 152:3633–3641

Maiti M, Chatterjee SN (1971) Characteristics of a group IV cholera phage. J Gen Virol 13:327–330

Maiti M, Sur P, Chatterjee SN (1977) Amino sugar contents and phage inactivating properties of lipopolysaccharide from cholera and El Tor vibrios. Ann Microbiol (Paris) 128A:35–39

Mandal D, Chatterjee SN (1987) Mitomycin-induced prophage induction in *Vibrio cholerae* cells. Indian J Biochem Biophys 24:305–307

Manning PA, Brown MH, Heuzenroeder MW (1984) Cloning of the structural gene (hly) for the haemolysin of *Vibrio cholerae* El Tor strain 017. Gene 31:225–231

Manning PA, Heuzenroeder MW, Yeadon J, Leavesley DI, Reeves PR, Rowley D (1986) Molecular cloning and expression in *Escherichia coli* K-12 of the O antigens of the Inaba and Ogawa serotypes of the *Vibrio cholerae* O1 lipopolysaccharides and their potential for vaccine development. Infect Immun 53:272–277

Manning PA, Stroeher UH, Karageorgos LE, Morona R (1995) Putative O-antigen transport genes within the rfb region of *Vibrio cholerae* O1 are homologous to those for capsule transport. Gene 158:1–7

Manning PA, Stroeher UH, Morona R (1994) Molecular basis for O-antigen biosynthesis in *Vibrio cholerae* O1: Ogawa–Inaba switching. In: Wachsmuth IK, Blake PA, and Olsvik O (eds) *Vibrio cholerae* and Cholera: Molecular to Global Perspectives. ASM Press, Washington, DC, pp 77–94

Marinaro M, Di Tommaso A, Uzzau S, Fasano A, De Magistris MT (1999) Zonula occludens toxin is a powerful mucosal adjuvant for intranasally delivered antigens. Infect Immun 67:1287–1291

Marinaro M, Fasano A, De Magistris MT (2003) Zonula occludens toxin acts as an adjuvant through different mucosal routes and induces protective immune responses. Infect Immun 71:1897–1902

Marsh JW, Taylor RK (1998) Identification of the *Vibrio cholerae* type 4 prepilin peptidase required for cholera toxin secretion and pilus formation. Mol Microbiol 29:1481–1492

Massol RH, Larsen JE, Fujinaga Y, Lencer WI, Kirchhausen T (2004) Cholera toxin toxicity does not require functional Arf6- and dynamin-dependent endocytic pathways. Mol Biol Cell 15:3631–3641

Mathur J, Waldor MK (2004) The *Vibrio cholerae* ToxR-regulated porin OmpU confers resistance to antimicrobial peptides. Infect Immun 72:3577–3583

Matlack KE, Mothes W, Rapoport TA (1998) Protein translocation: Tunnel vision. Cell 92:381–390

Matson JS, DiRita VJ (2005) Degradation of the membrane-localized virulence activator TcpP by the YaeL protease in *Vibrio cholerae*. Proc Natl Acad Sci USA 102:16403–16408

Mazel D, Dychinco B, Webb VA, Davies J (1998) A distinctive class of integron in the *Vibrio cholerae* genome. Science 280:605–608

McBroom AJ, Johnson AP, Vemulapalli S, Kuehn MJ (2006) Outer membrane vesicle production by *Escherichia coli* is independent of membrane instability. J Bacteriol 188:5385–5392

McBroom AJ, Kuehn MJ (2007) Release of outer membrane vesicles by Gram-negative bacteria is a novel envelope stress response. Mol Microbiol 63(2):545–558

McCardell BA, Kothary MH, Hall RH, Sathyamoorthy V (2000) Identification of a CHO cell-elongating factor produced by *Vibrio cholerae* O1. Microb Pathog 29:1–8

McCardell BA, Sathyamoorthy V, Michalski J, Lavu S, Kothary M et al. (2002) Cloning, expression and characterization of the CHO cell elongating factor (Cef) from *Vibrio cholerae* O1. Microb Pathog 32:165–172

McIntire FC, Hargie MP, Schenck JR, Finley RA, Sievert HW, Rietschel ET, Rosenstreich DL (1976) Biologic properties of nontoxic derivatives of a lipopolysaccharide from *Escherichia coli* K235. J Immunol 117:674–678

McIntyre OR, Feeley JC, Greenough WB 3rd, Benenson AS, Hassan SI, Saad A (1965) Diarrhea caused by non-cholera vibrios. Am J Trop Med Hyg 14:412–418

McLeod SM, Waldor MK (2004) Characterization of XerC- and XerD-dependent CTX phage integration in *Vibrio cholerae*. Mol Microbiol 54:935–947

Meeks MD, Saksena R, Ma X, Wade TK, Taylor RK, Kovac P, Wade WF (2004) Synthetic fragments of *Vibrio cholerae* O1 Inaba O-specific polysaccharide bound to a protein carrier are immunogenic in mice but do not induce protective antibodies. Infect Immun 72:4090–4101

Mekalanos JJ (1983) Duplication and amplification of toxin genes in *Vibrio cholerae*. Cell 35:253–263

Mekalanos JJ (1985) Cholera toxin: Genetic analysis, regulation, and role in pathogenesis. Curr Top Microbiol Immunol 118:97–118

Mekalanos JJ, Swartz DJ, Pearson GD, Harford N, Groyne F, de Wilde M (1983) Cholera toxin genes: Nucleotide sequence, deletion analysis and vaccine development. Nature 306:551–557

Mekalanos JJ, Waldor MK, Gardel CL, Coster TS, Kenner J et al. (1995) Live cholera vaccines: perspectives on their construction and safety. Bull Inst Pasteur 93:255–262

Meno Y, Waldor MK, Mekalanos JJ, Amako K (1998) Morphological and physical characterization of the capsular layer of *Vibrio cholerae* O139. Arch Microbiol 170:339–344

Merrell DS, Hava DL, Camilli A (2002) Identification of novel factors involved in colonization and acid tolerance of *Vibrio cholerae*. Mol Microbiol 43:1471–1491

Merritt EA, Hol WG (1995) AB5 toxins. Curr Opin Struct Biol 5:165–171

Merritt EA, Kuhn P, Sarfaty S, Erbe JL, Holmes RK, Hol WG (1998) The 1.25 A resolution refinement of the cholera toxin B-pentamer: Evidence of peptide backbone strain at the receptor-binding site. J Mol Biol 282:1043–1059

Merritt EA, Sarfaty S, Jobling MG, Chang T, Holmes RK, Hirst TR, Hol WG (1997) Structural studies of receptor binding by cholera toxin mutants. Protein Sci 6:1516–1528

Merritt EA, Sarfaty S, van den Akker F, L'Hoir C, Martial JA, Hol WG (1994a) Crystal structure of cholera toxin B-pentamer bound to receptor GM1 pentasaccharide. Protein Sci 3:166–175

Merritt EA, Sixma TK, Kalk KH, van Zanten BA, Hol WG (1994b) Galactose-binding site in *Escherichia coli* heat-labile enterotoxin (LT) and cholera toxin (CT). Mol Microbiol 13:745–753

Miller VL, DiRita VJ, Mekalanos JJ (1989) Identification of toxS, a regulatory gene whose product enhances toxR-mediated activation of the cholera toxin promoter. J Bacteriol 171:1288–1293

Miller VL, Mekalanos JJ (1984) Synthesis of cholera toxin is positively regulated at the transcriptional level by toxR. Proc Natl Acad Sci USA 81:3471–3475

Miller VL, Mekalanos JJ (1985) Genetic analysis of the cholera toxin-positive regulatory gene toxR. J Bacteriol 163:580–585

Miller VL, Mekalanos JJ (1988) A novel suicide vector and its use in construction of insertion mutations: Osmoregulation of outer membrane proteins and virulence determinants in *Vibrio cholerae* requires toxR. J Bacteriol 170:2575–2583

Miller VL, Taylor RK, Mekalanos JJ (1987) Cholera toxin transcriptional activator toxR is a transmembrane DNA binding protein. Cell 48:271–279

Minami A, Hashimoto S, Abe H, Arita M, Taniguchi T et al. (1991) Cholera enterotoxin production in *Vibrio cholerae* O1 strains isolated from the environment and from humans in Japan. Appl Environ Microbiol 57:2152–2157

Mitra R, Basu A, Dutta D, Nair GB, Takeda Y (1996) Resurgence of *Vibrio cholerae* O139 Bengal with altered antibiogram in Calcutta, India. Lancet 348:1181

Miyamoto Y, Kato T, Obara Y, Akiyama S, Takizawa K, Yamai S (1969) In vitro hemolytic characteristic of *Vibrio parahaemolyticus*: its close correlation with human pathogenicity. J Bacteriol 100:1147–1149

Mizunoe Y, Wai SN, Takade A, Yoshida SI (1999) Isolation and characterization of rugose form of *Vibrio cholerae* O139 strain MO10. Infect Immun 67:958–963

Mooi FR, Bik EM (1997) The evolution of epidemic *Vibrio cholerae* strains. Trends Microbiol 5:161–165

Mori H, Ito K (2001) The Sec protein-translocation pathway. Trends Microbiol 9:494–500

Morris JG (1994) Cholera and other vibrios. In: Hoeprich PD, Jordan MC, and Ronald AR (eds) Infectious Diseases: A Treatise of Infectious Processes. J.B. Lippincott, Philadelphia, PA, pp 753–762

Morris JG, Jr., Losonsky GE, Johnson JA, Tacket CO, Nataro JP, Panigrahi P, Levin MM (1995) Clinical and immunologic characteristics of *Vibrio cholerae* O139 Bengal infection in North American volunteers. J Infect Dis 171:903–908

Morris JG, Jr., Sztein MB, Rice EW, Nataro JP, Losonsky GA et al. (1996) *Vibrio cholerae* O1 can assume a chlorine-resistant rugose survival form that is virulent for humans. J Infect Dis 174:1364–1368

Moschioni M, Tombola F, de Bernard M, Coelho A, Zitzer A, Zoratti M, Montecucco C (2002) The *Vibrio cholerae* haemolysin anion channel is required for cell vacuolation and death. Cell Microbiol 4:397–409

Moseley SL, Echeverria P, Seriwatana J, Tirapat C, Chaicumpa W, Sakuldaipeara T, Falkow S (1982) Identification of enterotoxigenic *Escherichia coli* by colony hybridization using three enterotoxin gene probes. J Infect Dis 145:863–869

Mosley WH (1969) The role of immunity in cholera: A review of epidemiological and serological studies. Tex Rep BiolMed 27:227–241

Mosley WH, McCormack WM, Ahmed A, Chowdhury AK, Barui RK (1969) Report of the 1966–67 cholera vaccine field trial in rural East Pakistan. 2. Results of the serological surveys in the study population: the relationship of case rate to antibody titre and an estimate of the inapparent infection rate with *Vibrio cholerae*. Bull World Health Organ 40:187–197

Moss J, Fishman PH, Richards RL, Alving CR, Vaughan M, Brady RO (1976) Choleragen-mediated release of trapped glucose from liposomes containing ganglioside GM1. Proc Natl Acad Sci USA 73:3480–3483

Moss J, Richardson SH (1978) Activation of adenylate cyclase by heat-labile *Escherichia coli* enterotoxin. Evidence for ADP-ribosyltransferase activity similar to that of choleragen. J Clin Invest 62:281–285

Moss J, Vaughan M (1977) Mechanism of action of choleragen. Evidence for ADP-ribosyltransferase activity with arginine as an acceptor. J Biol Chem 252:2455–2457

Moss J, Vaughan M (1991) Activation of cholera toxin and *Escherichia coli* heat-labile enterotoxins by ADP-ribosylation factors, a family of 20 kDa guanine nucleotide-binding proteins. Mol Microbiol 5:2621–2627

Mosser G, Mallouh V, Brisson A (1992) A 9 A two-dimensional projected structure of cholera toxin B-subunit-GM1 complexes determined by electron crystallography. J Mol Biol 226:23–28

Moyer KE, Kimsey HH, Waldor MK (2001) Evidence for a rolling-circle mechanism of phage DNA synthesis from both replicative and integrated forms of CTXphi. Mol Microbiol 41:311–323

Muhlardt PF, Wray V, Lehmann V (1977) A 31P-nuclear magnetic resonance study of the phosphate groups in lipopolysaccharide and lipid A from *Salmonella*. Eur J Biochem 81:193–203

Mukerjee S (1963a) The bacteriophage susceptibility test in differentiating *Vibrio cholerae* and Vibrio El Tor. Bull World Health Organ 28:333–336
Mukerjee S (1963b) Bacteriopahge typing of cholera. Bull World Health Organ 28:337–345
Mukerjee S, Takeya K (1974) Vibriophages. In: Barua D and Burrows W (eds) Cholera. Saunders, Philadelphia, PA, pp 61–84
Mukhopadhyay AK, Basu A, Garg P, Bag PK, Ghosh A et al. (1998) Molecular epidemiology of reemergent *Vibrio cholerae* O139 Bengal in India. J Clin Microbiol 36:2149–2152
Mukhopadhyay AK, Chakraborty S, Takeda Y, Nair GB, Berg DE (2001) Characterization of VPI pathogenicity island and CTXphi prophage in environmental strains of *Vibrio cholerae*. J Bacteriol 183:4737–4746
Mukhopadhyay S, Nandi B, Ghose AC (2000) Antibodies (IgG) to lipopolysaccharide of *Vibrio cholerae* O1 mediate protection through inhibition of intestinal adherence and colonisation in a mouse model. FEMS Microbiol Lett 185:29–35
Muller M, Klosgen RB (2005) The Tat pathway in bacteria and chloroplasts (review). Mol Membr Biol 22:113–121
Munro S (2003) Lipid rafts: Elusive or illusive? Cell 115:377–388
Murley YM, Carroll PA, Skorupski K, Taylor RK, Calderwood SB (1999) Differential transcription of the tcpPH operon confers biotype-specific control of the *Vibrio cholerae* ToxR virulence regulon. Infect Immun 67:5117–5123
Murphy RA, Boyd EF (2008) Three pathogenicity islands of *Vibrio cholerae* can excise from the chromosome and form circular intermediates. J Bacteriol 190:636–647
Nag S, Das S, Chaudhuri K (2005) In vivo induced clpB1 gene of *Vibrio cholerae* is involved in different stress responses and affects in vivo cholera toxin production. Biochem Biophys Res Commun 331:1365–1373
Nagamune K, Yamamoto K, Naka A, Matsuyama J, Miwatani T, Honda T (1996) In vitro proteolytic processing and activation of the recombinant precursor of El Tor cytolysin/hemolysin (pro-HlyA) of *Vibrio cholerae* by soluble hemagglutinin/protease of V. cholerae, trypsin, and other proteases. Infect Immun 64:4655–4658
Nagayama K, Oguchi T, Arita M, Honda T (1994) Correlation between cell-associated mannose-sensitive hemagglutination by *Vibrio parahaemolyticus* and adherence to a human colonic cell line Caco-2. FEMS Microbiol Lett 120:207–210
Nair GB, Bhattacharya SK, Takeda T (1993) Identification of heat-stable enterotoxin-producing strains of *Yersinia enterocolitica* and *Vibrio cholerae* non-O1 by a monoclonal antibody-based enzyme-linked immunosorbent assay. Microbiol Immunol 37:181–186
Nair GB, Faruque SM, Bhuiyan NA, Kamruzzaman M, Siddique AK, Sack DA (2002) New variants of *Vibrio cholerae* O1 biotype El Tor with attributes of the classical biotype from hospitalized patients with acute diarrhea in Bangladesh. J Clin Microbiol 40:3296–3299
Nair GB, Oku Y, Takeda Y, Ghosh A, Ghosh RK et al. (1988) Toxin profiles of *Vibrio cholerae* non-O1 from environmental sources in Calcutta, India. Appl Environ Microbiol 54:3180–3182
Nambiar MP, Oda T, Chen C, Kuwazuru Y, Wu HC (1993) Involvement of the Golgi region in the intracellular trafficking of cholera toxin. J Cell Physiol 154:222–228
Nandi B, Nandy RK, Vicente AC, Ghose AC (2000) Molecular characterization of a new variant of toxin-coregulated pilus protein (TcpA) in a toxigenic non-O1/Non-O139 strain of *Vibrio cholerae*. Infect Immun 68:948–952
Nandi S, Maiti D, Saha A, Bhadra RK (2003) Genesis of variants of *Vibrio cholerae* O1 biotype El Tor: Role of the CTXphi array and its position in the genome. Microbiology 149:89–97
Nandy RK, Albert MJ, Ghose AC (1996) Serum antibacterial and antitoxin responses in clinical cholera caused by *Vibrio cholerae* O139 Bengal and evaluation of their importance in protection. Vaccine 14:1137–1142
Nato F, Boutonnier A, Rajerison M, Grosjean P, Dartevelle S et al. (2003) One-step immunochromatographic dipstick tests for rapid detection of *Vibrio cholerae* O1 and O139 in stool samples. Clin Diagn Lab Immunol 10:476–478
Neoh SH, Rowley D (1970) The antigens of *Vibrio cholerae* involved in the vibriocidal action of antibody and complement. J Infect Dis 121:505–513

Nesper J, Blass J, Fountoulakis M, Reidl J (1999) Characterization of the major control region of *Vibrio cholerae* bacteriophage K139: immunity, exclusion, and integration. J Bacteriol 181:2902–2913

Nesper J, Kapfhammer D, Klose KE, Merkert H, Reidl J (2000) Characterization of *Vibrio cholerae* O1 antigen as the bacteriophage K139 receptor and identification of IS1004 insertions aborting O1 antigen biosynthesis. J Bacteriol 182:5097–5104

Nesper J, Kraiss A, Schild S, Blass J, Klose KE, Bockemuhl J, Reidl J (2002a) Comparative and genetic analyses of the putative *Vibrio cholerae* lipopolysaccharide core oligosaccharide biosynthesis (wav) gene cluster. Infect Immun 70:2419–2433

Nesper J, Lauriano CM, Klose KE, Kapfhammer D, Kraiss A, Reidl J (2001) Characterization of *Vibrio cholerae* O1 El tor galU and galE mutants: influence on lipopolysaccharide structure, colonization, and biofilm formation. Infect Immun 69:435–445

Nesper J, Schild S, Lauriano CM, Kraiss A, Klose KE, Reidl J (2002b) Role of *Vibrio cholerae* O139 surface polysaccharides in intestinal colonization. Infect Immun 70:5990–5996

Nichols B (2003) Caveosomes and endocytosis of lipid rafts. J Cell Sci 116:4707–4714

Nichols BJ, Kenworthy AK, Polishchuk RS, Lodge R, Roberts TH et al. (2001) Rapid cycling of lipid raft markers between the cell surface and Golgi complex. J Cell Biol 153:529–541

Nishibuchi M, Fasano A, Russell RG, Kaper JB (1992) Enterotoxigenicity of *Vibrio parahaemolyticus* with and without genes encoding thermostable direct hemolysin. Infect Immun 60:3539–3545

Nishibuchi M, Kaper JB (1995) Thermostable direct hemolysin gene of *Vibrio parahaemolyticus*: A virulence gene acquired by a marine bacterium. Infect Immun 63:2093–2099

Northrup RS, Fauci AS (1972) Adjuvant effect of cholera enterotoxin on the immune response of the mouse to sheep red blood cells. J Infect Dis 125:672–673

Nye MB, Pfau JD, Skorupski K, Taylor RK (2000) *Vibrio cholerae* H-NS silences virulence gene expression at multiple steps in the ToxR regulatory cascade. J Bacteriol 182:4295–4303

O'Brien AD, Chen ME, Holmes RK, Kaper J, Levine MM (1984) Environmental and human isolates of *Vibrio cholerae* and *Vibrio parahaemolyticus* produce a *Shigella dysenteriae* 1 (Shiga)-like cytotoxin. Lancet 1:77–78

O'Brien AD, Holmes RK (1987) Shiga and Shiga-like toxins. Microbiol Rev 51:206–220

Ogawa A, Kato J, Watanabe H, Nair BG, Takeda T (1990) Cloning and nucleotide sequence of a heat-stable enterotoxin gene from *Vibrio cholerae* non-O1 isolated from a patient with traveler's diarrhea. Infect Immun 58:3325–3329

Ogawa A, Takeda T (1993) The gene encoding the heat-stable enterotoxin of *Vibrio cholerae* is flanked by 123-base pair direct repeats. Microbiol Immunol 37:607–616

Ogawa Y, Lei PS, Kovac P (1996) Synthesis of eight glycosides of hexasaccharide fragments representing the terminus of the O-polysaccharide of *Vibrio cholerae* O:1, serotype Inaba and Ogawa, bearing aglycons suitable for linking to proteins. Carbohydr Res 293:173–194

Ogierman MA, Fallarino A, Riess T, Williams SG, Attridge SR, Manning PA (1997) Characterization of the *Vibrio cholerae* El Tor lipase operon lipAB and a protease gene downstream of the hly region. J Bacteriol 179:7072–7080

Oku Y, Uesaka Y, Hirayama T, Takeda Y (1988) Development of a highly sensitive bead-ELISA to detect bacterial protein toxins. Microbiol Immunol 32:807–816

Olson R, Gouaux E (2005) Crystal structure of the *Vibrio cholerae* cytolysin (VCC) pro-toxin and its assembly into a heptameric transmembrane pore. J Mol Biol 350:997–1016

Olsvik O, Wahlberg J, Petterson B, Uhlen M, Popovic T, Wachsmuth IK, Fields PI (1993) Use of automated sequencing of polymerase chain reaction-generated amplicons to identify three types of cholera toxin subunit B in *Vibrio cholerae* O1 strains. J Clin Microbiol 31:22–25

O'Neal CJ, Amaya EI, Jobling MG, Holmes RK, Hol WG (2004) Crystal structures of an intrinsically active cholera toxin mutant yield insight into the toxin activation mechanism. Biochemistry 43:3772–3782

O'Neal CJ, Jobling MG, Holmes RK, Hol WG (2005) Structural basis for the activation of cholera toxin by human ARF6-GTP. Science 309:1093–1096

Orlandi PA (1997) Protein-disulfide isomerase-mediated reduction of the A subunit of cholera toxin in a human intestinal cell line. J Biol Chem 272:4591–4599

Orlandi PA, Curran PK, Fishman PH (1993) Brefeldin A blocks the response of cultured cells to cholera toxin. Implications for intracellular trafficking in toxin action. J Biol Chem 268:12010–12016

Osek J, Jonson G, Svennerholm AM, Holmgren J (1994) Role of antibodies against biotype-specific *Vibrio cholerae* pili in protection against experimental classical and El Tor cholera. Infect Immun 62:2901–2907

Osek J, Svennerholm AM, Holmgren J (1992) Protection against *Vibrio cholerae* El Tor infection by specific antibodies against mannose-binding hemagglutinin pili. Infect Immun 60:4961–4964

O'Shaughnessy WB (1832) Report on the clinical pathology of the malignant cholera. Lancet i 929–936

O'Shea YA, Boyd EF (2002) Mobilization of the Vibrio pathogenicity island between *Vibrio cholerae* isolates mediated by CP-T1 generalized transduction. FEMS Microbiol Lett 214:153–157

O'Shea YA, Finnan S, Reen FJ, Morrissey JP, O'Gara F, Boyd EF (2004) The Vibrio seventh pandemic island-II is a 26.9 kb genomic island present in *Vibrio cholerae* El Tor and O139 serogroup isolates that shows homology to a 43.4 kb genomic island in *V. vulnificus*. Microbiology 150:4053–4063

Overbye LJ, Sandkvist M, Bagdasarian M (1993) Genes required for extracellular secretion of enterotoxin are clustered in *Vibrio cholerae*. Gene 132:101–106

Pang H, Le PU, Nabi IR (2004) Ganglioside GM1 levels are a determinant of the extent of caveolae/raft-dependent endocytosis of cholera toxin to the Golgi apparatus. J Cell Sci 117:1421–1430

Panigrahi P, Srinivas S, Johnson JA, DeTolla LJ (1992) Modulation of immunity in non-O1 *Vibrio cholerae*. 92nd Annu Meet Am Soc Microbiol Abstr B-316

Parker C, Gauthier D, Tate A, Richardson K, Romig WR (1979) Expanded linkage map of *Vibrio cholerae*. Genetics 91:191–214

Parsot C, Mekalanos JJ (1990) Expression of ToxR, the transcriptional activator of the virulence factors in *Vibrio cholerae*, is modulated by the heat shock response. Proc Natl Acad Sci USA 87:9898–9902

Paul S, Sen AK, Banerjee N, Chatterjee AN, Das J (1990) Lipid A mutants of *Vibrio cholerae*: isolation and partial characterization. Biochem Biophys Res Commun 169:116–122

Pearson GD, DiRita VJ, Goldberg MB, Boyko SA, Calderwood SB, Mekalanos JJ (1990) New attenuated derivatives of *Vibrio cholerae*. Res Microbiol 141:893–899

Pearson GD, Mekalanos JJ (1982) Molecular cloning of *Vibrio cholerae* enterotoxin genes in *Escherichia coli* K-12. Proc Natl Acad Sci USA 79:2976–2980

Pearson GD, Woods A, Chiang SL, Mekalanos JJ (1993) CTX genetic element encodes a site-specific recombination system and an intestinal colonization factor. Proc Natl Acad Sci USA 90:3750–3754

Peek JA, Taylor RK (1992) Characterization of a periplasmic thiol:disulfide interchange protein required for the functional maturation of secreted virulence factors of *Vibrio cholerae*. Proc Natl Acad Sci USA 89:6210–6214

Peterson KM (2002) Expression of *Vibrio cholerae* virulence genes in response to environmental signals. Curr Issues Intest Microbiol 3:29–38

Peterson KM, Mekalanos JJ (1988) Characterization of the *Vibrio cholerae* ToxR regulon: Identification of novel genes involved in intestinal colonization. Infect Immun 56: 2822–2829

Pfau JD, Taylor RK (1998) Mutations in toxR and toxS that separate transcriptional activation from DNA binding at the cholera toxin gene promoter. J Bacteriol 180:4724–4733

Phillips RA (1964) Water and electrolyte losses in cholera. Fed Proc 23:705–712

Pichel M, Rivas M, Chinen I, Martin F, Ibarra C, Binsztein N (2003) Genetic diversity of *Vibrio cholerae* O1 in Argentina and emergence of a new variant. J Clin Microbiol 41:124–134

Pierce NF, Cray WC Jr, Kaper JB, Mekalanos JJ (1988) Determinants of immunogenicity and mechanisms of protection by virulent and mutant *Vibrio cholerae* O1 in rabbits. Infect Immun 56:142–148

Pierce NF, Cray WC, Jr., Sacci JB, Jr. (1982) Oral immunization of dogs with purified cholera toxin, crude cholera toxin, or B subunit: Evidence for synergistic protection by antitoxic and antibacterial mechanisms. Infect Immun 37:687–694

Pike LJ (2003) Lipid rafts: Bringing order to chaos. J Lipid Res 44:655–667

Pizza M, Scarlato V, Masignani V, Giuliani MM, Arico B et al. (2000) Identification of vaccine candidates against serogroup B meningococcus by whole-genome sequencing. Science 287:1816–1820

Poirot E, Zhang X, Whittaker NF, Kovac P (2001) Syntheses of the L-manno and some other analogs of the terminal determinants of the O-PS of *Vibrio cholerae* O:1. Carbohydr Res 330:7–20

Politzer R (1959) Cholera. World Health Organization, Geneva

Pollard TD, Borisy GG (2003) Cellular motility driven by assembly and disassembly of actin filaments. Cell 112:453–465

Popovic T, Bopp C, Olsvik O, Wachsmuth K (1993) Epidemiologic application of a standardized ribotype scheme for *Vibrio cholerae* O1. J Clin Microbiol 31:2474–2482

Popovic T, Fields PI, Olsvik O, Wells JG, Evins GM et al. (1995) Molecular subtyping of toxigenic *Vibrio cholerae* O139 causing epidemic cholera in India and Bangladesh, 1992–1993. J Infect Dis 171:122–127

Prescott LM, Datta A, Datta GC (1968) R-factors in Calcutta strains of *Vibrio cholerae* and members of the Enterobacteriaceae. Bull World Health Organ 39:971–973

Preston LM, Xu Q, Johnson JA, Joseph A, Maneval DR Jr et al. (1995) Preliminary structure determination of the capsular polysaccharide of *Vibrio cholerae* O139 Bengal Al1837. J Bacteriol 177:835–838

Proctor RA (1985) Effects of endotoxins on neutrophils. In: Berry LJ (ed) Handbook of Endotoxins. Elsevier, Amsterdam, pp 244–259

Prouty MG, Osorio CR, Klose KE (2005) Characterization of functional domains of the *Vibrio cholerae* virulence regulator ToxT. Mol Microbiol 58:1143–1156

Provenzano D, Klose KE (2000) Altered expression of the ToxR-regulated porins OmpU and OmpT diminishes *Vibrio cholerae* bile resistance, virulence factor expression, and intestinal colonization. Proc Natl Acad Sci USA 97:10220–10224

Provenzano D, Kovac P, Wade WF (2006) The ABCs (Antibody, B cells, and Carbohydrate epitopes) of cholera immunity: considerations for an improved vaccine. Microbiol Immunol 50:899–927

Provenzano D, Schuhmacher DA, Barker JL, Klose KE (2000) The virulence regulatory protein ToxR mediates enhanced bile resistance in *Vibrio cholerae* and other pathogenic *Vibrio* species. Infect Immun 68:1491–1497

Pugsley AP (1993) The complete general secretory pathway in Gram-negative bacteria. Microbiol Rev 57:50–108

Py B, Loiseau L, Barras F (2001) An inner membrane platform in the type II secretion machinery of Gram-negative bacteria. EMBO Rep 2:244–248

Qadri F, Ahmed F, Karim MM, Wenneras C, Begum YA et al. (1999) Lipopolysaccharide- and cholera toxin-specific subclass distribution of B-cell responses in cholera. Clin Diagn Lab Immunol 6:812–818

Qadri F, Mohi G, Hossain J, Azim T, Khan AM et al. (1995a) Comparison of the vibriocidal antibody response in cholera due to *Vibrio cholerae* O139 Bengal with the response in cholera due to *Vibrio cholerae* O1. Clin Diagn Lab Immunol 2:685–688

Qadri F, Hasan JA, Hossain J, Chowdhury A, Begum YA et al. (1995b) Evaluation of the monoclonal antibody-based kit Bengal SMART for rapid detection of *Vibrio cholerae* O139 synonym Bengal in stool samples. J Clin Microbiol 33:732–734

Qadri F, Raqib R, Ahmed F, Rahman T, Wenneras C et al. (2002) Increased levels of inflammatory mediators in children and adults infected with *Vibrio cholerae* O1 and O139. Clin Diagn Lab Immunol 9:221–229

Qadri F, Wenneras C, Albert MJ, Hossain J, Mannoor K et al. (1997) Comparison of immune responses in patients infected with *Vibrio cholerae* O139 and O1. Infect Immun 65:3571–3576

Quinones M, Kimsey HH, Waldor MK (2005) LexA cleavage is required for CTX prophage induction. Mol Cell 17:291–300

Rader AE, Murphy JR (1988) Nucleotide sequences and comparison of the hemolysin determinants of *Vibrio cholerae* El Tor RV79(Hly+) and RV79(Hly–) and classical 569B(Hly–). Infect Immun 56:1414–1419

Raetz CR (1998) Enzymes of lipid A biosynthesis: Targets for the design of new antibiotics. Prog Clin Biol Res 397:1–14

Raetz CR, Whitfield C (2002) Lipopolysaccharide endotoxins. Annu Rev Biochem 71:635–700

Raetz CRH (1996) Bacterial lipopolysaccharides: A remarkable family of bioactive macroamphiphiles. In: Neidhardt FC, Curtiss III R, Ingraham JL, Lin ECC, Low KB, Megasanik B, Reznikoff WS, Riley M, Schaechter M, and Umberger HE (eds) *Escherichia coli* and *Salmonella*: Cellular and Molecular Biology, 2nd edn. ASM Press, Washington, DC, pp 1035–1063

Rahman MS, Pal AK, Chatterjee SN (1993) Induction of SOS like responses by nitrofurantoin in *Vibrio cholerae* El Tor cells. Arch Microbiol 159:98–100

Rajanna C, Wang J, Zhang D, Xu Z, Ali A, Hou YM, Karaolis DK (2003) The vibrio pathogenicity island of epidemic *Vibrio cholerae* forms precise extrachromosomal circular excision products. J Bacteriol 185:6893–6901

Ramamurthy T, Garg S, Sharma R, Bhattacharya SK, Nair GB et al. (1993) Emergence of novel strain of *Vibrio cholerae* with epidemic potential in southern and eastern India. Lancet 341:703–704

Ramamurthy T, Yamasaki S, Takeda Y, Nair GB (2003) *Vibrio cholerae* O139 Bengal: Odyssey of a fortuitous variant. Microbes Infect 5:329–344

Raychaudhuri C, Chatterjee SN, Maiti M (1970) Effects of furazolidone on the macromolecular synthesis and morphology of *Vibrio cholerae* cells. Biochim Biophys Acta 222:637–646

Raziuddin S (1977) Characterization of lipid A and polysaccharide moieties of the lipopolysaccharides from *Vibrio cholerae*. Biochem J 167:147–154

Raziuddin S (1978) Toxic and immunological properties of the lipopolysaccharides (O-antigens) from Vibrio el-tor. Immunochemistry 15:611–614

Raziuddin S (1979) Structure–function relationship: Biological activities of the lipopolysaccharides and lipid A from *Vibrio cholerae*. J Infect Dis 140:590–595

Raziuddin S (1980a) Biological activities of chemically modified endotoxins from *Vibrio cholerae*. Biochim Biophys Acta 620:193–204

Raziuddin S (1980b) Immunochemical studies of the lipopolysaccharides of *Vibrio cholerae*: Constitution of O specific side chain and core polysaccharide. Infect Immun 27:211–215

Raziuddin S, Kawasaki T (1976) Biochemical studies on the cell wall lipopolysaccharides (O-antigens) of *Vibrio cholerae* 569 B (Inaba) and El-tor (Inaba). Biochim Biophys Acta 431:116–126

Recchia GD, Hall RM (1995) Gene cassettes: A new class of mobile element. Microbiology 141(Pt 12):3015–3027

Redmond JW (1978) The 4-amino sugars present in the lipopolysaccharides of *Vibrio cholerae* and related vibrios. Biochim Biophys Acta 542:378–384

Redmond JW (1979) The structure of the O-antigenic side chain of the lipopolysaccharide of *Vibrio cholerae* 569B (Inaba). Biochim Biophys Acta 584:346–352

Redmond JW, Korsch MJ, Jackson GD (1973) Immunochemical studies to the O-antigens of *Vibrio cholerae*. Partial characterization of an acid-labile antigenic determinant. Aust J Exp Biol Med Sci 51:229–235

Reed RA, Mattai J, Shipley GG (1987) Interaction of cholera toxin with ganglioside GM1 receptors in supported lipid monolayers. Biochemistry 26:824–832

Reeves PR, Hobbs M, Valvano MA, Skurnik M, Whitfield C et al. (1996) Bacterial polysaccharide synthesis and gene nomenclature. Trends Microbiol 4:495–503

Reich KA, Schoolnik GK (1994) The light organ symbiont *Vibrio fischeri* possesses a homolog of the *Vibrio cholerae* transmembrane transcriptional activator ToxR. J Bacteriol 176:3085–3088

Reidl J, Mekalanos JJ (1995) Characterization of *Vibrio cholerae* bacteriophage K139 and use of a novel mini-transposon to identify a phage-encoded virulence factor. Mol Microbiol 18:685–701

Rhine JA, Taylor RK (1994) TcpA pilin sequences and colonization requirements for O1 and O139 vibrio cholerae. Mol Microbiol 13:1013–1020

Ribi E, Anacker RL, Brown R, Haskins WT, Malmgren B, Milner KC, Rudbach JA (1966) Reaction of endotoxin and surfactants. I. Physical and biological properties of endotoxin treated with sodium deoxycholate. J Bacteriol 92:1493–1509

Ribi HO, Ludwig DS, Mercer KL, Schoolnik GK, Kornberg RD (1988) Three-dimensional structure of cholera toxin penetrating a lipid membrane. Science 239:1272–1276

Richards AA, Stang E, Pepperkok R, Parton RG (2002) Inhibitors of COP-mediated transport and cholera toxin action inhibit simian virus 40 infection. Mol Biol Cell 13:1750–1764

Richardson SH (1969) Factors influencing in vitro skin permeability factor production by *Vibrio cholerae*. J Bacteriol 100:27–34

Rietschel ET (1976) Absolute configuration of 3-hydroxy fatty acids present in lipopolysaccharides from various bacterial groups. Eur J Biochem 64:423–428

Rietschel ET, Galanos C, Tanaka A, Ruschmann E, Luderitz O, Westphal O (1971) Biological activities of chemically modified endotoxins. Eur J Biochem 22:218–224

Rietschel ET, Luderitz O, Volk WA (1975) Nature, type of linkage, and absolute configuration of (hydroxy) fatty acids in lipopolysaccharides from *Xanthomonas sinensis* and related strains. J Bacteriol 122:1180–1188

Rietschel ET, Palin WJ, Watson DW (1973) Nature and linkages of the fatty acids present in lipopolysaccharides from *Vibrio metchnikovii* and *Vibrio parahemolyticus*. Eur J Biochem 37:116–120

Rietschel ET, Wollenweaber HW, Brade H, Zahringer U, Linder B et al. (1984) Structure and conformation of the lipid A component of lipopolysaccharides. In: Proctor RA (ed) Handbook of Endotoxins. Elsevier, Amsterdam, pp 187–220

Riordan JR, Rommens JM, Kerem B, Alon N, Rozmahel R et al. (1989) Identification of the cystic fibrosis gene: Cloning and characterization of complementary DNA. Science 245:1066–1073

Rivera IN, Chun J, Huq A, Sack RB, Colwell RR (2001) Genotypes associated with virulence in environmental isolates of *Vibrio cholerae*. Appl Environ Microbiol 67:2421–2429

Roantree R (1957) Salmonella O-antigen and virulence. Ann Rev Microbiol 21:443–466

Robbins JB, Schneerson R, Szu SC (1995) Perspective: Hypothesis: Serum IgG antibody is sufficient to confer protection against infectious diseases by inactivating the inoculum. J Infect Dis 171:1387–1398

Robert A, Silva A, Benitez JA, Rodriguez BL, Fando R et al. (1996) Tagging a *Vibrio cholerae* El Tor candidate vaccine strain by disruption of its hemagglutinin/protease gene using a novel reporter enzyme: *Clostridium thermocellum* endoglucanase A. Vaccine 14:1517–1522

Robert-Pillot A, Baron S, Lesne J, Fournier J-M, Quilici ML (2002) Improved specific detection of *Vibrio cholerae* in environmental water samples by culture on selective medium and colony hybridization assay with an oligonucleotide probe. FEMS Microbiol Ecol 40:39–46

Robien MA, Krumm BE, Sandkvist M, Hol WG (2003) Crystal structure of the extracellular protein secretion NTPase EpsE of *Vibrio cholerae*. J Mol Biol 333:657–674

Rocchetta HL, Burrows LL, Lam JS (1999) Genetics of O-antigen biosynthesis in *Pseudomonas aeruginosa*. Microbiol Mol Biol Rev 63:523–553

Rodighiero C, Aman AT, Kenny MJ, Moss J, Lencer WI, Hirst TR (1999) Structural basis for the differential toxicity of cholera toxin and *Escherichia coli* heat-labile enterotoxin. Construction of hybrid toxins identifies the A2-domain as the determinant of differential toxicity. J Biol Chem 274:3962–3969

Rodighiero C, Fujinaga Y, Hirst TR, Lencer WI (2001) A cholera toxin B-subunit variant that binds ganglioside G(M1) but fails to induce toxicity. J Biol Chem 276:36939–36945

Rodighiero C, Tsai B, Rapoport TA, Lencer WI (2002) Role of ubiquitination in retro-translocation of cholera toxin and escape of cytosolic degradation. EMBO Rep 3:1222–1227

Rosner M, Tang J, Barzilay I, Khorana HG (1979) Structure of the lipopolysaccharide from an *Escherichia coli* heptose-less mutant: 1. Chemical degradation and identification of products. J Biol Chem 254:5906–5917

Rowe-Taitt CA, Cras JJ, Patterson CH, Golden JP, Ligler FS (2000) A ganglioside-based assay for cholera toxin using an array biosensor. Anal Biochem 281:123–133

Rubin EJ, Lin W, Mekalanos JJ, Waldor MK (1998) Replication and integration of a *Vibrio cholerae* cryptic plasmid linked to the CTX prophage. Mol Microbiol 28:1247–1254

Ruddock LW, Coen JJ, Cheesman C, Freedman RB, Hirst TR (1996) Assembly of the B subunit pentamer of *Escherichia coli* heat-labile enterotoxin. Kinetics and molecular basis of rate-limiting steps in vitro. J Biol Chem 271:19118–19123

Ruddock LW, Ruston SP, Kelly SM, Price NC, Freedman RB, Hirst TR (1995) Kinetics of acid-mediated disassembly of the B subunit pentamer of *Escherichia coli* heat-labile enterotoxin. Molecular basis of pH stability. J Biol Chem 270:29953–29958

Russel M (1995) Moving through the membrane with filamentous phages. Trends Microbiol 3:223–228

Russel M (1998) Macromolecular assembly and secretion across the bacterial cell envelope: Type II protein secretion systems. J Mol Biol 279:485–499

Sack DA, Clemens JD, Huda S, Harris JR, Khan MR et al. (1991) Antibody responses after immunization with killed oral cholera vaccines during the 1985 vaccine field trial in Bangladesh. J Infect Dis 164:407–411

Sack DA, Huda S, Neogi PK, Daniel RR, Spira WM (1980) Microtiter ganglioside enzyme-linked immunosorbent assay for vibrio and *Escherichia coli* heat-labile enterotoxins and antitoxin. J Clin Microbiol 11:35–40

Sack RB, Cassells J, Mitra R, Merritt C, Butler T et al. (1970) The use of oral replacement solutions in the treatment of choleraand other severe diarrhoeal disorders. Bull World Health Organ 43:351–360

Safrin S, Morris JG, Adams M, Pons V, Jacob R, Conte JE (1987) Non-O1 *Vibrio cholerae* bacteremia: A case report and review. Rev Infect Dis 10:1012–1017

Saha S, Sanyal SC (1989) Immunobiological relationships among new cholera toxins produced by CT gene-negative strains of *Vibrio cholerae* O1. J Med Microbiol 28:33–37

Saha S, Sanyal SC (1990) Immunobiological relationships of the enterotoxins produced by cholera toxin gene-positive (CT+) and -negative (CT–) strains of *Vibrio cholerae* O1. J Med Microbiol 32:33–37

Saka HA, Bidinost C, Sola C, Carranza P, Collino C et al. (2008) *Vibrio cholerae* cytolysin is essential for high enterotoxicity and apoptosis induction produced by a cholera toxin gene-negative *V. cholerae* non-O1, non-O139 strain. Microb Pathog 44:118–128

Sakazaki R (1968) Proposal of *Vibrio alginolyticus* for the biotype 2 of *Vibrio parahaemolyticus*. Jpn J Med Sci Biol 21:359–362

Sakazaki R (1992) Bacteriology of *Vibrio* and related organisms. In: Barua D and Greenough WB III (eds) Cholera. Plenum, New York, pp 37–55

Sakazaki R, Shimada T (1977) Serovars of *Vibrio cholerae*. Jpn J Med Sci Biol 30:279–282

Sakazaki R, Tamura K (1971) Somatic antigen variation in *Vibrio cholerae*. Jpn J Med Sci Biol 24:93–100

Saksena R, Chernyak A, Karavanov A, Kovac P (2003) Conjugating low molecular mass carbohydrates to proteins. 1. Monitoring the progress of conjugation. Methods Enzymol 362:125–139

Saksena R, Ma X, Wade TK, Kovac P, Wade WF (2006) Length of the linker and the interval between immunizations influences the efficacy of *Vibrio cholerae* O1, Ogawa hexasaccharide neoglycoconjugates. FEMS Immunol Med Microbiol 47:116–128

Salama NN, Eddington ND, Fasano A (2006) Tight junction modulation and its relationship to drug delivery. Adv Drug Deliv Rev 58:15–28

Samad SA, Bhattacharyya SC, Chatterjee SN (1987) Ultraviolet inactivation and photoreactivation of the cholera phage "kappa." Radiat Environ Biophys 26:295–300

Sambrook J, Fritsch EF, Maniatis T (1989) Molecular Cloning: A Laboratory Manual. Cold Spring Harbor Laboratory Press, Cold Spring Harbor, NY
Sandkvist M (2001a) Biology of type II secretion. Mol Microbiol 40:271–283
Sandkvist M (2001b) Type II secretion and pathogenesis. Infect Immun 69:3523–3535
Sandkvist M, Bagdasarian M, Howard SP, DiRita VJ (1995) Interaction between the autokinase EpsE and EpsL in the cytoplasmic membrane is required for extracellular secretion in *Vibrio cholerae*. Embo J 14:1664–1673
Sandkvist M, Keith JM, Bagdasarian M, Howard SP (2000) Two regions of EpsL involved in species-specific protein–protein interactions with EpsE and EpsM of the general secretion pathway in *Vibrio cholerae*. J Bacteriol 182:742–748
Sandkvist M, Michel LO, Hough LP, Morales VM, Bagdasarian M et al. (1997) General secretion pathway (eps) genes required for toxin secretion and outer membrane biogenesis in *Vibrio cholerae*. J Bacteriol 179:6994–7003
Sandkvist M, Morales V, Bagdasarian M (1993) A protein required for secretion of cholera toxin through the outer membrane of *Vibrio cholerae*. Gene 123:81–86
Sandvig K, Grimmer S, Lauvrak SU, Torgersen ML, Skretting G, van Deurs B, Iversen TG (2002) Pathways followed by ricin and Shiga toxin into cells. Histochem Cell Biol 117:131–141
Sandvig K, van Deurs B (1994) Endocytosis without clathrin. Trends Cell Biol 4:275–277
Sano Y, Kondo S, Isshiki Y, Shimada T, Hisatsune K (1996) An *N*-[(*R*)-(–)-2-hydroxypropionyl]-alpha-L-perosamine homopolymer constitutes the O-polysaccharide chain of the lipopolysaccharide from *Vibrio cholerae* O144 which has antigenic factor(s) in common with *V. cholerae* O76. Microbiol Immunol 40:735–741
Sanyal SC, Alam K, Neogi PK, Huq MI, Al-Mahmud KA (1983) A new cholera toxin. Lancet 1:1337
Sargent F, Gohlke U, De Leeuw E, Stanley NR, Palmer T, Saibil HR, Berks BC (2001) Purified components of the *Escherichia coli* Tat protein transport system form a double-layered ring structure. Eur J Biochem 268:3361–3367
Sathyamoorthy V, Hall RH, McCardell BA, Kothary MH, Ahn SJ, Ratnayake S (2000) Purification and characterization of a cytotonic protein expressed In vitro by the live cholera vaccine candidate CVD 103-HgR. Infect Immun 68:6062–6065
Sauvonnet N, Poquet I, Pugsley AP (1995) Extracellular secretion of pullulanase is unaffected by minor sequence changes but is usually prevented by adding reporter proteins to its N- or C-terminal end. J Bacteriol 177:5238–5246
Schengrund CL, Ringler NJ (1989) Binding of *Vibrio cholera* toxin and the heat-labile enterotoxin of *Escherichia coli* to GM1, derivatives of GM1, and nonlipid oligosaccharide polyvalent ligands. J Biol Chem 264:13233–13237
Schmitz A, Herrgen H, Winkeler A, Herzog V (2000) Cholera toxin is exported from microsomes by the Sec61p complex. J Cell Biol 148:1203–1212
Schofield CL, Field RA, Russell DA (2007) Glyconanoparticles for the colorimetric detection of cholera toxin. Anal Chem 79:1356–1361
Schon A, Freire E (1989) Thermodynamics of intersubunit interactions in cholera toxin upon binding to the oligosaccharide portion of its cell surface receptor, ganglioside GM1. Biochemistry 28:5019–5024
Schoolnik GK, Voskuil MI, Schnappinger D, Yildiz FH, Meibom K et al. (2001) Whole genome DNA microarray expression analysis of biofilm development by *Vibrio cholerae* O1 E1 Tor. Methods Enzymol 336:3–18
Schuhmacher DA, Klose KE (1999) Environmental signals modulate ToxT-dependent virulence factor expression in *Vibrio cholerae*. J Bacteriol 181:1508–1514
Scott ME, Dossani ZY, Sandkvist M (2001) Directed polar secretion of protease from single cells of *Vibrio cholerae* via the type II secretion pathway. Proc Natl Acad Sci USA 98:13978–13983
Seifert PS, Haeffner-Cavaillon N, Appay MD, Kazatchkine MD (1991) Bacterial lipopolysaccharides alter human endothelial cell morphology in vitro independent of cytokine secretion. J Lab Clin Med 118:563–569

Selander RK, Caugant DA, Ochman H, Musser JM, Gilmour MN, Whittam TS (1986) Methods of multilocus enzyme electrophoresis for bacterial population genetics and systematics. Appl Environ Microbiol 51:873–884

Sen AK, Mukherjee AK (1978) Structural studies of a specific polysaccharide isolated from non-agglutinable *Vibrio*. Carbohydr Res 64:215–223

Sen AK, Mukherjee AK, Guhathakurta B, Dutta A, Sasmal D (1979) Structural investigations of the lipopolysaccharide isolated from *Vibrio cholera*, Inaba 569 B. Carbohydr Res 72: 191–199

Senchenkova SN, Zatonsky GV, Shashkov AS, Knirel YA, Jansson PE, Weintraub A, Albert MJ (1998) Structure of the O-antigen of *Vibrio cholerae* O155 that shares a putative D-galactose 4,6-cyclophosphate-associated epitope with *V. cholerae* O139 Bengal. Eur J Biochem 254:58–62

Shakhnovich EA, Hung DT, Pierson E, Lee K, Mekalanos JJ (2007) Virstatin inhibits dimerization of the transcriptional activator ToxT. Proc Natl Acad Sci USA 104:2372–2377

Sharma C, Nair GB, Mukhopadhyay AK, Bhattacharya SK, Ghosh RK, Ghosh A (1997) Molecular characterization of *Vibrio cholerae* O1 biotype El Tor strains isolated between 1992 and 1995 in Calcutta, India: Evidence for the emergence of a new clone of the El Tor biotype. J Infect Dis 175:1134–1141

Sharma DP, Stroeher UH, Thomas CJ, Manning PA, Attridge SR (1989) The toxin-coregulated pilus (TCP) of *Vibrio cholerae*: Molecular cloning of genes involved in pilus biosynthesis and evaluation of TCP as a protective antigen in the infant mouse model. Microb Pathog 7:437–448

Sheahan KL, Cordero CL, Satchell KJ (2004) Identification of a domain within the multifunctional *Vibrio cholerae* RTX toxin that covalently cross-links actin. Proc Natl Acad Sci USA 101:9798–9803

Sheahan KL, Satchell KJ (2007) Inactivation of small Rho GTPases by the multifunctional RTX toxin from *Vibrio cholerae*. Cell Microbiol 9:1324–1335

Shields JM, Haston WS (1985) Behaviour of neutrophil leucocytes in uniform concentrations of chemotactic factors: Contraction waves, cell polarity and persistence. J Cell Sci 74:75–93

Shimamura T, Watanabe S, Sasaki S (1985) Enhancement of enterotoxin production by carbon dioxide in *Vibrio cholerae*. Infect Immun 49:455–456

Shogomori H, Futerman AH (2001) Cholera toxin is found in detergent-insoluble rafts/domains at the cell surface of hippocampal neurons but is internalized via a raft-independent mechanism. J Biol Chem 276:9182–9188

Shull GE, Greeb J (1988) Molecular cloning of two isoforms of the plasma membrane Ca^{2+}-transporting ATPase from rat brain. Structural and functional domains exhibit similarity to Na^+, K^+ and other cation transport ATPases. J Biol Chem 263:8646–8657

Sigel SP, Payne SM (1982) Effect of iron limitation on growth, siderophore production, and expression of outer membrane proteins of *Vibrio cholerae*. J Bacteriol 150:148–155

Sigler PB, Dryan ME, Kiuefer HC, Finkelstein RA (1977) Cholera toxin crystals suitable for X-ray diffraction. Science 197:1277–1279

Sillerud LO, Prestegard JH, Yu RK, Konigsberg WH, Schafer DE (1981) Observation by 13C NMR of interactions between cholera toxin and the oligosaccharide of ganglioside GM1. J Biol Chem 256:1094–1097

Silva AJ, Eko FO, Benitez JA (2008) Exploiting cholera vaccines as a versatile antigen delivery platform. Biotechnol Lett 30:571–579

Silva TM, Schleupner MA, Tacket CO, Steiner TS, Kaper JB, Edelman R, Guerrant R (1996) New evidence for an inflammatory component in diarrhea caused by selected new, live attenuated cholera vaccines and by El Tor and Q139 *Vibrio cholerae*. Infect Immun 64:2362–2364

Simons K, Ikonen E (1997) Functional rafts in cell membranes. Nature 387:569–572

Singh RD, Puri V, Valiyaveettil JT, Marks DL, Bittman R, Pagano RE (2003) Selective caveolin-1-dependent endocytosis of glycosphingolipids. Mol Biol Cell 14:3254–3265

Sixma TK, Kalk KH, van Zanten BA, Dauter Z, Kingma J, Witholt B, Hol WG (1993) Refined structure of *Escherichia coli* heat-labile enterotoxin, a close relative of cholera toxin. J Mol Biol 230:890–918

Sixma TK, Pronk SE, Kalk KH, van Zanten BA, Berghuis AM, Hol WG (1992) Lactose binding to heat-labile enterotoxin revealed by X-ray crystallography. Nature 355:561–564

Sixma TK, Pronk SE, Kalk KH, Wartna ES, van Zanten BA, Witholt B, Hol WG (1991) Crystal structure of a cholera toxin-related heat-labile enterotoxin from *E. coli*. Nature 351:371–377

Skorupski K, Taylor RK (1997) Control of the ToxR virulence regulon in *Vibrio cholerae* by environmental stimuli. Mol Microbiol 25:1003–1009

Skorupski K, Taylor RK (1999) A new level in the *Vibrio cholerae* ToxR virulence cascade: AphA is required for transcriptional activation of the tcpPH operon. Mol Microbiol 31:763–771

Smith DJ, King WF, Barnes LA, Trantolo D, Wise DL, Taubman MA (2001) Facilitated intranasal induction of mucosal and systemic immunity to mutans streptococcal glucosyltransferase peptide vaccines. Infect Immun 69:4767–4773

Smith HR (1984) Genetics of enterotoxin production in *Escherichia coli*. Biochem Soc Trans 12:187–189

Sommerfelt H, Svennerholm AM, Kalland KH, Haukanes BI, Bjorvatn B (1988) Comparative study of colony hybridization with synthetic oligonucleotide probes and enzyme-linked immunosorbent assay for identification of enterotoxigenic *Escherichia coli*. J Clin Microbiol 26:530–534

Song X, Shi J, Swanson B (2000) Flow cytometry-based biosensor for detection of multivalent proteins. Anal Biochem 284:35–41

Sozhamannan S, Deng YK, Li M, Sulakvelidze A, Kaper JB et al. (1999) Cloning and sequencing of the genes downstream of the wbf gene cluster of *Vibrio cholerae* serogroup O139 and analysis of the junction genes in other serogroups. Infect Immun 67:5033–5040

Spangler BD (1992) Structure and function of cholera toxin and the related *Escherichia coli* heat-labile enterotoxin. Microbiol Rev 56:622–647

Spangler BD, Westbrook EM (1989) Crystallization of isoelectrically homogeneous cholera toxin. Biochemistry 28:1333–1340

Sperandio V, Giron JA, Silveira WD, Kaper JB (1995) The OmpU outer membrane protein, a potential adherence factor of *Vibrio cholerae*. Infect Immun 63:4433–4438

Steele EJ, Chaicumpa W, Rowley D (1974) Isolation and biological properties of three classes of rabbit antibody to *Vibrio cholerae*. J Infect Dis 130:93–103

Steele EJ, Chaicumpa W, Rowley D (1975) Further evidence for cross-linking as a protective factor in experimental cholera: Properties of antibody fragments. J Infect Dis 132:175–180

Steele-Mortimer O, Knodler LA, Finlay BB (2000) Poisons, ruffles and rockets: Bacterial pathogens and the host cell cytoskeleton. Traffic 1:107–118

Stroeher UH, Karageorgos LE, Morona R, Manning PA (1992) Serotype conversion in *Vibrio cholerae* O1. Proc Natl Acad Sci USA 89:2566–2570

Stroeher UH, Karageorgos LE, Brown MH, Morona R, Manning PA (1995a) A putative pathway for perosamine biosynthesis is the first function encoded within the rfb region of *Vibrio cholerae* O1. Gene 166:33–42

Stroeher UH, Jedani KE, Dredge BK, Morona R, Brown MH et al. (1995b) Genetic rearrangements in the rfb regions of *Vibrio cholerae* O1 and O139. Proc Natl Acad Sci USA 92:10374–10378

Stroeher UH, Karageorgos LE, Morona R, Manning PA (1995c) In *Vibrio cholerae* serogroup O1, rfaD is closely linked to the rfb operon. Gene 155:67–72

Stroeher UH, Manning PA (1997a) *Vibrio cholerae* serotype O139: swapping genes for surface polysaccharide biosynthesis. Trends Microbiol 5:178–180

Stroeher UH, Parasivam G, Dredge BK, Manning PA (1997b) Novel *Vibrio cholerae* O139 genes involved in lipopolysaccharide biosynthesis. J Bacteriol 179:2740–2747

Stroeher UH, Jedani KE, Manning PA (1998) Genetic organization of the regions associated with surface polysaccharide synthesis in *Vibrio cholerae* O1, O139 and *Vibrio anguillarum* O1 and O2: A review. Gene 223:269–282

Strom MS, Lory S (1993) Structure–function and biogenesis of the type IV pili. Annu Rev Microbiol 47:565–596

Sun DX, Mekalanos JJ, Taylor RK (1990) Antibodies directed against the toxin-coregulated pilus isolated from *Vibrio cholerae* provide protection in the infant mouse experimental cholera model. J Infect Dis 161:1231–1236

Sun DX, Seyer JM, Kovari I, Sumrada RA, Taylor RK (1991) Localization of protective epitopes within the pilin subunit of the *Vibrio cholerae* toxin-coregulated pilus. Infect Immun 59:114–118

Sur P, Maiti M, Chatterjee SN (1974) Physico-chemical studies of *Vibrio cholerae* cell envelope fraction. Indian J Exp Biol 12:479–483

Sveen K (1978) The importance of C5 and the role of the alternative complement pathway in leukocyte chemotaxis induced in vivo and in vitro by *Bacteroides fragilis* lipopolysaccharide. Acta Pathol Microbiol Scand B 86:93–100

Sveneerholm AM, Wikstrom M, Lindholm L, Holmgren J (1986) Monoclonal antibodies and immunodetection methods for *Vibrio cholerae* and *Escherichia coli* enterotoxins. In: Macario AJ and deMacario EC (eds) Monoclonal Antibodies Against Bacteria. Academic, New York, pp 77–97

Svennerholm AM (1975) Experimental studies on cholera immunization. 4. The antibody response to formalinized *Vibrio cholerae* and purified endotoxin with special reference to protective capacity. Int Arch Allergy Appl Immunol 49:434–452

Svennerholm AM, Holmgren J (1976) Synergistic protective effect in rabbits of immunization with *Vibrio cholerae* lipopolysaccharide and toxin/toxoid. Infect Immun 13:735–740

Svennerholm AM, Jertborn M, Gothefors L, Karim AM, Sack DA, Holmgren J (1984) Mucosal antitoxic and antibacterial immunity after cholera disease and after immunization with a combined B subunit-whole cell vaccine. J Infect Dis 149:884–893

Svennerholm AM, Sack DA, Holmgren J, Bardhan PK (1982) Intestinal antibody responses after immunisation with cholera B subunit. Lancet 1:305–308

Szu SC, Gupta R, Robbins JB (1994) Induction of serum vibriocidal antibodies by O-specific polysaccharide-protein conjugate vaccines for prevention of cholera. In: Wachsmuth IK, Blake PA, and Olsvik O (eds) *Vibrio cholerae* and Cholera. American Society for Microbiology, Washington, DC, pp 381–394

Tacket CO, Forrest B, Morona R, Attridge SR, LaBrooy J et al. (1990) Safety, immunogenicity, and efficacy against cholera challenge in humans of a typhoid-cholera hybrid vaccine derived from *Salmonella typhi* Ty21a. Infect Immun 58:1620–1627

Tacket CO, Losonsky G, Nataro JP, Comstock L, Michalski J et al. (1995a) Initial clinical studies of CVD 112 *Vibrio cholerae* O139 live oral vaccine: Safety and efficacy against experimental challenge. J Infect Dis 172:883–886

Tacket CO, Losonsky G, Nataro JP, Wasserman SS, Cryz SJ, Edelman R, Levine MM (1995b) Extension of the volunteer challenge model to study South American cholera in a population of volunteers predominantly with blood group antigen O. Trans R Soc Trop Med Hyg 89:75–77

Takada H, Kotani S (1992) In: Morrison DC and Ryan JL (eds) Bacterial Endotoxic Lipopolysaccharides. CRC Press, Boca Raton, FL, pp 107–130

Takao T, Hitouji T, Aimoto S, Shimonishi Y, Hara S et al. (1983) Amino acid sequence of a heat-stable enterotoxin isolated from enterotoxigenic *Escherichia coli* strain 18D. FEBS Lett 152:1–5

Takao T, Shimonishi Y, Kobayashi M, Nishimura O, Arita M et al. (1985) Amino acid sequence of heat-stable enterotoxin produced by *Vibrio cholerae* non-01. FEBS Lett 193:250–254

Takao T, Tominaga N, Shimonishi Y, Hara S, Inoue T, Miyama A (1984) Primary structure of heat-stable enterotoxin produced by *Yersinia enterocolitica*. Biochem Biophys Res Commun 125:845–851

Takeda T, Peina Y, Ogawa A, Dohi S, Abe H, Nair GB, Pal SC (1991) Detection of heat-stable enterotoxin in a cholera toxin gene-positive strain of *Vibrio cholerae* O1. FEMS Microbiol Lett 64:23–27

Takeya K, Shimodori S (1963) "Prophage-typing" of El Tor vibrios. J Bacteriol 85:957–958

Tamura S, Yamanaka A, Shimohara M, Tomita T, Komase K et al. (1994) Synergistic action of cholera toxin B subunit (and *Escherichia coli* heat-labile toxin B subunit) and a trace amount of cholera whole toxin as an adjuvant for nasal influenza vaccine. Vaccine 12:419–426

Taylor RK, Miller VL, Furlong DB, Mekalanos JJ (1987) Use of phoA gene fusions to identify a pilus colonization factor coordinately regulated with cholera toxin. Proc Natl Acad Sci USA 84:2833–2837

Tebbey PW, Scheuer CA, Peek JA, Zhu D, LaPierre NA et al. (2000) Effective mucosal immunization against respiratory syncytial virus using purified F protein and a genetically detoxified cholera holotoxin, CT-E29H. Vaccine 18:2723–2734

Terai A, Baba K, Shirai H, Yoshida O, Takeda Y, Nishibuchi M (1991) Evidence for insertion sequence-mediated spread of the thermostable direct hemolysin gene among *Vibrio* species. J Bacteriol 173:5036–5046

Teter K, Allyn RL, Jobling MG, Holmes RK (2002) Transfer of the cholera toxin A1 polypeptide from the endoplasmic reticulum to the cytosol is a rapid process facilitated by the endoplasmic reticulum-associated degradation pathway. Infect Immun 70:6166–6171

Teter K, Holmes RK (2002) Inhibition of endoplasmic reticulum-associated degradation in CHO cells resistant to cholera toxin, *Pseudomonas aeruginosa* exotoxin A, and ricin. Infect Immun 70:6172–6179

Teter K, Jobling MG, Holmes RK (2003) A class of mutant CHO cells resistant to cholera toxin rapidly degrades the catalytic polypeptide of cholera toxin and exhibits increased endoplasmic reticulum-associated degradation. Traffic 4:232–242

Teter K, Jobling MG, Sentz D, Holmes RK (2006) The cholera toxin A1(3) subdomain is essential for interaction with ADP-ribosylation factor 6 and full toxic activity but is not required for translocation from the endoplasmic reticulum to the cytosol. Infect Immun 74:2259–2267

Thungapathra M, Sharma C, Gupta N, Ghosh RK, Mukhopadhyay A et al. (1999) Construction of a recombinant live oral vaccine from a non-toxigenic strain of *Vibrio cholerae* O1 serotype inaba biotype E1 Tor and assessment of its reactogenicity and immunogenicity in the rabbit model. Immunol Lett 68:219–227

Tikoo A, Singh DV, Shukla BN, Sanyal SC (1996) Development of an improved synthetic medium for a better production of the new cholera toxin and its immunological relationship with the toxin produced by *Vibrio cholerae* O139 strains. FEMS Immunol Med Microbiol 14:67–72

Tinker JK, Erbe JL, Holmes RK (2005) Characterization of fluorescent chimeras of cholera toxin and *Escherichia coli* heat-labile enterotoxins produced by use of the twin arginine translocation system. Infect Immun 73:3627–3635

Tischler AD, Camilli A (2004) Cyclic diguanylate (c-di-GMP) regulates *Vibrio cholerae* biofilm formation. Mol Microbiol 53:857–869

Tischler AD, Camilli A (2005) Cyclic diguanylate regulates *Vibrio cholerae* virulence gene expression. Infect Immun 73:5873–5882

Toida N, Hajishengallis G, Wu HY, Russell MW (1997) Oral immunization with the saliva-binding region of *Streptococcus mutans* AgI/II genetically coupled to the cholera toxin B subunit elicits T-helper-cell responses in gut-associated lymphoid tissues. Infect Immun 65:909–915

Trucksis M, Conn TL, Fasano A, Kaper JB (1997) Production of *Vibrio cholerae* accessory cholera enterotoxin (Ace) in the yeast *Pichia pastoris*. Infect Immun 65:4984–4988

Trucksis M, Conn TL, Wasserman SS, Sears CL (2000) *Vibrio cholerae* ACE stimulates Ca(2+)-dependent Cl(–)/HCO(3)(–) secretion in T84 cells in vitro. Am J Physiol Cell Physiol 279:C567–577

Trucksis M, Galen JE, Michalski J, Fasano A, Kaper JB (1993) Accessory cholera enterotoxin (Ace), the third toxin of a *Vibrio cholerae* virulence cassette. Proc Natl Acad Sci USA 90:5267–5271

Trucksis M, Michalski J, Deng YK, Kaper JB (1998) The *Vibrio cholerae* genome contains two unique circular chromosomes. Proc Natl Acad Sci USA 95:14464–14469

Tsai B, Rapoport TA (2002) Unfolded cholera toxin is transferred to the ER membrane and released from protein disulfide isomerase upon oxidation by Ero1. J Cell Biol 159:207–216

Tsai B, Rodighiero C, Lencer WI, Rapoport TA (2001) Protein disulfide isomerase acts as a redox-dependent chaperone to unfold cholera toxin. Cell 104:937–948

Tsai B, Ye Y, Rapoport TA (2002) Retro-translocation of proteins from the endoplasmic reticulum into the cytosol. Nat Rev Mol Cell Biol 3:246–255

Tweedy JM, Park RW, Hodgkiss W (1968) Evidence for the presence of fimbriae (pili) on vibrio species. J Gen Microbiol 51:235–244

Ujiye A, Kobari K (1970) Protective effect on infections with *Vibrio cholerae* in suckling mice caused by the passive immunization with milk of immune mothers. J Infect Dis 121(Suppl 121):150+

Uzzau S, Lu R, Wang W, Fiore C, Fasano A (2001) Purification and preliminary characterization of the zonula occludens toxin receptor from human (CaCo2) and murine (IEC6) intestinal cell lines. FEMS Microbiol Lett 194:1–5

Vajdy M, Lycke N (1993) Stimulation of antigen-specific T- and B-cell memory in local as well as systemic lymphoid tissues following oral immunization with cholera toxin adjuvant. Immunology 80:197–203

Vajdy M, Lycke NY (1992) Cholera toxin adjuvant promotes long-term immunological memory in the gut mucosa to unrelated immunogens after oral immunization. Immunology 75:488–492

Val ME, Bouvier M, Campos J, Sherratt D, Cornet F, Mazel D, Barre FX (2005) The single-stranded genome of phage CTX is the form used for integration into the genome of *Vibrio cholerae*. Mol Cell 19:559–566

van den Akker F, Merritt EA, Pizza M, Domenighini M, Rappuoli R, Hol WG (1995) The Arg7Lys mutant of heat-labile enterotoxin exhibits great flexibility of active site loop 47-56 of the A subunit. Biochemistry 34:10996–11004

Van der Heijden PJ, Bianchi AT, Dol M, Pals JW, Stok W, Bokhout BA (1991) Manipulation of intestinal immune responses against ovalbumin by cholera toxin and its B subunit in mice. Immunology 72:89–93

van Ginkel FW, Jackson RJ, Yuki Y, McGhee JR (2000) Cutting edge: The mucosal adjuvant cholera toxin redirects vaccine proteins into olfactory tissues. J Immunol 165:4778–4782

van Heyningen S (1976) The subunits of cholera toxin: Structure, stoichiometry, and function. J Infect Dis 133(Suppl):5–13

Van Heyningen WE, Carpenter CC, Pierce NF, Greenough WB 3rd (1971) Deactivation of cholera toxin by ganglioside. J Infect Dis 124:415–418

van Wezenbeek PM, Hulsebos TJ, Schoenmakers JG (1980) Nucleotide sequence of the filamentous bacteriophage M13 DNA genome: comparison with phage fd. Gene 11:129–148

Verma AK, Filoteo AG, Stanford DR, Wieben ED, Penniston JT et al. (1988) Complete primary structure of a human plasma membrane Ca^{2+} pump. J Biol Chem 263:14152–14159

Vicente AC, Coelho AM, Salles CA (1997) Detection of *Vibrio cholerae* and *V. mimicus* heat-stable toxin gene sequence by PCR. J Med Microbiol 46:398–402

Vignon G, Kohler R, Larquet E, Giroux S, Prevost MC, Roux P, Pugsley AP (2003) Type IV-like pili formed by the type II secreton: Specificity, composition, bundling, polar localization, and surface presentation of peptides. J Bacteriol 185:3416–3428

Villeneuve S, Boutonnier A, Mulard LA, Fournier JM (1999) Immunochemical characterization of an Ogawa-Inaba common antigenic determinant of *Vibrio cholerae* O1. Microbiology 145(Pt 9):2477–2484

Villeneuve S, Souchon H, Riottot MM, Mazie JC, Lei P et al. (2000) Crystal structure of an anti-carbohydrate antibody directed against *Vibrio cholerae* O1 in complex with antigen: Molecular basis for serotype specificity. Proc Natl Acad Sci USA 97:8433–8438

Vimont S, Dumontier S, Escuyer V, Berche P (1997) The rfaD locus: A region of rearrangement in *Vibrio cholerae* O139. Gene 185:43–47

Vinogradov EV, Bock K, Holst O, Brade H (1995) The structure of the lipid A-core region of the lipopolysaccharides from *Vibrio cholerae* O1 smooth strain 569B (Inaba) and rough mutant strain 95R (Ogawa). Eur J Biochem 233:152–158

Vinogradov EV, Stuike-Prill R, Bock K, Holst O, Brade H (1993) The structure of the carbohydrate backbone of the core-lipid-A region of the lipopolysaccharide from *Vibrio cholerae* strain H11 (non-O1). Eur J Biochem 218:543–554

Viswanathan S, Wu LC, Huang MR, Ho JA (2006) Electrochemical immunosensor for cholera toxin using liposomes and poly(3,4-ethylenedioxythiophene)-coated carbon nanotubes. Anal Chem 78:1115–1121

Voss E, Manning PA, Attridge SR (1996) The toxin-coregulated pilus is a colonization factor and protective antigen of *Vibrio cholerae* El Tor. Microb Pathog 20:141–153

Wachsmuth I, Blake P, Olsvik O (1994) *Vibrio cholerae* and Cholera, Molecular to Global Prespectives. ASM Press, Washington, DC

Wade WF (2006) B-cell responses to lipopolysaccharide epitopes: Who sees what first—does it matter? Am J Reprod Immunol 56:329–336

Wai SN, Mizunoe Y, Takade A, Kawabata SI, Yoshida SI (1998) *Vibrio cholerae* O1 strain TSI-4 produces the exopolysaccharide materials that determine colony morphology, stress resistance, and biofilm formation. Appl Environ Microbiol 64:3648–3655

Waldor MK, Colwell R, Mekalanos JJ (1994) The *Vibrio cholerae* O139 serogroup antigen includes an O-antigen capsule and lipopolysaccharide virulence determinants. Proc Natl Acad Sci USA 91:11388–11392

Waldor MK, Mekalanos JJ (1994) Emergence of a new cholera pandemic: molecular analysis of virulence determinants in *Vibrio cholerae* O139 and development of a live vaccine prototype. J Infect Dis 170:278–283

Waldor MK, Mekalanos JJ (1996) Lysogenic conversion by a filamentous phage encoding cholera toxin. Science 272:1910–1914

Waldor MK, Rubin EJ, Pearson GD, Kimsey H, Mekalanos JJ (1997) Regulation, replication, and integration functions of the *Vibrio cholerae* CTXphi are encoded by region RS2. Mol Microbiol 24:917–926

Walia K, Ghosh S, Singh H, Nair GB, Ghosh A et al. (1999) Purification and characterization of novel toxin produced by *Vibrio cholerae* O1. Infect Immun 67:5215–5222

Wandersman C, Delepelaire P (1990) TolC, an *Escherichia coli* outer membrane protein required for hemolysin secretion. Proc Natl Acad Sci USA 87:4776–4780

Wang J, Villeneuve S, Zhang J, Lei P, Miller CE et al. (1998) On the antigenic determinants of the lipopolysaccharides of *Vibrio cholerae* O:1, serotypes Ogawa and Inaba. J Biol Chem 273:2777–2783

Wang W, Uzzau S, Goldblum SE, Fasano A (2000) Human zonulin, a potential modulator of intestinal tight junctions. J Cell Sci 113 Pt 24:4435–4440

Ward HM, Manning PA (1989) Mapping of chromosomal loci associated with lipopolysaccharide synthesis and serotype specificity in *Vibrio cholerae* 01 by transposon mutagenesis using Tn5 and Tn2680. Mol Gen Genet 218:367–370

Ward HM, Morelli G, Kamke M, Morona R, Yeadon J, Hackett JA, Manning PA (1987) A physical map of the chromosomal region determining O-antigen biosynthesis in *Vibrio cholerae* O1. Gene 55:197–204

Wassermann SS, Losonsky GA, Noriega F, Tacket CO, Castaned E, Levine MM (1994) Kinetics of the vibriocidal antibody response to live oral cholera vaccines. Vaccine 12:1000–1003

Watnick PI, Fullner KJ, Kolter R (1999) A role for the mannose-sensitive hemagglutinin in biofilm formation by *Vibrio cholerae* El Tor. J Bacteriol 181:3606–3609

Watnick PI, Kolter R (1999) Steps in the development of a *Vibrio cholerae* El Tor biofilm. Mol Microbiol 34:586–595

Watnick PI, Lauriano CM, Klose KE, Croal L, Kolter R (2001) The absence of a flagellum leads to altered colony morphology, biofilm development and virulence in *Vibrio cholerae* O139. Mol Microbiol 39:223–235

Weintraub A, Widmalm G, Jansson PE, Jansson M, Hultenby K, Albert MJ (1994) *Vibrio cholerae* O139 Bengal possesses a capsular polysaccharide which may confer increased virulence. Microb Pathog 16:235–241

Weissbach A, Hurwitz J (1959) The formation of 2-keto-3-deoxyheptonic acid in extracts of *Escherichia coli* B. I. Identification. J Biol Chem 234:705–709

Welch RA (2001) RTX toxin structure and function: A story of numerous anomalies and few analogies in toxin biology. Curr Top Microbiol Immunol 257:85–111

Welsh CF, Moss J, Vaughan M (1994) ADP-ribosylation factors: A family of approximately 20-kDa guanine nucleotide-binding proteins that activate cholera toxin. Mol Cell Biochem 138:157–166

Westphal O, Luderitz O (1954) Chemilische erforschung von lipopolysacchariden gram-negativen bacterien. Angew Chem 66:407–417

Westphal O, Luderitz O, Bister F (1952) Uber die extraction von bakterien mit phenol/wasser. Z Naturforsch 7b:148–155

Wilkinson SG (1977) Composition and structure of bacterial lipopolysaccharides. In: Sutherland IW (ed) Surface Carbohydrates of the Prokaryotic Cell. Academic, New York, pp 97–175

Wilkinson SG (1996) Bacterial lipopolysaccharides: Themes and variations. Prog Lipid Res 35:283–343

Winner L 3rd, Mack J, Weltzin R, Mekalanos JJ, Kraehenbuhl JP, Neutra MR (1991) New model for analysis of mucosal immunity: Intestinal secretion of specific monoclonal immunoglobulin A from hybridoma tumors protects against *Vibrio cholerae* infection. Infect Immun 59:977–982

Withey JH, Dirita VJ (2005a) *Vibrio cholerae* ToxT independently activates the divergently transcribed *aldA* and *tagA* genes. J Bacteriol 187:7890–7900

Withey JH, DiRita VJ (2005b) Activation of both *acfA* and *acfD* transcription by *Vibrio cholerae* ToxT requires binding to two centrally located DNA sites in an inverted repeat conformation. Mol Microbiol 56:1062–1077

Withey JH, DiRita VJ (2006) The toxbox: Specific DNA sequence requirements for activation of *Vibrio cholerae* virulence genes by ToxT. Mol Microbiol 59:1779–1789

Wolf AA, Fujinaga Y, Lencer WI (2002) Uncoupling of the cholera toxin-G(M1) ganglioside receptor complex from endocytosis, retrograde Golgi trafficking, and downstream signal transduction by depletion of membrane cholesterol. J Biol Chem 277:16249–16256

Wolf AA, Jobling MG, Wimer-Mackin S, Ferguson-Maltzman M, Madara JL, Holmes RK, Lencer WI (1998) Ganglioside structure dictates signal transduction by cholera toxin and association with caveolae-like membrane domains in polarized epithelia. J Cell Biol 141:917–927

Wright AC, Guo Y, Johnson JA, Nataro JP, Morris JG Jr (1992) Development and testing of a nonradioactive DNA oligonucleotide probe that is specific for *Vibrio cholerae* cholera toxin. J Clin Microbiol 30:2302–2306

Xiang SH, Hobbs M, Reeves PR (1994) Molecular analysis of the rfb gene cluster of a group D2 *Salmonella enterica* strain: Evidence for its origin from an insertion sequence-mediated recombination event between group E and D1 strains. J Bacteriol 176:4357–4365

Xu-Amano J, Jackson RJ, Fujihashi K, Kiyono H, Staats HF, McGhee JR (1994) Helper Th1 and Th2 cell responses following mucosal or systemic immunization with cholera toxin. Vaccine 12:903–911

Xu-Amano J, Kiyono H, Jackson RJ, Staats HF, Fujihashi K et al. (1993) Helper T cell subsets for immunoglobulin A responses: oral immunization with tetanus toxoid and cholera toxin as adjuvant selectively induces Th2 cells in mucosa associated tissues. J Exp Med 178:1309–1320

Yam WC, Lung ML, Ng KY, Ng MH (1989) Molecular epidemiology of *Vibrio cholerae* in Hong Kong. J Clin Microbiol 27:1900–1902

Yamamoto K, Al-Omani M, Honda T, Takeda Y, Miwatani T (1984) Non-O1 *Vibrio cholerae* hemolysin: Purification, partial characterization, and immunological relatedness to El Tor hemolysin. Infect Immun 45:192–196

Yamamoto K, Ichinose Y, Shinagawa H, Makino K, Nakata A et al. (1990) Two-step processing for activation of the cytolysin/hemolysin of *Vibrio cholerae* O1 biotype El Tor: Nucleotide sequence of the structural gene (hlyA) and characterization of the processed products. Infect Immun 58:4106–4116

Yamamoto S, Takeda Y, Yamamoto M, Kurazono H, Imaoka K et al. (1997) Mutants in the ADP-ribosyltransferase cleft of cholera toxin lack diarrheagenicity but retain adjuvanticity. J Exp Med 185:1203–1210

Yamasaki S, Shimizu T, Hoshino K, Ho ST, Shimada T, Nair GB, Takeda Y (1999) The genes responsible for O-antigen synthesis of *Vibrio cholerae* O139 are closely related to those of *Vibrio cholerae* O22. Gene 237:321–332

Yanez ME, Korotkov KV, Abendroth J, Hol WG (2008a) Structure of the minor pseudopilin EpsH from the Type 2 secretion system of *Vibrio cholerae*. J Mol Biol 377:91–103

Yanez ME, Korotkov KV, Abendroth J, Hol WG (2008b) The crystal structure of a binary complex of two pseudopilins: EpsI and EpsJ from the type 2 secretion system of *Vibrio vulnificus*. J Mol Biol 375:471–486

Yang J, Tamm LK, Tillack TW, Shao Z (1993) New approach for atomic force microscopy of membrane proteins. The imaging of cholera toxin. J Mol Biol 229:286–290

Yildiz FH, Dolganov NA, Schoolnik GK (2001) VpsR, a member of the response regulators of the two-component regulatory systems, is required for expression of vps biosynthesis genes and EPS(ETr)-associated phenotypes in *Vibrio cholerae* O1 El Tor. J Bacteriol 183:1716–1726

Yildiz FH, Schoolnik GK (1999) *Vibrio cholerae* O1 El Tor: Identification of a gene cluster required for the rugose colony type, exopolysaccharide production, chlorine resistance, and biofilm formation. Proc Natl Acad Sci USA 96:4028–4033

Yoh M, Honda T, Miwatani T (1986) Purification and partial characterization of a non-O1 *Vibrio cholerae* hemolysin that cross-reacts with thermostable direct hemolysin of *Vibrio parahaemolyticus*. Infect Immun 52:319–322

Yoh M, Miyagi K, Matsumoto Y, Hayashi K, Takarada Y, Yamamoto K, Honda T (1993) Development of an enzyme-labeled oligonucleotide probe for the cholera toxin gene. J Clin Microbiol 31:1312–1314

Yoon SS, Mekalanos JJ (2008) Decreased potency of the *Vibrio cholerae* sheathed flagellum to trigger host innate immunity. Infect Immun 76:1282–1288

Yoshino K, Miyachi M, Takao T, Bag PK, Huang X et al. (1993) Purification and sequence determination of heat-stable enterotoxin elaborated by a cholera toxin-producing strain of *Vibrio cholerae* O1. FEBS Lett 326:83–86

Yoshiyama Y, Brown WR (1987) Specific antibodies to cholera toxin in rabbit milk are protective against *Vibrio cholerae*-induced intestinal secretion. Immunology 61:543–547

Yu J, McLaughlin S, Freedman RB, Hirst TR (1993) Cloning and active site mutagenesis of *Vibrio cholerae* DsbA, a periplasmic enzyme that catalyzes disulfide bond formation. J Biol Chem 268:4326–4330

Yu J, Webb H, Hirst TR (1992) A homologue of the *Escherichia coli* DsbA protein involved in disulphide bond formation is required for enterotoxin biogenesis in *Vibrio cholerae*. Mol Microbiol 6:1949–1958

Yu RR, DiRita VJ (2002) Regulation of gene expression in *Vibrio cholerae* by ToxT involves both antirepression and RNA polymerase stimulation. Mol Microbiol 43:119–134

Zhang D, Takahashi J, Seno T, Tani Y, Honda T (1999) Analysis of receptor for *Vibrio cholerae* El Tor hemolysin with a monoclonal antibody that recognizes glycophorin B of human erythrocyte membrane. Infect Immun 67:5332–5337

Zhang L, Radziejewska-Lebrecht J, Krajewska-Pietrasik D, Toivanen P, Skurnik M (1997) Molecular and chemical characterization of the lipopolysaccharide O-antigen and its role in the virulence of *Yersinia enterocolitica* serotype O:8. Mol Microbiol 23:63–76

Zhang RG, Scott DL, Westbrook ML, Nance S, Spangler BD, Shipley GG, Westbrook EM (1995a) The three-dimensional crystal structure of cholera toxin. J Mol Biol 251:563–573

Zhang RG, Westbrook ML, Westbrook EM, Scott DL, Otwinowski Z et al. (1995b) The 2.4 A crystal structure of cholera toxin B subunit pentamer: choleragenoid. J Mol Biol 251:550–562

Zhao S, Sandt CH, Feulner G, Vlazny DA, Gray JA, Hill CW (1993) Rhs elements of *Escherichia coli* K-12: Complex composites of shared and unique components that have different evolutionary histories. J Bacteriol 175:2799–2808

Zhou DX, Massenet O, Quigley F, Marion MJ, Moneger F, Huber P, Mache R (1988) Characterization of a large inversion in the spinach chloroplast genome relative to *Marchantia*: A possible transposon-mediated origin. Curr Genet 13:433–439

Zhu J, Miller MB, Vance RE, Dziejman M, Bassler BL, Mekalanos JJ (2002) Quorum-sensing regulators control virulence gene expression in *Vibrio cholerae*. Proc Natl Acad Sci USA 99:3129–3134

Zitzer A, Wassenaar TM, Walev I, Bhakdi S (1997a) Potent membrane-permeabilizing and cytocidal action of *Vibrio cholerae* cytolysin on human intestinal cells. Infect Immun 65:1293–1298

Zitzer A, Palmer M, Weller U, Wassenaar T, Biermann C, Tranum-Jensen J, Bhakdi S (1997b) Mode of primary binding to target membranes and pore formation induced by *Vibrio cholerae* cytolysin (hemolysin). Eur J Biochem 247:209–216

Zitzer A, Walev I, Palmer M, Bhakdi S (1995) Characterization of *Vibrio cholerae* El Tor cytolysin as an oligomerizing pore-forming toxin. Med Microbiol Immunol 184:37–44

Zitzer A, Zitzer O, Bhakdi S, Palmer M (1999) Oligomerization of *Vibrio cholerae* cytolysin yields a pentameric pore and has a dual specificity for cholesterol and sphingolipids in the target membrane. J Biol Chem 274:1375–1380

Zughaier SM, Zimmer SM, Datta A, Carlson RW, Stephens DS (2005) Differential induction of the toll-like receptor 4-MyD88-dependent and -independent signaling pathways by endotoxins. Infect Immun 73:2940–2950

Index

Printing: Krips bv, Meppel, The Netherlands
Binding: Stürtz, Würzburg, Germany